Elementary St

Elementary
Statistical Methods

Third edition

G. BARRIE WETHERILL
Professor of Statistics
University of Kent at Canterbury

with a contribution to
Appendix One by
E. E. Bassett
University of Kent

LONDON   NEW YORK
CHAPMAN AND HALL

First published 1967 by
Methuen and Co. Ltd
Second edition 1972
Published by Chapman and Hall Ltd
11 New Fetter Lane, London EC4P 4EE
Third edition 1982
Published in the USA by
Chapman and Hall
in association with Methuen, Inc.
733 Third Avenue, New York NY 10017

Printed in Great Britain by
Redwood Burn Limited
Trowbridge, Wiltshire

ISBN 0 412 24000 9

British Library Cataloguing in Publication Data

Wetherill, G. Barrie
    Elementary statistical methods.—3rd ed.—(Science paperbacks)
    1. Mathematical statistics
    I. Title
    519.5   QA276.AI
    ISBN 0–412–24000–9

Preface

This book is mainly based on lectures given by Professor D. R. Cox and myself at Birkbeck College over a period of eight to nine years. It began as a joint venture, but pressure of other work made it necessary for Professor Cox to withdraw early on. I have throughout received much valuable advice and encouragement from Professor Cox, but of course, I am solely responsible for the text, and any errors remaining in it.

The book is intended as a first course on statistical methods, and there is a liberal supply of exercises. Although the mathematical level of the book is low, I have tried to explain carefully the logical reasoning behind the use of the methods discussed. Some of the exercises which require more difficult mathematics are marked with an asterisk, and these may be omitted. In this way, I hope that the book will satisfy the needs for a course on statistical methods at a range of mathematical levels.

It is essential for the reader to work through the numerical exercises, for only in this way can he grasp the full meaning and usefulness of the statistical techniques, and gain practice in the interpretation of the results.

Chapters 7 and 8 discuss methods appropriate for use on ranked or discrete data, and Chapters 9–12 do not depend on these chapters. Chapters 7 and 8 may therefore be omitted, if desired.

Examples, tables and figures are numbered serially within chapters, and referred to as, for example, Example 5.3, Table 1.12, etc., the first figure denoting the chapter. To refer to exercises, such as to Exercise 3 at the end of § 8.8, I use the notation Exercise 8.8.3.

I am grateful to a number of people for detailed criticisms of earlier drafts of the book, and apart from Professor Cox, I mention particularly Mr G. E. G. Campling, Dr W. G. Gilchrist, and Mrs E. J. Snell.

G. B. W.

University of Kent at Canterbury

Preface to the Third Edition

I am grateful for the kind reception the book has had, and I am indebted to a number of people for helpful comments. The most substantial revision is to Appendix One, which has been carried out by my colleague, Dr E. Eryl Bassett. The BASIC programs given there should assist in using the book. Programs, some very substantial, have been written in other languages in the Statistics Department of the University of Kent, and in particular in APL, and readers should write to us if they are interested.

G . B . W .

Contents

Preface	*page vii*
Preface to the Third Edition	*viii*
Acknowledgements	*xiii*

1 Introduction

1.1 Examples of random variation	15
1.2 One-dimensional frequency distributions	19
1.3 Summarizing quantities	26
1.4 Frequency distributions in two or more dimensions	33
1.5 Some illustrative examples	38
1.6 Populations, samples and probability	44

2 Probability and Probability Distributions

2.1 Probability	47
2.2 Addition law of probability	48
2.3 Conditional probability and statistical independence	51
2.4 Examples	55
2.5 Discrete random variables	58
2.6 Continuous random variables	63
2.7 Several random variables	75

3 Expectation and its Applications

3.1 Expectation	79
3.2 Variance	84
3.3 Higher moments	87
3.4 Dependence and covariance	90
3.5 Normal models	93

4 Sampling Distributions and Statistical Inference

4.1 Statistical inference	101
4.2 Pseudo random deviates	105

4.3 A sampling experiment   110
4.4 Estimation   117
4.5 Significance tests   124

5 Single Sample Problems

5.1 Introduction   129
5.2 Point estimates of μ and σ^2 130
5.3 Interval estimates for μ (σ^2 unknown)   133
5.4 Interval estimates for σ^2 136
5.5 Significance test for a mean   139
5.6 Significance test for a variance   141
5.7 Departures from assumptions   143

6 Two Sample Problems

6.1 Introduction   146
6.2 The comparison of two independent sample means   147
6.3 The comparison of two independent sample variances   150
6.4 Analysis of paired samples   152
6.5 An example   155
6.6 Departures from assumptions   160

7 Non-parametric Tests

7.1 Introduction   162
7.2 Normal approximation to the binomial distribution   162
7.3 The sign test   165
7.4 The signed rank (Wilcoxon one sample) test   169
7.5 Two sample rank (Wilcoxon) test   173
7.6 Discussion   175

8 The Analysis of Discrete Data

8.1 Introduction   177
8.2 Distributions and approximations   178
8.3 Inference about a single Poisson mean   182
8.4 Inference about a single binomial probability   185
8.5 The comparison of two Poisson variates   186
8.6 The comparison of two binomial variates   188
8.7 Comparison of proportions in matched pairs   193
8.8 Examination of Poisson frequency table   195
8.9 Examination of binomial frequency tables   198
8.10 Comparison of observed and expected frequencies   201
8.11 Contingency tables   206
8.12 A tasting experiment   210

9 Statistical Models and Least Squares

9.1 General points   216
9.2 An example   218
9.3 Least squares   221

10 Linear Regression

10.1 Introduction   224
10.2 Least squares estimates   227
10.3 Properties of $\hat{\alpha}$ and $\hat{\beta}$ 230
10.4 Predictions from regressions   235
10.5 Comparison of two regression lines   236
10.6 Equally spaced x-values   238
10.7 Use of residuals   239
10.8 Discussion of models   243

11 Multiple Regression

11.1 Introduction   245
11.2 Theory for two explanatory variables only   249
11.3 Analysis of Example 11.2   253
11.4 Discussion   259

12 Analysis of Variance

12.1 The problem   263
12.2 Theory of one-way analysis of variance   267
12.3 Procedure for analysis   270
12.4 Two-way analysis of variance   276
12.5 Linear contrasts   280
12.6 Randomized blocks   285
12.7 Components of variance   289
12.8 Departures from assumptions   294

Miscellaneous Exercises   298

Appendix One   Notes on calculation and computing   307

Appendix Two   Statistical tables   324

Appendix Three   Hints to the solution of selected exercises   331

References   348

Author Index   352

Subject Index   354

Acknowledgements

I am indebted to the following sources and persons for permission to publish:

To the American Society for Testing Materials for the data used in Example 3.15. This was selected from data published in 'The Effect of Surface Preparation and Roughness on Rockwell Hardness Readings', prepared by a Task Group of Subcommittee VI of ASTM Committee E-1 on Methods of Testing, *Materials Research and Standards*, Vol. 4, No. 12, December 1964.

To Bell Telephone Laboratories, Murray Hill, N.J., for the data used in Example 1.8.

To Dr N. T. Gridgeman of the National Research Council, Ottowa, Canada, for the data used in Example 8.12, Table 8.15.

To Imperial Chemical Industries Ltd, Mond Division, for the data used in Example 11.2.

To the Cambridge University Press, the Clarendon Press, Oxford and to the University of London for permission to publish various Oxford, Cambridge and London University examination questions. Of course, the responsibility for any discussion or answers to these questions is the author's.

To the Biometrika Trustees, for permission to publish extracts from *Biometrika Tables for Statisticians*.

G. B. W.

CHAPTER ONE

Introduction

1.1 Examples of random variation

This book is concerned with situations in which there is appreciable random variation. Such situations occur in many fields and we begin with a few examples.

Example 1.1. *Testing of seeds.* In the routine testing of consignments of seed a procedure similar to the following is used. From a consignment, 100 seeds are selected by a well-defined method of sampling and kept under carefully standardized conditions favourable to germination. After 14 days the number of seeds that germinate is counted. The resulting observation will be one of the integers 0, 1, . . ., 100. □ □ □

The whole procedure can be repeated many times. For each repetition a fresh selection of seeds is made from the same consignment, and the conditions of testing are kept as constant as possible. It is a well-established fact that repeat observations vary haphazardly within quite wide limits, and Table 1.1 gives some typical results. The smallest observation is 84 and the largest is 95. A good summary of the variation is given by a table of the observed frequencies for all possible outcomes, and this is called a *frequency table*. A frequency table for the 80 trials recorded in Table 1.1 is shown in Table 1.2, and presented graphically in Figure 1.1. Thus the percentage frequency of 10 for the value 89 simply means that the value 89 is observed 8 times in the 80

Table 1.1 *A series of* 80 *repetitions of Example* 1.1
Data record the number of seed germinating

86	95	92	89	92	92	91	88	88	94	90	88	93	92
89	90	86	88	93	92	91	87	91	94	92	94	92	92
91	92	88	94	93	91	92	91	92	90	89	86	93	91
89	90	90	89	84	89	88	93	92	95	86	93	93	94
91	90	91	89	89	95	92	94	94	90	86	93	88	93
86	94	90	87	93	91	90	92	93	94				

Table 1.2   *Frequency table for the results shown in Table 1.1*

No. germinating	84	85	86	87	88	89	90
Frequency	1	0	6	2	7	8	9
Relative frequency	1·3	0	7·5	2·5	8·7	10·0	11·3

No. germinating	91	92	93	94	95	Total
Frequency	10	14	11	9	3	80
Relative frequency	12·5	17·5	13·7	11·3	3·7	100%

trials recorded in Table 1.1. The frequencies vary reasonably smoothly. It is found in practice that the larger the number of observations, the more nearly does the relative frequency diagram analogous to Figure 1.1 approach a stable set of values which vary smoothly. It is this tendency for the frequencies to behave in an orderly way that will enable us to make statements about the probability that the number of seeds germinating is such and such, even though we cannot say what any particular observation will turn out to be.

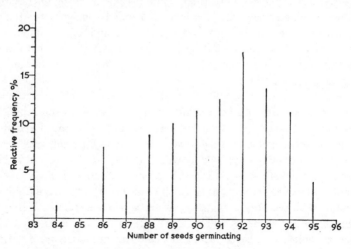

Figure 1.1 *Relative frequency diagram for Example 1.1*

It is convenient to introduce some general terms. *A trial* is the making of one observation. The resulting observation is sometimes called the outcome of the trial. The set of all possible values that the observation could conceivably take is called the *set of possible outcomes*. In the present example the set of possible outcomes is the set of integers 0, 1, . . ., 100. If we think of a geometrical representation of the set of possible outcomes, we have a *sample space*; examples will be given later. The whole system will be called a system with random variation, or more briefly, a *random system*. The main things determining this particular random system are the consignment of seeds, the method of sampling and the conditions of testing. If these change in an important respect the random system will be changed.

Example 1.2. *Counting radioactive emissions*. Suppose that a trial consists in counting the number of emissions from a stable radioactive source in a 5-second period. Again, the observation can be repeated many times and again very variable results will be obtained. Thus a typical short run of observations is '5,'8, 3, 5, 9, 6, . . .'. Here the set of possible outcomes is the set of all non-negative integers 0, 1, 2, etc. A frequency table for a series of 200 observations is shown in Table 1.3. A frequency table preserves all the

Table 1.3 *Frequency table of the number of emissions from a radioactive source in a 5-second period*

No. of emissions	0	1	2	3	4	5	6	7
Frequency	0	2	4	13	17	29	24	25
Relative frequency	0·000	0·010	0·020	0·065	0·085	0·145	0·120	0·125

No. of emissions	8	9	10	11	12	13	14	Total
Frequency	31	27	13	5	5	4	1	200
Relative frequency	0·155	0·135	0·065	0·025	0·025	0·020	0·005	1·000

information on the magnitude of the outcomes, losing only the information about the order in which they occur.   □ □ □

These examples illustrate the following important points about random variation.

(*a*) We cannot predict the outcome of a particular trial.

(*b*) There is a long run regularity in that, to take Example 1.2, the proportion of outcomes greater than say 10 appears to tend to a limit as the number of trials increases.

(*c*) The series of results are *random* in that sub-series have the same properties. For example, if in Example 1.2 we select a sub-series containing (*i*) every other term, or (*ii*) every term following a 5, or any such similar rule, then a frequency table made from the sub-series appears to tend to the same limit as a frequency table made from the whole data as the number of observations increases.

These properties can be checked empirically from the data; see Exercise 1.1.1.

Example 1.3. *Production of plastic pipe*. In part of an experiment by Gill concerned with the strength of plastic conduit pipe used for domestic water supply, successive 8-ft lengths produced by the extruder were cut, measured for length, and weighed. The weight per unit length gave a series of observations such as the following: 1·377, 1·365, 1·330, 1·319, 1·300, 1·290, 1·310, 1·295, 1·319, 1·305, 1·302, 1·324, 1·326, 1·292, 1·326, 1·326, 1·321, 1·318. In this experiment the outcomes are not restricted to integers; and in principle the observations could be recorded to as many decimal places as we like to name, although in practice we may be content with two or three decimal places. Thus the observations in this experiment give rise to a continuous set of possible outcomes, as do many observations made in the quantitative sciences and technologies.   □ □ □

In Example 1.3 a record was also kept of the thickness of each 8-ft length of pipe, and measurements were made which described the strength of the pieces of pipe. In this example, therefore, the outcome is not merely a single number but a triplet describing the weight per unit length, the thickness and the strength of each 8-ft length of pipe. The observation made on each trial

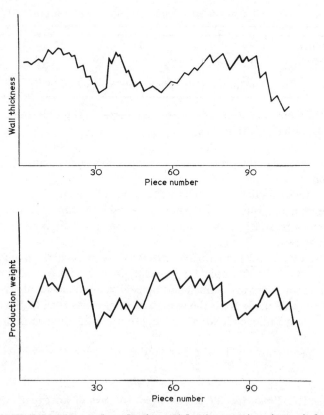

Figure 1.2 *Wall thickness and production weight of successive pieces of plastic pipe*

is said to be *multivariate*. The situation is further complicated by the presence of trends in the characteristics of the pipe, which are superimposed on the random variation. This can be seen clearly from Figure 1.2, which shows traces of the wall thickness and of the weight per unit length of successive pieces of pipe produced on one extruder from the same batch of material. The data shown in Exercise 1.3 demonstrate the presence of local random variation but this does not overrule the presence of systematic trends. An analysis of the data might proceed by attempting to separate the local random variation and the systematic trends. In this example, therefore, it would be inadvisable to pool the data to get an overall picture of the random variation.

The traces shown in Figure 1.2 are examples of what are called time series, and are composed of systematic trends, cyclic fluctuations and random variation. Methods of analysing time series will not be discussed in this book and the reader is referred to treatises such as Kendall and Stuart (1966). However, graphs such as Figure 1.2 can easily be made to check whether such effects are present.

These examples of random variation will suffice for the present. The task of the statistician is the interpretation of observations which contain appreciable random variation, and this leads to certain implications for experimental design. In the remainder of this chapter we consider some simple situations, and show how graphical and tabular methods can be used in the analysis.

Exercises 1.1

1. Obtain the frequencies for the data of Table 1.1. and hence check Figure 1.1. Form and compare similar tables using (a) the first 20 observations only, and (b) the first 50 observations only.

2. Using the calculations of Exercise 1.1.1, plot diagrams similar to Figure 1.1, (a) plotting all frequencies separately, (b) combining frequencies for pairs of cells, 84–85, 86–87, 88–89, . . ., 94–95, and plotting only one vertical 'spoke' for each of the six groups, (c) combining frequencies of four successive cells, 84–87, 88–91, 92–95.

1.2 One-dimensional frequency distributions

If in Example 1.1 the order of the results is totally irrelevant, then Table 1.2 or Figure 1.1 summarizes all the information in the data. The results of Exercise 1.1.1 show clearly that the relative frequencies of the outcomes fluctuate wildly until a sufficient number of trials has been conducted. The aim of computing a frequency table, whether presented in tabular or graphical form, is to exhibit as much as possible of the shape of the underlying frequency distribution, without at the same time preserving merely random fluctuations in the relative frequencies. Exercise 1.1.2 shows that if only a small number of trials have been made, a better picture of the distribution is obtained by combining cells, but this example also shows that combining cells unnecessarily can mask information available in the data. Therefore we have to strike a balance between a desire to keep the outcomes distinct so as to exhibit more features of the distribution, and a desire to combine outcomes so as to obtain more stable relative frequencies. In practice it will often be necessary to find the best grouping interval by trial and error. The tail of a frequency distribution will nearly always be ragged, unless there are physical restrictions on the range of the outcomes, and it will often be better to combine frequencies for tail outcomes, unless the tail is of special interest. Thus in Table 1.2, it might be better to group outcomes 84–87 together. Very little is gained by presenting a ragged tail.

In order to treat continuous data similarly we must divide the range into intervals or groups, and count the number of observations in each group. This involves loss of information concerning where an observation came in its group, but this will not seriously matter if a sensible grouping has been chosen. Consider the following example.

Example 1.4. *Measuring blood pressure.* Suppose that we are interested in the systolic blood pressure of normal British males, say in the age range 30 to 60 years. We first specify as precisely as possible what we mean by normal. This will mean excluding men with certain diseases. A trial consists of selecting a normal man and measuring his systolic blood pressure by a standardized technique. The procedure used to select men for measurement is important and will be discussed briefly when we deal with methods of sampling. A typical set of results recorded to one decimal place is shown in Table 1.4.

□ □ □

The procedure for calculating a frequency distribution for this data and plotting it, is illustrated in Table 1.5. The steps are:

(*i*) Decide upon a grouping interval to try. This will be discussed further below, but we have to take the same kind of considerations into account as in deciding whether to combine outcomes in the discrete case. It depends on the range of the variation, and the number of observations but it is better to try first an interval which is too fine, since frequencies are readily combined, if a coarser interval turns out to be necessary.

(*ii*) Uniquely define the groups. That is, something like 1·0–1·5, 1·5–2·0, . . ., is to be avoided because of the doubt about which cell 1·50 goes into.

(*iii*) Count the number of observations in each cell as indicated below, and calculate the relative frequencies.

Table 1.4 *Measurements of systolic blood pressure on 70 normal British males*
Each observation is the average of several readings. The units are mm of mercury.*

151·7	123·4	120·6	123·8	96·3
110·9	115·7	117·6	112·3	127·9
120·0	122·8	121·9	117·0	123·3
110·1	136·6	108·6	124·6	146·8
110·6	115·7	155·5	130·1	161·8
116·2	121·1	120·8	117·6	106·4
99·0	137·2	133·7	127·9	121·7
131·3	110·4	109·6	138·5	120·5
143·7	163·1	186·3	145·1	130·5
148·5	105·7	116·0	153·7	168·4
123·2	118·6	133·9	136·7	120·0
138·6	113·8	107·3	118·8	152·0
153·5	177·8	125·3	132·1	144·2
123·6	117·5	128·0	101·5	147·7

* Fictitious data, based on a survey by Armitage *et al.*, 1966.

(*iv*) When plotting the data, plot (relative frequency) divided by (width of grouping interval). This is to ensure that the diagram has a consistent meaning in terms of area; see below.

Table 1.5 *Calculations for a histogram of Table* 1.4 *data*

95·0– 99·9	11	2
100·0–104·9	1	1
105·0–109·9	╫╫	5
110·0–114·9	╫╫ 1	6
115·0–119·9	╫╫ ╫╫	10
120·0–124·9	╫╫ ╫╫ ╫╫	15
125·0–129·9	1111	4
130·0–134·9	╫╫ 1	6
135·0–139·9	╫╫	5
140·0–144·9	11	2
145·0–149·9	1111	4
150·0–154·9	1111	4
155·0–159·9	1	1
160·0–164·9	11	2
165·0–169·9	1	1
170·0–174·9		
175·0–179·9	1	1
180·0–184·9		
185·0–189·9	1	1
	Total	70

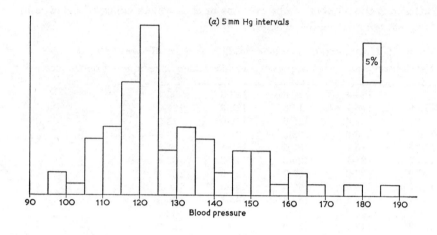

(a) 5 mm Hg intervals

5%

Blood pressure

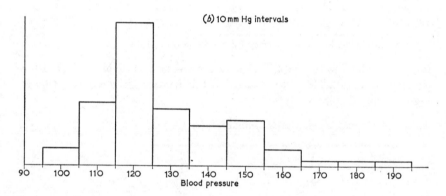

(b) 10 mm Hg intervals

Blood pressure

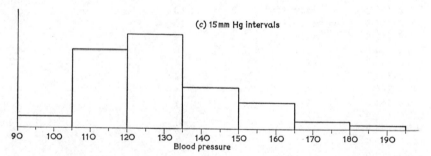

(c) 15 mm Hg intervals

Blood pressure

Figure 1.3 *Histograms of Table* 1.4 *data, for three different grouping intervals*

The diagrams shown in Figure 1.3 are called *histograms*. The interpretation of a histogram is that the areas of the cells above each grouping interval are equal to the relative frequencies with which observations in each particular group occur. More generally, if x and y are the lowest points of any two intervals then the area under the histogram between x and y is the percentage frequency with which results fall between x and y. The total area under the histogram is unity, or 100%.

If all cells of a frequency table have the same width, then it is not necessary to divide by the width of the interval before plotting the histogram, but it is nevertheless advisable to do so, so that histograms constructed on different grouping intervals will correspond. However, suppose we calculate a histogram for Table 1.4 data as in Table 1.5, but combining the tail frequencies before plotting, then Figure 1.4 shows what happens if we do not divide

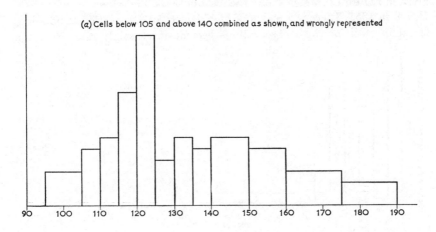

(a) Cells below 105 and above 140 combined as shown, and wrongly represented

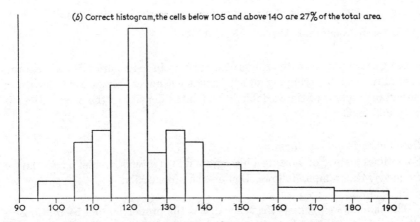

(b) Correct histogram, the cells below 105 and above 140 are 27% of the total area

Figure 1.4 *The effect of not dividing by width of cell on a histogram*

before drawing the histogram. In Figure 1.4(*a*), the importance of the top
and bottom cells is made to appear disproportionately large, since they make
up only 27% of the observations together. Figure 1.4(*b*) is the correct graph,
and preserves the interpretation in terms of area described above, and the
top and bottom cells make together the correct proportion 27% of the total
area under the histogram.

The choice of the grouping interval is basically a matter of judgment. It
depends on the number and range of the observations, the shape of the distribution, and the parts of it which are of interest. However, with some practice,
the reader will have little difficulty in choosing a suitable grouping interval.
Figure 1.4(*b*) appears to be the best histogram for presenting Table 1.4 data.

Frequently one grouping interval is satisfactory for most, if not all, of one
histogram, but there are some cases where this is not so. An illustration is
given in Figure 1.5, which shows the distribution of individual income in the
U.S.A. in 1957. There is a region of high density of observations, and a long
straggling tail in which observations become increasingly rare. No single

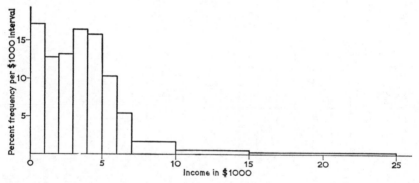

Figure 1.5 *Money income of persons with income, males over 14 years*

(Data taken from Stats. Abs. of U.S.A., 1957)

spacing could be used without loss of sensitivity in some part of the histogram.
A need for variable grouping interval arises whenever there are extensive and
important regions of the variable in which the density of the observations is
very different.

Cumulative frequency diagrams

Sometimes instead of plotting a histogram it is of interest to plot a cumulative
frequency diagram, as follows. Denote the observations $x_1, \ldots, x_n$, then we
define cumulative frequency:

The *cumulative* (*relative*) *frequency* of x is the proportion of observations in
the set $x_1, \ldots, x_n$ which are less than or equal to x.

Suppose we have the following data, 1·6, 1·9, 0·8, 0·6, 0·1, 0·2, then the first step in obtaining the cumulative frequency diagram is to put the observations in order, and we then have

$$0·1, 0·2, 0·6, 0·8, 1·6, 1·9.$$

For $x < 0·1$, the proportion of observations less than or equal to x is zero. For $0·1 \leqslant x < 0·2$, the proportion of observations less than or equal to x is $1/6$, etc.

In general, when we have observations $x_1, \ldots, x_n$, our first step is to put them in order. The ordered observations are usually denoted $x_{(i)}$, so that $x_{(1)} \leqslant x_{(2)} \leqslant \ldots \leqslant x_{(n)}$, and these correspond to the set $0·1, 0·2, \ldots, 1·9$, given above.

Then the cumulative (relative) frequency is 0 for $x < x_{(1)}$, $1/n$ for $x_{(1)} \leqslant x < x_{(2)}$, $2/n$ for $x_{(2)} \leqslant x < x_{(3)}$, etc., and 1·0 for $x \geqslant x_{(n)}$. Thus cumulative frequency is a step function, which rises from 0 to 1·0 with equal steps of $1/n$ if all the observations are distinct, and steps which are multiples of $1/n$ when several observations are equal. When this step function is plotted out we call it a *cumulative frequency diagram*. Such a diagram for the radioactive emissions experiment is shown in Figure 1.6.

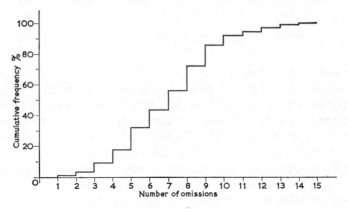

Figure 1.6 *Cumulative frequency diagram for the radioactive emissions experiment*

Unless a computer is available, it may be very laborious to calculate and plot the exact cumulative frequency diagram in any particular case. It is often sufficient to calculate the cumulative frequencies at a number of suitable points x, such as a set of equally spaced points. Thus the discussion given above on grouping intervals for histograms applies here for the intervals used in plotting a cumulative frequency diagram.

There are two situations in which a cumulative frequency diagram is of interest. Firstly when we are interested in quantities such as the proportion of samples below a certain standard. The seed germination experiment,

Example 1.1, comes into this category. A cumulative frequency diagram for this data would plot the proportion of samples of 100 seeds in which the number germinating was less than or equal to x, for $0 \leqslant x \leqslant 100$. The second situation requiring a cumulative frequency diagram is when we want to check that the distribution follows some particular mathematical form. For any specified mathematical form for the distribution. the percentage frequency scale can be transformed so that a cumulative frequency diagram should be exactly a straight line, if the specified form is exactly followed. This will be discussed again later, see § 2.6(v).

Exercises 1.2

1. Describe a random system with which you are familiar. Obtain a number of observations, and form them into a frequency histogram. (Suitable examples can be obtained from experimental work, from the weather, from sport, etc. Where possible, it is best for the reader to generate his own data by running simple experiments of the type described in Jowett and Davies (1960).)

1.3 Summarizing quantities

A need for summarizing quantities arises in three situations. Firstly, the comparison of two or more different frequency distributions is facilitated if the important features of the frequency distributions can be summarized in a few numbers or *statistics*. Secondly, frequency distributions are useful only if large numbers of observations are available. With smaller samples we shall have to be satisfied with much cruder information about the underlying frequency distribution. Thirdly, the summarizing statistics sometimes have a practical meaning and are of interest in themselves.

One way of describing a frequency distribution is to use the *percentiles*. The Pth percentile is the value which is exceeded by $P\%$ of the observations. Thus the 50th percentile is the value which is exceeded by half of the observations; this is of especial interest and is called the *median*. The use of percentiles is illustrated in Table 1.6, which is taken from Lydall (1959). This summarizes the distribution of British income for four separate years. We notice that the 1st percentile increased from 1140 in 1938 to 2450 in 1957, while the median increased from about 110 to 520.

The distribution of income is not a symmetrical shape and is said to be *skew*. It has a long tail to the right. The tendency over the years 1938 to 1957 has been for the tail to become shortened and the distribution of incomes

Table 1.6 *Percentiles of British income before tax* (in £)*

Percentile	1938	1949	1954	1957
1st	1140	1860	2210	2450
5th	393	765	995	1180
10th	266	565	795	940
20th	185	430	635	792
50th	110	261	382	512

* From Lydall, 1959, *J. R. Statist. Soc. A*, **122**, 7.

more closely bunched. This example illustrates the first reason given above for the need for summarizing statistics.

Measures of location

An obvious and very important summarizing statistic is the *mean*. This is simply the average of all the observations. If this is calculated from a frequency table then we merely add together relative frequency times centre of group interval for all the groups. There is some error involved in using this formula because of the grouping of the observations. See Appendix I for details of calculations.

The mean is a useful measure of location of a frequency distribution for frequency distributions which are unimodal and bell-shaped. For bimodal

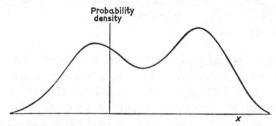

Figure 1.7 *A bimodal distribution*

distributions, Figure 1.7, or highly skew distributions, Figure 1.8, the mean is of less interest unless it has a practical meaning in terms of the problem being discussed.

An important point which bears on our choice of summarizing statistics is that practical data often contain stray or erroneous observations. If we suspect the presence of wild observations then instead of using the mean we may prefer to use one of the following:

The *P% truncated mean* is obtained by rejecting the highest and lowest $(P/2)\%$ of the observations and taking the average of the remainder.

The *P% Winsorized mean* is obtained as follows. We reject the $(P/2)\%$ highest observations and count them as if they had occurred at the highest observation still included. We now do the same thing for the $(P/2)\%$ lowest observations, and then average the result.

There are many other statistics which could be used for measuring the location of a distribution. For example, we could use the sample median or we could use the geometric mean of the observations, etc. The choice between these alternatives depends to some extent on what we know about the shape

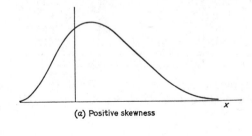

(a) Positive skewness

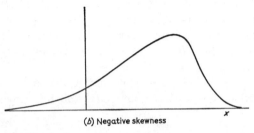

(b) Negative skewness

Figure 1.8 *Skew distributions*

of the underlying frequency distribution and also on whether wild observations are likely or not. A discussion of the choice between alternative statistics will be postponed until Chapter 4. Here we remark that the sample mean is one of the commonest measures of location used, although perhaps the truncated or Winsorized means ought to be used much more commonly than they are.

Measures of dispersion

We have now discussed possible measures of location for a frequency distribution and we turn next to consider possible measures of spread or dispersion The two discussed below are in common use.

(i) The *range* of a set of observations is the difference between the largest and the smallest observation. Clearly, the range of a sample of n independent observations increases with n, and this is particularly important at small n. The range can therefore be used most readily to compare the dispersion in samples of the same size. If the samples to be compared are of different sizes, then it is necessary to make some assumptions about the underlying frequency distribution and adjust for n; see Biometrika Tables, Pearson and Hartley (1966), Tables 20, 22, 23 and 27. It is intuitively clear that except in small samples, the range uses only part of the information about dispersion available in a sample; this point will be discussed in greater detail in Chapter 4. In practice the range is often used in samples of size 5–10, and if the sample size is larger, then either a different measure of dispersion is used or, less frequently, the sample is split up into groups of size 5–10 by some random procedure, and the average of the separate ranges for each group taken. The range is used commonly in industrial quality control, as illustrated in **Example 1.5** below.

(*ii*) The commonest measure of spread is the *standard deviation*. This is the square root of the *sample variance*, and is calculated from the formula

$$\text{sample variance} = s^2 = \frac{1}{n-1} \sum_{i=1}^{n} (x_i - \bar{x})^2$$

$$= \frac{1}{n-1} \left\{ \sum x_i^2 - \left(\sum x_i \right)^2 / n \right\} \qquad (1.1)$$

$$\text{standard deviation} = \sqrt{\{\text{sample variance}\}},$$

where $x_1, \ldots, x_n$ are the observations and $\bar{x} = \sum x_i / n$ is the sample mean. The terms inside the brackets of (1.1) are the sum of squared deviations from the sample mean, and although there are n observations, there are only $(n-1)$ independent deviations from the mean. Partly for this reason the divisor outside the brackets is $(n-1)$, and not n as might have been expected: a better justification of this divisor will be given later. The sample variance, or standard deviation, is the simplest measure of dispersion to manipulate mathematically, and has certain other theoretical advantages to be discussed later.

The calculation of the variance can often be made easier by noticing that if the observations $x_1, \ldots, x_n$ are transformed by

$$y = a + bx$$

for any constants a and b, then

$$s^2 = \frac{1}{n-1} \left\{ \sum x_i^2 - \left(\sum x_i \right)^2 / n \right\} = \frac{1}{b^2} \frac{1}{n-1} \left\{ \sum y_i^2 - \left(\sum y_i \right)^2 / n \right\}$$

That is,

$$\text{sample variance of } y\text{'s} = b^2 \times (\text{sample variance of } x\text{'s}). \qquad (1.2)$$

Therefore for the calculation of the variance, we can choose any convenient origin to reduce the size of the numbers, and this has no effect on the answer. If further, we remove fractions, etc., by multiplying up the x's by a constant b, then the sample variance will be multiplied by b^2. Computational routines for calculating the sample variance are described in Appendix I.

There are many other possibilities for measures of dispersion, such as the mean absolute deviation from the sample mean, the sample variance of a truncated or Winsorized sample, and the interquartile range, which is the difference between the 25th and 75th percentiles. However, the sample variance is the most commonly used statistic, except where it would be inconvenient or troublesome to calculate (1.1), in which case the range is used. All summarizing statistics suffer an inherent difficulty in that they highlight some features of a frequency distribution and ignore others. For

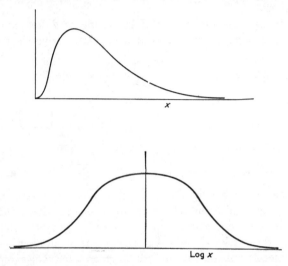

Figure 1.9 *Transformations to symmetry*

example the range ignores the shape of the middle of a distribution, the interquartile range ignores the tails, etc.

Throughout this discussion of summarizing statistics there has been the implicit assumption that the observations are a homogeneous set, and not, for example, a mixture of two or more quite different distributions. Further, most summarizing statistics are more meaningful when the frequency distribution is nearly symmetrical and unimodal except in special cases. If we have a unimodal distribution which is not symmetrical, then often a very simple transformation of the variable will give near symmetry. An example using the logarithmic transformation is shown in Figure 1.9: other simple transformations used are $\sqrt{x}$, x^2 and $1/x$.

An example

We close this section with an illustration of the use of summarizing statistics.

Example 1.5. *Quality control inspection.* A factory mass produces valves, and at the packing stage five valves are drawn at regular intervals from the production line. The valves selected are inspected for about a dozen variables, and some typical results for one variable are shown in Table 1.7 (unspecified units). If the process is working correctly, the mean measurement should all be close to 2·00 and the ranges fairly small. If either the means deviate too much from 2·00, or the ranges are too large, this will be an indication that

Table 1.7 *Typical results from quality control inspection*

| | Sample no. | | | | | |
	1	2	3	4	5	6
	2·4	1·4	1·0	0·6	1·6	1·7
	1·6	2·1	1·3	1·9	1·2	3·6
	3·4	1·9	1·4	4·1	1·4	2·9
	1·2	1·2	3·1	1·6	3·1	1·6
	2·4	2·1	1·2	1·3	1·9	1·7
Average	2·20	1·74	1·60	1·90	1·84	2·30
Range	2·2	0·9	2·1	3·5	1·9	2·0

| | Sample no. | | | | | |
	7	8	9	10	11	12
	0·3	1·1	1·4	2·2	0·5	1·4
	1·7	1·4	1·2	2·1	1·2	3·1
	2·7	1·1	4·9	2·4	1·9	1·4
	2·4	2·4	4·6	8·4	1·2	1·1
	0·9	3·4	2·9	2·0	1·9	1·9
Average	1·60	1·88	3·00	3·42	1·34	1·78
Range	2·4	2·3	3·7	6·4	1·4	2·0

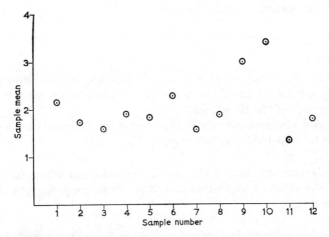

Figure 1.10 *X-chart for Table 1.7 data*

something has gone wrong with the process, and corrective action must be taken. One standard procedure for detecting these conditions is to plot successive means and ranges as in Figures 1.10 and 1.11. By looking at these

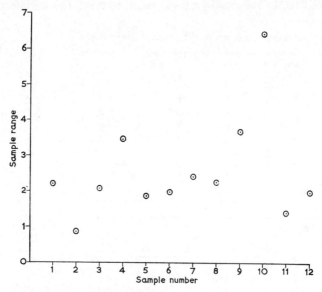

Figure 1.11 *Range chart for Table 1.7 data*

charts, it is possible to see at a glance whether a set of results indicates the need for corrective action or not. Usually, warning lines are drawn in on these charts, placed so that points will only rarely be beyond them if the process is working correctly. The theory for placing the warning lines will be discussed in Exercise 6.2.2.   □ □ □

Exercises 1.3

1. Calculate the sample mean and sample variance for the blood pressure data, Example 1.4, (*i*) directly from the data, (*ii*) from the grouped frequency tables, for each of the three group sizes.

Under what circumstances would the calculation of $\bar{x}$ and s^2 from a grouped frequency table be liable to lead to appreciable error?

2. What dangers are there in the use of truncated and Winsorized means? Consider the effect of truncation and Winsorizing frequency distributions which are skew.

3. With reference to quality control inspection, Example 1.5, a plot which is

more sensitive to slight changes of the mean from the target value is the *cumulative sum chart* (CUSUM chart). This is similar to Figure 1.10 except that the quantity plotted vertically is not the mean, but the sum of (observation — target value) over all observations. Plot a CUSUM chart for Example 1.5 data, assuming 2.0 as the target value.

1.4 Frequency distributions in two or more dimensions

Relationships between two variables
The concept of a frequency distribution is readily extended to situations where we are considering two or more variables at a time. For example, in Example 1.2 one very simple and obvious way of checking the independence between the number of emissions from the radioactive source in successive 5-second periods is as follows. Suppose the source is observed for a series of pairs of 5-second periods, the 5-second periods within each pair being successive, but the pairs being separated by a short interval. The data can be set out as in Table 1.8.

Table 1.8 *Numbers of emissions from a radioactive source in pairs of 5-second periods*

		No. of emissions in 2nd period														Marginal total
		1	2	3	4	5	6	7	8	9	10	11	12	13	14	
No. of	1	–	–	–	–	–	–	1	–	–	–	–	–	–	–	1
emissions	2	–	–	1	–	–	–	–	–	–	–	–	–	–	–	1
in 1st	3	–	–	–	2	–	1	3	1	–	–	–	1	–	–	8
period	4	–	1	–	–	1	–	–	2	1	–	–	–	1	–	6
	5	–	–	–	1	2	3	4	5	2	1	–	1	–	–	19
	6	–	1	–	3	3	3	1	–	–	–	1	–	–	–	12
	7	–	–	1	–	–	2	1	2	1	1	–	–	–	–	8
	8	–	–	–	2	2	–	1	2	5	1	1	–	–	–	14
	9	1	1	1	2	1	2	2	1	2	2	–	–	–	–	15
	10	–	–	–	1	1	–	3	2	–	–	–	–	–	–	7
	11	–	–	1	–	–	–	1	1	–	–	–	–	–	–	3
	12	–	–	1	–	–	–	–	1	–	–	–	–	–	–	2
	13	–	–	–	–	–	1	–	–	1	1	–	–	–	–	3
	14	–	–	–	–	–	–	–	–	–	–	–	1	–	–	1
Marginal total		1	3	5	11	10	12	17	17	12	6	2	3	1	0	100

There were no zero emissions for any of the 200 5-second periods observed, so that the zero column and row are omitted from the table. The entry two in column four, row three, means that there were two pairs of 5-second periods in which there were three emissions in the first period and four in the second, and the other entries have a similar meaning. The marginal totals are simply totals over rows or columns; for example, the marginal entry eight in row three means that there were in all eight pairs of periods in which the number of emissions in the first period of each pair was exactly three. Clearly, the marginal totals add to 100, which is the number of pairs of 5-second periods observed. The marginal totals for rows defines what we call the *marginal frequency distribution* for rows, and similarly for columns. As we would expect, these two frequency distributions are very similar.

Table 1.9 shows the frequency distributions of the number of emissions

Table 1.9 *Frequency distributions for groups of rows of Table 1.8 data*

		No. of emissions in 2nd period															
---	---	0	1	2	3	4	5	6	7	8	9	10	11	12	13	14	*Total*
		frequency															
1st emission	0 *to* 5	0	0	1	1	3	3	4	8	8	3	1	0	2	1	0	35
	6, 7 *or* 8	0	0	1	1	5	5	5	3	4	6	2	2	0	0	0	34
	9 *or more*	0	1	1	3	3	2	3	6	5	3	3	0	1	0	0	31

in the second 5-second period for a given numbers of emissions in the first: this table shows no evidence of any change in the distribution of emissions in the second 5-second period, whatever the emission in the first. In this example therefore, all information is summarized in the marginal distributions, which also appear to be identical. This data shows up in sharp contrast to the next example.

Example 1.6. Pearson and Lee, see also Fisher 1958, reported data on the

Table 1.10 *Heights of fathers and daughters*[*]

	Fathers height in inches								
Daughters height in inches	58	60	62	64	66	68	70	72	74	76
52	O	O	1	O	O	O	O	O	O
54	O	O	O	1	O	O	O	O	O
56	1	2	7	4	3	1	O	O	O
58	2	3	13	14	23	6	2	O	O
60	2	6	36	68	75	40	11	2	O
62	1	7	31	87	135	97	43	4	O
64	O	3	7	56	76	124	74	15	4
66	O	O	2	17	36	80	62	20	2
68	O	O	1	1	5	13	20	11	5
70	O	O	O	O	O	3	6	4	O
72	O	O	O	O	O	1	O	O	O
74									

[*] Extracted from Fisher (1958) and slightly modified.

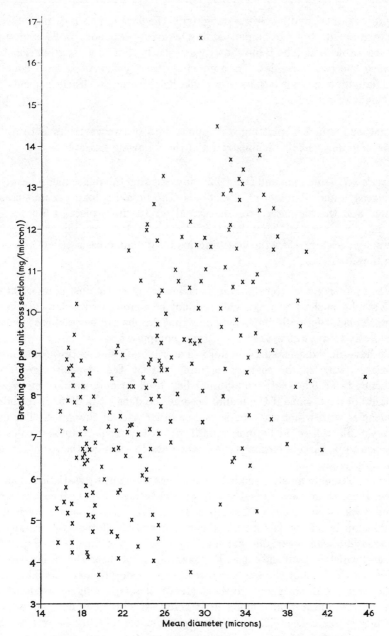

Figure 1.12 *Scatter diagram of breaking load and mean diameter for wool fibres*

heights of fathers and heights of daughters. The data have a continuous scale of measurement, but can be plotted in a bivariate frequency table by grouping, see Table 1.10. The table shows very clearly that there is a relationship between the two variables. The frequency distribution along each column – for daughters having fathers in a given height group – clearly depends on the height of the father.   □ □ □

Another method of plotting continuous data in two variables is to use a *scatter diagram*, and this is illustrated in the following example.

Example 1.7. Anderson and Cox (1950) investigated the relationship between the strength and diameter of wool fibres. The breaking load per unit cross-section, and the diameter, were measured for 185 fibre pieces 1·5 cm long. The data are shown in Figure 1.12, and despite the large scatter, the diagram shows very clearly that the breaking load per unit cross-section increases with fibre diameter.   □ □ □

When plotting a scatter diagram, it is better not to use just points, but to use a symbol such as $\times$ or $\odot$, which cannot be confused with stray marks or points. This is especially important now that graphs can be produced photographically from plots made by electronic computers.

We have now discussed two alternative ways of representing continuous bivariate data, by the scatter diagram, and by the bivariate (grouped) frequency table. The scatter diagram is best used for only small or moderate amounts of data, since if the points become too dense it is difficult to judge frequencies with which they lie in one or other part of the graph. The frequency table will often be more useful, if for any purpose further calculations are to be carried out on parts of the data to examine the form of relationship present.

Some comments must be made here on what is meant by statistical dependence. It must not be confused in any way with causal dependence, although in many problems, such as Examples 1.6 and 1.7, an observed statistical relationship is due in fact to an underlying causal relationship. However, there may be a third variable which is related to both of the observed quantities, and which is itself the cause. For example, it has been shown beyond all reasonable doubt that there is a *statistical* relationship between smoking and lung cancer, see Berkson and Elveback (1960); that is, among heavy smokers, more die of cancer than among light smokers. This does not prove that smoking causes cancer. The relationship could conceivably be due to the presence of a class of people of a nervous disposition, who, because of this, are prone to both cancer and cigarette smoking. This theory could, of course, be examined statistically, but one can never be sure that all such possibilities have been eliminated.

Relationships between three variables

For three variables, the methods used are just an extension of the methods used for two variables. If we have discrete data, or if we group the data in all three variables, we can construct a three-dimensional frequency table. This will simply be a series of tables like Table 1.10 – one for each grouping interval of the third variable.

For continuous data, a series of scatter diagrams can be produced if the data are grouped in one dimension, and a separate scatter diagram made for each grouping interval. This method is very suitable for use on electronic computers. The computer can be programmed to produce a filmstrip (or a series of slides) and then as the filmstrip is projected the change in the pattern of the points in the two projected dimensions will be seen. Thus time is used for plotting the third dimension. This method does not treat variables symmetrically of course, and it is probably better to use time for the least important variable.

In some circumstances the entire range of one variable can be represented by three or four groups without much loss of information. This will be possible whenever the change in the distribution of the points in this direction is small relative to the range covered. In these circumstances the data can be plotted in one scatter diagram by using different colours, or different kinds of point, such as:

An alternative method, due to E. A. Anderson (1954), is to use a circle with a prong of variable length, this 'prong' representing the value of the third variable. A simple extension of this idea is to have a number of prongs, each

for a different variable, such as Figure 1.13(*a*), with examples of points of this system in Figure 1.13(*b*). In this way, seven or eight variables can be plotted in one graph, provided that all but two of them can be reduced to a few classes.

Figure 1.13 *Anderson's method of plotting multivariate observations*

A good summary and discussion of graphical methods and their uses is given by Pearson (1956), and this paper includes a number of interesting examples which are discussed in some detail. For a manual on data analysis methods, including computer programs in FORTRAN and APL, see McNeil (1977).

Exercises 1.4

1. In studying a certain fatal disease, the following observations were obtained, where y is the logarithm of months to death, x_1 is an objective measure of blood properties, and x_2 is the doctor's subjective prognosis, ranked on a five-point scale (1 good; 5 very bad).

y	x_1	x_2	y	x_1	x_2
0·51	2·0	3	0·72	2·9	5
0·42	2·1	4	1·21	2·9	3
0·48	2·3	5	1·12	3·0	2
0·39	2·3	4	0·76	3·1	3
0·51	2·5	4	0·77	3·1	2
0·79	2·6	3	1·31	3·2	2
0·58	2·8	3	1·02	3·3	2
1·10	2·8	4	1·21	3·5	2
0·59	2·9	4	1·32	3·6	1

Plot a scatter diagram of the data, representing the three variables. Write a brief report describing your findings.

2. Calculate the sample means and variances for the three frequency distributions of Table 1.9. Comment on what further light this throws on any relationship between the numbers of emissions in pairs of 5-second periods.

1.5 Some illustrative examples
In § 1.1 we have defined the role of the statistician as the interpretation of experimental results in which there is random variation, and we have discussed some of the tools at his disposal. In this section we shall discuss the detailed results of experiments, and show how far we can progress in the interpretation of the results with the simple tools mentioned, and then consider the need for more advanced and more precise techniques.

Example 1.8. Snoke (1956) has described experiments for the bioassay of wood preservatives. Blocks of $\frac{3}{4}$-in cubes of wood were prepared, impregnated with a preservative, and then exposed to a particular strain of fungus for 90 days. After this the fungal growth (if any) was carefully brushed off and the weight loss noted. The amount of preservative in a block of wood is calculated on a scale called the retention, and it was not possible accurately to attain desired retention values. Some typical results are are shown in Table 1.11 on the following page.

Table 1.11 *Typical results for bioassay of wood preservatives*

| Group | | | | | | | | | | | |
| 1 | | 2 | | 3 | | 4 | | 5 | | 6 | |
R	W	R	W	R	W	R	W	R	W	R	W
1·99	19·95	3·01	2·50	4·16	1·61	5·04	2·01	5·86	0·22	7·06	−0·22
1·96	25·42	2·85	5·56	4·03	1·14	4·94	0·22	6·06	0·45	7·23	0·22
1·96	22·33	3·03	8·82	4·01	0·69	5·13	0·44	6·09	0	7·04	0
1·98	25·32	3·04	4·11	4·10	0·47	5·06	0·45	6·15	0·22	7·19	0·44
1·98	14·81	3·05	5·53	4·03	2·06	5·12	−0·22	6·04	0·44	7·25	−0·22
2·00	20·88	3·09	8·20	4·19	0·69	5·15	−0·22	6·06	0·22	6·96	0·44
1·97	9·05	2·98	0·69	4·09	0·23	4·97	0	6·04	0·67	7·15	0·22
1·85	11·49	3·05	7·67	4·07	0·91	4·93	1·11	6·22	0·23	7·27	0·44
2·01	12·30	3·09	7·60	4·12	2·28	4·99	0·91	5·88	0·22	7·00	0
2·02	9·31	3·04	7·05	4·01	0·69	4·96	0·67	5·86	0·45	7·21	−0·22

R = Retention level. W = % weight loss

The data exhibit rather a large amount of random variation, but in spite of this the main trend is clear. A scatter diagram of the results is shown in Figure 1.14. The relationship between percentage weight loss and retention

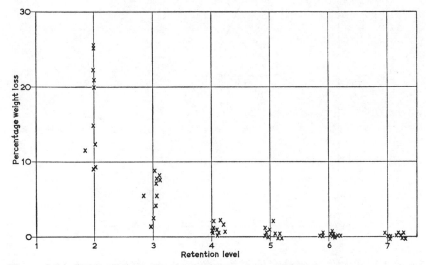

Figure 1.14 *Scatter diagram for wood preservative experiment*

appears to be approximately negative exponential, which suggests that we plot log (percentage weight loss). The extent of the random variation also shows a clear pattern, as can be seen either from Figure 1.14, or from the following table of the range of each group of results against the mean percentage weight loss for the group. This table has to be interpreted with care, since the retention levels in the groups are not identical. However, the range of results in any group seems to be approximately proportional to the mean percentage weight loss. A much simpler pattern emerges when log (percentage weight loss) is plotted as in Figure 1.15. The scatter of the points is now

Table 1.12 *The range and average % weight losses for each group of retention level*

Mean W	17·09	5·77	1·08	0·54	0·31	0·11
Range of W	16·37	8·13	2·05	2·23	0·67	0·66

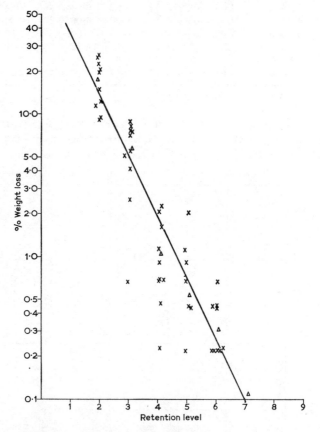

Figure 1.15 *Scatter diagram of log (% weight loss) and retention level.*
[Obs at retention level 7 not plotted because of negative observations].
× = individual points   △ = group means.

about the same in each group, and the mean percentage weight loss seems to fall on a straight line. The line shown in Figure 1.15 was drawn by eye, and is clearly satisfactory for most purposes. For example, from the line we can estimate the retention necessary to give any desired level of protection, such as the retention corresponding to a 1% weight loss. If we wish to give such an estimate a measure of error, or if we wish to compare the 1% retention levels for two different preservatives, some more elaborate theoretical structure is needed. The problem of comparison between characteristics of two fitted lines also occurs in the next example.   □ □ □

Example 1.9 (B.Sc. Special, London, 1958). Two pieces of the same type of apparatus, A and B, for measuring electrically the moisture content of tobacco were calibrated by dividing each of fifteen samples of tobacco into three sub-samples, and measuring the moisture content of each sub-sample by one of these methods: using A, using B, and using a direct chemical method. The results are given in Table 1.13. The instruments do not give measurements on a scale purporting to represent moisture directly, and it is required to calibrate them, so as to convert the scale readings to readings of true moisture content. It is also required to decide if one piece of apparatus is preferable for use.

Table 1.13 *Results of tobacco moisture content experiment*

Direct	A	B	Direct	A	B
11·0	12·0	10·1	20·2	43·1	46·7
11·1	12·1	13·5	19·1	38·2	38·3
7·2	7·5	8·5	25·0	69·0	64·8
8·3	8·0	9·6	10·2	11·8	12·0
12·4	16·0	16·8	13·3	20·0	17·5
14·7	24·5	23·6	23·6	57·6	55·2
5·1	5·0	4·9	12·0	15·0	14·8
21·7	47·9	47·8			

We shall denote the direct measurements in Example 1.9 as x, and the scale readings on A and B as y_A, y_B respectively. Calibration of the pieces of apparatus means estimating the relationship between the y's and x. It is therefore natural to begin by plotting the data, and Figure 1.16 shows the plot of y_A versus x. There appears to be a relationship of the form $y \propto x^2$ or $y \propto e^x$, so that we are led to try plotting $\sqrt{y}$ or log y against x, in the hope of obtaining a straight line. Figure 1.17 shows the plot of $\sqrt{y}$ against x, and an approximately linear relationship holds. The graph of $\sqrt{y}$ against x for machine B is left to the reader, see Exercise 1.5.2 (log y could be used instead of $\sqrt{y}$).

The relationships of $\sqrt{y}$ against x appear to be very nearly linear, and this is so clear that lines drawn by eye will differ little from lines fitted by much more advanced techniques. These lines can be used to calibrate the apparatuses, so that for any given value of y, the values of x are read from the lines.

The graphs of $\sqrt{y}$ against x are a very good summary of the data, for they

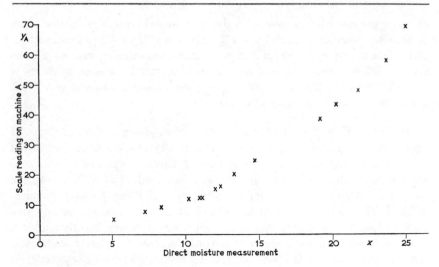

Figure 1.16 *Graph of the scale reading on apparatus* A, y_A, *against the direct chemical measurement of moisture, x*

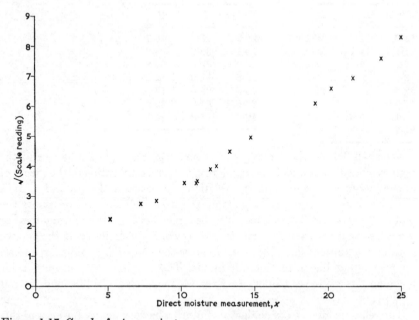

Figure 1.17 *Graph of* $\sqrt{y_A}$ *against x*

clearly illustrate the main information contained in it, and the scatter of the points about a fitted line is a measure of the error of estimation when using the particular apparatus. It would appear from the graphs that apparatus A has rather less scatter than apparatus B, though in fact this impression is mainly due to two wild observations on B. It could be that such observations were in error due to some extraneous factor, and the particular observations should be checked back for this. Apart from these observations, methods A and B appear to work equally well, and we might conclude that there may be some slight evidence that apparatus A is preferable.

To proceed further and compare the two sets of apparatuses, we could examine the frequency tables of the deviations in the y direction of the points, from the lines. However, these tables would doubtless show up the rather poor method of fitting the lines – by eye – and would not provide safe grounds for a comparison. Thus again, although the main conclusions can be drawn from the data by very simple graphical or tabular devices, any questions involving *precision* lead to a need for more elaborate techniques. □ □ □

Example 1.10. Cameron (1951) has discussed the sampling of a shipment of wool, the shipment being made up of a large number of bales. The price paid for the shipment, and the amount of duty payable, depends on the average clean content, that is on the ratio of the weight after cleaning to the raw weight. It is impracticable to determine this quantity for the entire shipment at the dockside, so that sampling must be employed. The design of the sampling plan is complicated by the presence of two different sources of variation. Firstly, the average clean content of the bales will vary from bale to bale; this variation may in part be due to the bales being drawn from a large number of farms over a large area. Secondly, the average clean content of small amounts of wool taken from any bale will vary about the average for the bale; this variation may in part be due to the variation in the amounts of dirt on different sheep, from one part to another of the same sheep, and different sheep of the same flock.

The basic form of the sampling plan was as follows: select k bales from the shipment in such a way that all bales are equally likely to be chosen, and then select n cores of wool (drawn from a tube) from each bale chosen. Clearly, a given amount of effort could be put to selecting say, 50 bales and 2 cores from each, or 25 bales and 4 cores from each, etc. It can be shown that the precision* of the final estimate of average clean content often varies very greatly in accordance with the particular sampling plan chosen, and this depends on the amount of variation in the two sources mentioned above. Therefore a pilot sample of 7 bales, and 4 cores from each were selected, in order to assess the amount of variation in average clean content between and within bales.

This information was then used to design a better sampling plan for the main sample. The data are given in Table 1.14.

* Not yet explicitly defined. See Chapter 4.

Table 1.14 *Percentage clean content of four cores of wool from each of seven bales**

	Bale 1	2	3	4	5	6	7
Core 1	52·33	56·99	54·64	54·90	59·89	57·76	60·27
2	56·26	58·69	57·48	60·08	57·76	59·68	60·30
3	62·86	58·20	59·29	58·72	60·26	59·58	61·09
4	50·46	57·35	57·41	55·61	57·53	58·08	61·45
Average	55·48	57·81	57·21	57·33	58·86	58·78	60·78

* From Cameron, 1951, **7**, *Biometrics*, **7**, 84.

We cannot proceed with this example at present, and we do not even have the necessary concepts to formulate the problem properly. A number of examples of this type could be presented and we must develop a theory to help us formulate the problems and discuss possible solutions. □ □ □

Exercises 1.5

1. Obtain a book catalogue from a large scientific publisher. Examine the data as best you can, using frequency tables, scatter diagrams, means, percentiles, etc., to study questions such as the following.
(*i*)   What relationship holds between the size and price of a book?
(*ii*) What changes in price have occurred over the years? (The catalogue should contain books published in a number of recent years.)
(*iii*) What changes in size and price distribution of books have occurred over the years?
Critically examine the conclusions you make from your analysis, noting any shortcomings of the data, the need for further data, etc.
2. Draw a graph of $\sqrt{y_B}$ against x.

1.6 Populations, samples and probability
In this section we introduce some important basic concepts used in theoretical statistics and we shall therefore limit ourselves to simple situations.
Consider again the seed-testing example, Example 1.1. The whole consignment of seed consists of a very large number N of seeds. Let p be the proportion which would germinate under the carefully standardized conditions used in the testing. The aim of the testing is to determine the value of p, and in order to do this precisely we would have to test the whole consignment. We refer to the consignment as the *population* under study. Practical testing proceeds by selecting a *sample* from this population. These terms sample and population are introduced by analogy with sampling human populations, as for example in carrying out a ballot poll in an electoral ward. We may define:

Population: The whole set or collection of items about which we want information.

Sample: An arbitrary subset of the population. That is, any collection of individuals from the population.

In the seed-testing example the population is both real and finite. This is not the most common situation, and we shall more often deal with populations which are hypothetical and infinite. For example, consider the emission of radioactive particles, discussed in Example 1.2. Here the population would be a collection of counts of the numbers of particles emitted in 5-second periods by the same source, at the same strength, for a very large – infinite – number of periods. This would in any case be impossible to carry out since radioactive sources are continuously decaying. The total observed time must be small compared to the half-life of the source. Although this population is hypothetical and infinite, we shall find that it is a very useful concept to work with. We say that Table 1.2 presents the results of a sample of 200 from this population.

For one more example, consider the strength testing of wool fibres example, Example 1.7. If we consider the strength measurement alone, the population can be considered as the strength measurements on the collection of all possible fibres grown on similar sheep in similar conditions. The measurement here is continuous, and the population is again hypothetical and infinite. If both strength and fibre diameter measurements are considered, the population is the set of couplets (strength, fibre diameter) on the same hypothetical and infinite collection of wool fibres.

Having defined a population, we can ask what kind of a sample we need in order to be able to draw valid inferences. Statistical methods are based on the assumption that we have *random samples*.

Random sample: A sample of n individuals from the population chosen in such a way that all possible sets of n individuals are equally likely to occur.

Random sampling contrasts with a method sometimes employed, by which an 'expert' purposely chooses out a sample he 'knows' to be representative. The reason why random sampling is essential to statistics will become more clear in Chapter 4, but briefly, it enables us to compare an observed sample with a reference set of other samples which are just as likely to occur.

Sometimes it is possible to subdivide a population into several smaller ones, and then randomly sample from these. When this can be done, a great increase in precision often results. Thus in terms of the original population, we can sometimes use restricted random sampling.

One further concept has been implied in this discussion – the concept of *probability*. Suppose the number of emissions from a radioactive source in a 5-second period is counted for a very large number of periods, and the frequency table Table 1.2 recalculated. As the number of observations increases, the proportion of trials in which any particular result appears seems to tend to a limit, and the frequency histogram tends to a definite shape. The same holds true for the other examples mentioned. For example, the proportion of wool fibres having strength between 5 and 6 mg/micron appears to tend to a limit as the number of observations is increased; the proportion of seeds germinating appears to tend to a limit as the number of seeds tested is increased. This is the empirical fact upon which probability is based, and we assume that the limiting proportions exist, and we call them *probabilities*. It can never be proved that the limit exists, since this would require an infinite number of trials at constant conditions. However, there is abundant evidence that the assumption that the limit exists is a fruitful one, and we may safely proceed on this assumption.

Exercises 1.6

1. Obtain a small can and a penny. Stand at a marked distance away from the can and aim the penny to drop into it. After practising this for a short period, carry out a series of 50 trials and note your results.

    Examine your data to see if there is evidence that your probability of success is increasing.

Probability and Probability Distributions

2.1 Probability

In Chapter 1 we showed that the science of statistics is concerned with the interpretation of experimental results from random systems. One of the most important tools used in the development of the subject is a branch of mathematics called the Theory of Probability. The reader who is merely interested in understanding how and when to use statistical methods will need very little indeed of this vast branch of mathematics, and the sketch given in this chapter will suffice. For others who wish to pursue a study of the Theory of Probability to a deeper level, there are some excellent books available, such as Feller (1957), Gnedenko (1962), Loéve (1960) and Parzen (1960).

We start for simplicity by considering Example 1.1, in which there are only two possible outcomes: each seed either germinates, or it does not, and these two outcomes can be denoted a and b respectively. It is found that as more and more observations from a single consignment are considered, the proportion of trials with outcome a appears to settle down towards a limiting value. This leads us to associate with the outcome a, a number $\Pr(a)$, called the probability of a, measuring this idealized limiting proportion. Similarly we associate with b a number $\Pr(b)$. Furthermore, proportions lie in the range zero to unity, and also the proportions of trials with outcomes a and b must add to unity since one of the events a or b must occur at every trial; therefore $\Pr(a)$ and $\Pr(b)$, which are limiting proportions, must obey the same rules,

$$0 \leqslant \Pr(a) \leqslant 1 \qquad 0 \leqslant \Pr(b) \leqslant 1 \tag{2.1}$$

$$\Pr(a) + \Pr(b) = 1. \tag{2.2}$$

Similar statements can be made about all the random systems described in § 1.1, but a continuation of the discussion of Example 1.1 will provide a sufficient illustration. Boxes of 100 seeds were tested for germination, and the observations were the numbers of seeds germinating in each box, so that the set of possible outcomes is one of the integers $\{0, 1, \ldots, 100\}$, see § 1.1 discussion. Let us suppose that results are available from a large number of

47

boxes. We can associate with each possible outcome $\{0, 1, \ldots, 100\}$ numbers $\{Pr(0), Pr(1), \ldots, Pr(100)\}$, which represent the limiting proportions of occurrence of the outcomes in a large series of trials. Again, we must have

$$0 \leqslant Pr(0) \leqslant 1, \qquad 0 \leqslant Pr(1) \leqslant 1, \quad \text{etc.}, \tag{2.3}$$

and

$$Pr(0) + Pr(1) + \ldots + Pr(100) = 1, \tag{2.4}$$

since probabilities are limiting proportions, and one of the events $\{0, 1, \ldots, 100\}$ must occur.

It is important to note that all probabilities considered in this book are to be found in principle by examining appropriate observations, such as counting the proportion of seeds germinating in Example 1.1. This means that in practice probabilities are only determined approximately, since experiments are necessarily of a limited size.

In general, for any random system, we proceed as follows:

(*a*) Compile a list of all possible distinct outcomes.

(*b*) Associate with each distinct outcome a number in the range $(0, 1)$, measuring the limiting proportion of trials on which these events occur. These numbers are probabilities.

(*c*) Since, by definition, one of the set of all possible outcomes must occur at each trial, the sum of the probabilities of all distinct outcomes must be unity. That is, an equation such as (2.2) or (2.4) holds.

Exercises 2.1

1. Give the set of all possible distinct outcomes in the following cases:

(*i*) The change in weight of an adult in a period of three months.

(*ii*) The results from throwing two six-sided dice.

(*iii*) The total score resulting from a throw of two six-sided dice.

The sets for (*ii*) and (*iii*) are different, and this emphasizes that there are often different ways of describing the outcome of any trial.

2. A particular experiment uses 6 mice, which all have different weights. The mice are divided at random into two groups of 3, and one group is used for the 'control' group and the other for the 'treatment' group. What is the probability that the control group contains the 3 heaviest mice?

2.2 Addition law of probability

So far we have defined probability only for single outcomes. We continue a discussion of Example 1.1, and suppose that we need the probability that 80 or more seeds germinate in a box of 100 seeds. Now for any number of trials the proportion in which 80 or more seeds germinate is

$$\begin{pmatrix} \text{proportion} \\ \text{of trials} \\ \text{in which} \\ 80 \\ \text{germinate} \end{pmatrix} + \begin{pmatrix} \\ 81 \\ \end{pmatrix} + \cdots + \begin{pmatrix} \\ 100 \\ \end{pmatrix}.$$

Therefore since we define probability as a limiting proportion we have

Pr(80 or more seeds germinate) =

$$\sum_{r=80}^{100} \text{Pr(exactly } r \text{ seeds germinate).} \quad (2.5)$$

We shall use capital letters $A, B, \ldots$, to denote that specified events have occurred. Thus in the example, we might use the symbol A to denote the event '80 or more seeds germinate'. Other events connected with this random system that we might have occasion to consider are

'the number of seeds germinating is 50 or more and less that 60',

'an even number of seeds germinate', etc.

Quite generally, an event is described by a subset of the collection of all possible outcomes, such as the subset $\{80, 81, \ldots, 100\}$ for the event A in the above example. The event occurs if and only if the outcome lies in the appropriate set.

Let us consider the following events

A '80 or more seeds germinate',

B '25 or less seeds germinate',

C 'the number of seeds germinating is 75 or more and less than 90'.

We shall denote by $(A \text{ or } B)$ the event that at a given trial, A occurs or that B occurs, or that both A and B occur. Also $(A \ \& \ B)$ will denote the event that both A and B occur. For the examples above, A and B can never occur together, so that $(A \ \& \ B)$ is an event which never occurs, whereas $(A \ \& \ C)$ occurs at any trial if the number of seeds germinating is 80, 81, $\ldots$, or 89, The event $(A \text{ or } C)$ occurs if the number of seeds germinating is 75 or more. and the event $(A \text{ or } B)$ occurs if the number of seeds germinating is one of the set $\{0, 1, \ldots, 25; 80, 81, \ldots, 100\}$. Therefore we have

$$\text{Pr}(B) = \sum_{r=0}^{25} \text{Pr(exactly } r \text{ seeds germinate),} \quad (2.6)$$

$$\text{Pr}(C) = \sum_{r=75}^{89} \text{Pr(exactly } r \text{ seeds germinate),} \quad (2.7)$$

similar to equation (2.5). For the compound events,

$$Pr(A \& B) = 0 \qquad (2.8)$$

since $(A \& B)$ never occurs, and

$$Pr(A \& C) = \sum_{r=80}^{89} Pr(\text{exactly } r \text{ seeds germinate}). \qquad (2.9)$$

$$Pr(A \text{ or } C) = \sum_{r=75}^{100} Pr(\text{exactly } r \text{ seeds germinate}). \qquad (2.10)$$

$$Pr(A \text{ or } B) = \left(\sum_{r=0}^{25} + \sum_{r=80}^{100} \right) Pr(\text{exactly } r \text{ seeds germinate}). \qquad (2.11)$$

The reader will now readily check the following result.

Addition law I. If A and B are two events which are mutually exclusive (i.e. they can never occur together) then

$$Pr(A \text{ or } B) = Pr(A) + Pr(B). \qquad (2.12)$$

For our example this law merely says that (2.11) is equal to (2.5) plus (2.6). We also see that the law cannot be used as it stands with events which can occur together. For example, with the above definitions of A and C,

$$Pr(A \text{ or } C) \neq Pr(A) + Pr(C)$$

since the left-hand side is (2.10), while the right-hand side is (2.5) + (2.7), which is

$$\sum_{r=75}^{100} Pr(\text{exactly } r \text{ seeds germinate}) + \sum_{r=80}^{89} Pr(\text{exactly } r \text{ seeds germinate})$$

$$= Pr(A \text{ or } C) + Pr(A \& C).$$

The following diagram represents the position. The probabilities of events A, C, A or C, and $A \& C$ are obtained by adding together the probabilities of the outcomes indicated.

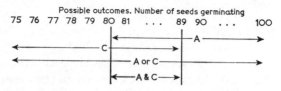

Figure 2.1 *The addition law of probabilities*

The reader will see that whenever any events A and B have outcomes in common, these outcomes are counted twice by the right-hand side of (2.12) We therefore have the following modified form of the addition law.

Addition law II. If A and B are any two events

$$Pr(A \text{ or } B) = Pr(A) + Pr(B) - Pr(A \& B). \qquad (2.13)$$

If A and B are mutually exclusive, $\Pr(A\ \&\ B)$ is zero, while if there are outcomes in common to A and B, we need to subtract $\Pr(A\ \&\ B)$ from $\Pr(A) + \Pr(B)$, to make this equal to $\Pr(A\text{ or }B)$.

Extensions and applications of these laws are given in the following exercises.

Exercises 2.2

1. The complementary event of A is the event that A does not occur, and it is written $\bar{A}$. Prove that

$$\Pr(A) + \Pr(\bar{A}) = 1.$$

2. Suppose that each trial consists of measuring the height of a man and recording the colour of his eyes. Let A be the event that the height exceeds 5 ft 6 in, and B the event that the eyes are brown. Suppose that $\Pr(A) = 0\cdot8$ and $\Pr(B) = 0\cdot7$. In order to calculate $\Pr(A\text{ or }B)$ someone calculates, incorrectly,

$$\Pr(A\text{ or }B) = \Pr(A) + \Pr(B) = 1\cdot5.$$

Why is this wrong?

3. A student at an evening class is due to arrive at 6 p.m. and leave at 9 p.m. Let A be the event that he arrives late, and B the event that he leaves early. Present graphically the events

$$(A\ \&\ B),\quad (A\text{ or }B),\quad (\bar{A}\ \&\ \bar{B}),\quad (\bar{A}\text{ or }\bar{B}).$$

2.3 Conditional probability and statistical independence

We now introduce through some examples the important idea of conditional probability.

Example 2.1. *Sexes of successive children in a family.* Consider the random system formed by recording the sex of a child, say as the birth is registered. If we ignore multiple births, each trial has two possible outcomes. Suppose further we restrict attention to cases where there is at least one previous child in the family and that we record the sex of that child too; the four possible outcomes are, in an obvious notation, (m_1, m_2); (m_1, f_2); (f_1, m_2); (f_1, f_2). A typical set of results is set out in Table 2.1 overleaf.

In Table 2.1, the proportion of male children in current births, among cases where the previous child was male, is $28/52 = 0\cdot54$, while for the cases where the previous birth was female we have $27/48 = 0\cdot56$. These proportions are almost identical, and there is no evidence from this data that the

Table 2.1 *Sexes of successive children*

Birth being registered	Previous child		Total
	m	f	
m	28	27	55
f	24	21	45
Totals	52	48	100

sex of a birth has any effect on the sex of the next birth. The next example stands out in sharp contrast to this conclusion.   ☐ ☐ ☐

Example 2.2. Heights of fathers and daughters. In Table 1.10, Example 1.6, we have some data on heights of fathers and daughters. If we present the heights in two categories, tall and short, dividing at 5 ft 8 in for fathers and 5 ft 4 in for daughters, we have the following 2 × 2 table.

Table 2.2 *Heights of fathers and daughters*

Fathers	Daughters		Totals
	short	tall	
short	522	204	726
tall	206	444	650
Totals	728	648	1376

The proportion of short daughters among short fathers is $522/726 = 0·72$, while among tall fathers we have $206/650 = 0·32$. The marked difference between these proportions indicates a connection between the heights of fathers and daughters, which is known to be partly genetic.   ☐ ☐ ☐

In these two examples we have been considering *conditional* proportions. For example, we examined the proportion of male births given the condition that the previous birth in the family was male. Now in Example 2.1 this proportion is

Proportion (male births|previous birth male)

$$= \left(\frac{\substack{\text{No. of trials} \\ \text{with outcome} \\ (m_1, m_2)}}{\substack{\text{No. of trials} \\ \text{with male at} \\ \text{previous birth}}} \right)$$

$$= \left(\begin{array}{c} \text{No. of trials} \\ \text{with outcome} \\ \underline{(m_1, m_2)} \\ \text{Total no. of} \\ \text{trials} \end{array} \right) \bigg/ \left(\begin{array}{c} \text{No. of trials} \\ \text{with male at} \\ \underline{\text{previous birth}} \\ \text{Total no. of} \\ \text{trials} \end{array} \right) \quad (2.14)$$

where the vertical rule means 'given that'. Now in § 2.1 we defined probability as a limiting proportion, so that if we imagine a large number of trials of Example 2.1, all the proportions in (2.14) become probabilities. We have

$$\Pr\{M_2 \mid M_1\} = \frac{\Pr(m_1, m_2)}{\Pr(M_1)}, \quad (2.15)$$

where M_1 denotes the event that the previous birth in the family was male and is a combination of the events (m_1, f_2) and (m_1, m_2). The probabilities are all defined in terms of the random system described in Example 2.1, so that $\Pr(m_1, m_2)$ is the limiting value of the proportion 28/100 shown in Table 2.1, and similarly $\Pr(M_1)$ is the limiting value of the proportion 52/100.

This discussion leads to the following general definition and result.

Definition. Given a random system and any two events A and B which can occur together, form a new random system by taking only those trials in which B occurs. The probability of A in this new random system is called the conditional probability of A given B, and is denoted $\Pr(A \mid B)$. If $\Pr(B) > 0$, then we have

$$\Pr(A \mid B) = \Pr(A \ \& \ B)/\Pr(B). \quad (2.16)$$

Example 2.3. Illustration of conditional probability.
(a) For Example 2.1, put A to be the event (m_1, m_2), and B to be the event M_1, then equation (2.16) is equation (2.15) and the definition merely summarizes the process of arriving at (2.15).
(b) For Example 2.2, let A be the event 'daughter is tall' and B the event 'father is tall', then (2.16) is the conditional probability that the daughter of a tall father is also tall, and it is the limiting value of the ratio 444/650 in Table 2.2.

The definition of conditional probability can, of course, be used with A and B in alternative roles. Thus we have

$$\Pr(B \mid A) = \Pr(A \ \& \ B)/\Pr(A). \quad (2.17)$$

It is important to realize that $\Pr(B \mid A)$ and $\Pr(A \mid B)$ have different meanings. In Example 2.3(b) for example, $\Pr(A \mid B)$ is the probability that the daughter is tall given that the father is tall, while $\Pr(B \mid A)$ is the probability that, given we have a tall daughter, her father is also tall.   □ □ □

In some situations we find that the fact that B occurs gives us no further information about whether A is likely to occur; that is, we have $\Pr(A \mid B)$

$= \Pr(A)$, and Example 2.3(a) is one such case where in general this holds to a very good approximation.

Definition. The event A is statistically independent of the event B if
$$\Pr(A \mid B) = \Pr(A) \tag{2.18}$$
so that
$$\Pr(A \ \& \ B) = \Pr(A).\Pr(B) \tag{2.19}$$
from (2.16).

The definition in the form (2.18) is unsymmetrical between A and B, but it is clear from (2.19) that if A is statistically independent of B, then B is statistically independent of A, provided $\Pr(A) > 0$. Often, when there is no danger of confusion with some other meaning of the word independent, we shall drop the adverb statistically, and simply say that A and B are independent.

Whether or not particular events are independent is in principle to be decided by examining appropriate observations. Often however there are physical reasons for expecting approximate independence. A clear case where even a small amount of data leads us to conclude that the events are not statistically independent is given by Example 2.2.

This definition of statistical independence is readily extended to deal with a series of events. Consider, for example, a series of throws of a six-sided die. It is very reasonable from physical reasons to expect that different throws are independent of each other, and the probability, say, of four sixes in four throws is
$$\Pr(6666) = \Pr(6).\Pr(6).\Pr(6).\Pr(6).$$
That is, where independence holds we simply multiply together the probabilities of the separate events. If independence does not hold we have a much more complicated situation, involving conditional probabilities.

The results of this section can now be summarized in the following theorem for multiplication of probabilities.

Multiplication law. If A and B are events
$$\Pr(A \ \& \ B) = \Pr(B) \Pr(A \mid B) \tag{2.20}$$
$$= \Pr(A) \Pr(B \mid A), \tag{2.21}$$
by (2.16) and (2.17), and if A and B are statistically independent
$$\Pr(A \ \& \ B) = \Pr(A) \Pr(B). \tag{2.22}$$

This law, together with the addition laws (2.12) and (2.13), are the fundamental laws of probability theory.

Exercises 2.3

1. Let A be the event that a man's left eye is brown, and B the event that a man's right eye is brown. Suppose $\Pr(A) = \Pr(B) = 0\cdot7$. Use the multi-

plication law to obtain the probability of the event $(A \ \& \ B)$, that both eyes are brown. Why is 0·49 incorrect?

2. A chain is formed from n links. The strengths of the links are mutually independent, and the probability that any one link fails under a specified load is θ. What is the probability that the chain fails under that load?

Hint: Pr(chain fails) $= 1 - $ Pr(all links hold).

3. A hand of 13 cards is dealt randomly from the pack of 52. What is the probability that the hand contains (i) no aces, (ii) just one ace?

2.4 Examples

Example 2.4. Geometric distribution. Suppose that the sexes of successive children in a family are independent, and that the size of a certain family is to be determined by the number of children required to ensure exactly one male child. Let the probability of a male child be θ. What is the probability that there are four children in the family?

Under the conditions stated, there will be four children in the family if the first three children were female and the fourth male. We want to find the probability Pr($FFFM$). By the multiplication law of probabilities for independent events this is

$$\mathrm{Pr}(FFFM) = \mathrm{Pr}(F)\,\mathrm{Pr}(F)\,\mathrm{Pr}(F)\,\mathrm{Pr}(M)$$
$$= (1 - \theta)(1 - \theta)(1 - \theta)\theta$$
$$= (1 - \theta)^3\theta.$$

We also see from this that the probability that there are n children in the family is

$$(1 - \theta)^{n-1}\theta, \quad \text{for } n = 1, 2, \ldots \tag{2.23}$$

□ □ □

Example 2.5. Binomial distribution. Let us refer back to Example 1.1 again, and suppose that we have a large consignment of seeds to test for germination. Let the proportion of seeds which would germinate under the conditions used be denoted θ. If 10 seeds are chosen at random and tested, what is the probability that exactly 8 germinate?

Denote the event that a seed germinates by S, and that a seed does not germinate by F. Then

$$\mathrm{Pr}(S) = \theta$$
$$\mathrm{Pr}(F) = 1 - \theta,$$

and if the first 8 seeds germinate,

$$\mathrm{Pr}(SSSSSSSSFF) = \mathrm{Pr}(S)\,\mathrm{Pr}(S) \ldots \mathrm{Pr}(S)\,\mathrm{Pr}(F)\,\mathrm{Pr}(F)$$
$$= \theta^8(1 - \theta)^2 \tag{2.24}$$

provided the trials are independent. But the particular sequence used in (2.24) is one of many in which 8 seeds germinate out of 10. The event that 8 seeds germinate out of 10 is the event

$$(SSSSSSSSFF) \quad \text{or} \quad (SSSSSSSFSF), \quad \text{or} \ldots \text{or} \quad (FFSSSSSSSS). \quad (2.25)$$

There are $^{10}C_8$ different sequences in (2.25), each of which has a probability given by (2.24). Therefore by the addition law of probabilities, the probability of (2.25) is

$$\text{Pr}(SSSSSSSSFF) + \text{Pr}(SSSSSSSFSF) + \ldots$$
$$= {}^{10}C_8 \theta^8 (1 - \theta)^2 = 45\theta^8 (1 - \theta)^2.$$

where $^nC_r = n!/r!(n - r)!$

We see from this that in general, if n seeds are tested, the probability that exactly r germinate is

$$^nC_r \theta^r (1 - \theta)^{n-r}, \quad \text{for } r = 0, 1, \ldots, n. \qquad (2.26)$$

This last result holds provided the trials are independent, and $\text{Pr}(S)$ is θ at each trial. For the example given, testing of seed, these conditions hold approximately provided the number n of seeds drawn is small relative to the number of seeds in the consignment; see Example 2.6.　□ □ □

Example 2.6. Industrial inspection. Batches of N items are presented for final inspection at a certain factory, before release to customers. The items can be classified as effective or defective, and items are drawn at random without replacement, for inspection. In the current batch there are D defectives. What is the probability that the first two items drawn are defectives? What is the probability that the third item is effective, when the first two items are defective?

Denote the events that the first and second items drawn are defectives by D_1 and D_2 respectively. For the first draw, any of the N items in the batch are equally likely to be drawn, while D of these are defectives, so that

$$\text{Pr}(D_1) = D/N.$$

There are now $(N - 1)$ items left in the batch, of which $(D - 1)$ are defective, so that the conditional probability that the second item drawn is defective given that the first item is defective is

$$\text{Pr}(D_2 \mid D_1) = \frac{D - 1}{N - 1}.$$

which is less than $\text{Pr}(D_1)$ provided $D < N$. Therefore by the multiplication law,

$$\text{Pr}(D_1 \ \& \ D_2) = \text{Pr}(D_1) \, \text{Pr}(D_2 \mid D_1)$$
$$= \frac{D}{N} \cdot \frac{D - 1}{N - 1}.$$

Given that the first two items are defective, the probability that the third item is an effective item is seen to be

$$\Pr(E_3 \mid D_1 \ \& \ D_2) = \frac{N - D}{N - 2},$$

in an obvious notation, since there are $(N - D)$ effective items in the batch. If we let $N \to \infty$ and $D \to \infty$ so that the ratio D/N is a constant and equal to θ, then

$$\Pr(E_3 \mid D_1 \ \& \ D_2) \to \frac{N - D}{N} = 1 - \theta,$$

and we revert to the situation of Example 2.5.   □ □ □

Exercises 2.4

1. A very large batch of items is submitted for inspection, and the items are classified as effective or defective. The proportion defective in the batch is θ. Items are sampled at random until exactly c defectives have been found. Show that the probability that this occurs at the nth item is

$$^{n-1}C_{c-1}\theta^c(1 - \theta)^{n-c}$$

for $n = c, c + 1, \ldots$ Hint: the last item sampled must be a defective.

2. In Example 2.4, the family size n must take one of the values $1, 2, 3, \ldots$ Hence the event

$$(n = 1) \quad \text{or} \quad (n = 2) \quad \text{or} \quad (n = 3) \ldots$$

is certain to occur, and its probability must be one. Prove that this is so.

3.* The following is a simplified model representing the spread of measles within a family of three susceptible children A, B, C. If A catches measles there is a period during which B may get the disease from him, and the chance that B does in fact succumb is θ. Similarly for C, and the probabilities for B and C are independent. Thus in the first stage none, one or two of the other children may get the disease. If the first or last event occurs, this completes the matter, but if B gets the disease and not C there is a further period during which C may, with probability θ, catch the disease from B.

Show that, given one child has measles, the chances that 0, 1, 2 of the other children getting it are $(1 - \theta)^2$, $2\theta(1 - \theta)^2$ and $\theta^2(3 - 2\theta)$.

(London, B.Sc. Gen., 1959)

2.5 Discrete random variables

(i) Generalities

When the outcome of a trial on a random system is a number, we say that there is defined a *random variable*. Most of the applications are either of this form, or else the outcome is easily made numerical by a scoring system. For example, a trial for germination on a single seed yields the outcomes 'seed germinates' or 'seed does not germinate', and these can be scored 1 and 0 respectively.

If the set of possible outcomes of a random system is not a continuous set, but is limited to a discrete set of numbers, such as $\{-1, 0, +1\}$, or $\{0, 1, 2, \ldots\}$, we say that we have a *discrete random variable*; the most common set of numbers involved is the set of non-negative integers $\{0, 1, 2, \ldots\}$, as in the radioactive emission experiment, Example 1.2.

The properties of a discrete random variable X are specified by the set of possible values, and by the probability attached to each such value. Such a specification is called a *probability distribution*, and it corresponds, for example, in the radioactive emission experiment, to a frequency table of the outcomes such as Table 1.3, if we could proceed to the limit and make counts for an infinite number of periods. One property of a probability distribution is that the sum of the probabilities over the set of all possible values must be one, since at least one of the set of all possible events must occur at each trial.

In practice we can estimate a probability distribution by making a frequency table, but we often attempt to summarize this in terms of a neat mathematical formula, containing a small number of parameters which we estimate or fit. This simplifies problems of inference to problems concerning the values of these parameters, provided we can assume the mathematical form of the distribution to be settled. This is discussed more in Chapter 4.

We now give a few simple examples of probability distributions.

(ii) Discrete rectangular distribution

This is the simplest possible probability distribution. The random variable takes the values $0, 1, \ldots, (M - 1)$, with equal probabilities $1/M$.

$$\Pr(X = i) = 1/M, \quad i = 0, 1, \ldots, (M - 1). \tag{2.27}$$

For example, random digits are random variables from this distribution with $M = 10$,

$$\Pr(0) = \Pr(1) = \ldots = \Pr(9) = 1/10. \tag{2.28}$$

Random digits can be thought of as being obtained by having equal numbers of cards in a hat with the digits $0, 1, \ldots, 9$, on them, and shuffling and drawing a card, replacing the card, shuffling, and drawing again, etc. Better methods of generating random deviates from a discrete rectangular distribution are discussed in § 4.2, where also the main use of such deviates is described, which is in connection with simulation studies.

Example 2.7. Sample surveys. Sample surveys are now used increasingly in the modern state, for example to estimate political allegiance, to estimate industrial production, etc. One essential step is frequently to draw a random sample from the designated population, and basically, this involves drawing random variables from the discrete rectangular distribution. □ □ □

(iii) Binomial distribution

In Example 2.5 we discussed an important type of random system. There are a number of independent trials, say n, and there are two possible outcomes to each trial, which can be designated 'success' or 'failure'. The probability of a success is assumed to be constant throughout the trials, and we denote this probability by θ. If we consider a random variable which we denote X_n, equal to the number of successes in n trials, irrespective of the order in which they occur, then we have a binomial distribution. An illustration of this situation is given in Example 2.5, which also derives the formula for the probability of r successes in n trials, which is

$$\Pr(X = r) = {}^nC_r\, \theta^r(1 - \theta)^{n-r}, \qquad (2.29)$$

for $r = 0, 1, \ldots, n$.

Now in any set of n trials, we must have either 0 or 1 or ... or n successes, so that the sum of the probabilities over X_n must add to unity,

$$\sum_{r=0}^{n} {}^nC_r\, \theta^r(1 - \theta)^{n-r} = 1. \qquad (2\ 30)$$

In fact all the terms of the probabilities are terms of the expansion of $\{\theta + (1 - \theta)\}^n$, which proves (2.30).

As an example, the binomial distribution for $n = 20$, $\theta = 0\cdot30$, is shown diagrammatically in Figure 2.2. The binomial distribution is skew for $\theta \neq 0\cdot5$ and small n, but becomes more symmetrical as n increases.

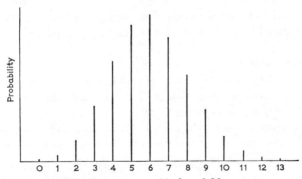

Figure 2.2 *The binomial distribution,* $n = 20$, $\theta = 0\cdot30$

Example 2.8. Industrial inspection. In a certain factory a batch of items is presented for final inspection before release to customers. The inspection is carried out by selecting a random sample of 50 items from the batch, and classifying the items as 'effective' or 'defective'. If not more than 2 defectives are found, the batch is passed.   □ □ □

In Example 2.6 we saw that if items are drawn successively, the probability of a defective at any draw is dependent on the results of previous draws. However, if the number of items in the batch is large compared with the number drawn, 50, this dependence of the probabilities can be ignored. Thus if there are, say, 5% of defective items in the batch, the system behaves as a system in which there are 50 independent trials, with a probability of 0·05 of a defective at each trial. This leads to a binomial distribution for the number of defectives, with $n = 50$, $\theta = 0·05$,

$$\Pr(r \text{ defectives}) = {}^{50}C_r(0·05)^r(0·95)^{50-r}$$

for $r = 0, 1, \ldots, 50$. The probability that a batch with 5% defectives is passed by the inspection is therefore

$$(0·95)^{50} + 50(0·95)^{49}(0·05) + \frac{50·49}{1·2}(0·95)^{48}(0·05)^2 = 0·54$$

which is the probability that the number of defectives is two or less.

Example 2.9. Multiple choice questions. A questionnaire containing 100 questions is given to students. To each question, three answers are provided, one and only one of which must be chosen. If the students guess the answers to all the questions, the distribution of the number of correct answers binomial with $n = 100$, $\theta = 1/3$.   □ □ □

When calculating terms of the binomial distribution, notice that the ratio of successive terms is

$$\frac{\Pr(x = r + 1)}{\Pr(x = r)} = \frac{{}^nC_{r+1}\, \theta^{r+1}(1 - \theta)^{n-r-1}}{{}^nC_r\, \theta^r(1 - \theta)^{n-r}} = \frac{(n - r)\theta}{(r + 1)(1 - \theta)}$$

Therefore if $(1 - \theta)^n$ is calculated first, successive terms of the distribution can be generated by multiplying by the appropriate factor.

(iv) Geometric distribution

In Example 2.4 we illustrated another type of discrete distribution. Again we have a series of independent trials, with two possible outcomes for each trial, which can be designated 'success' and 'failure', and the probability of a success is assumed constant throughout the trials and denoted θ. The distribution of the number of trials up to the first success is geometric,

$$\Pr(X = r) = (1 - \theta)^{r-1}\theta, \tag{2.31}$$

where $r = 1, 2, \ldots$

Example 2.10. *Process control.* An industrial process sometimes falls out of control, and if it does there is a constant probability θ that it will be detected in an hourly check-up. The number of hourly periods which pass between a failure and its detection has a geometric distribution.   □ □ □

(*v*) *Poisson distribution*
The Poisson distribution arises in two ways.

(*i*) As an approximation to the binomial distribution when θ is small and n is large, so that $n\theta = \mu$ is moderate.

(*ii*) For events distributed independently of one another in time (or space), the distribution of the number of events occurring in a fixed time is Poisson.

The distribution can be derived from the binomial distribution by putting $n\theta = \mu$ and letting n tend to infinity; see Exercise 2.5.6. The form of the distribution is

$$\Pr(X = r) = e^{-\mu}\mu^r/r!, \tag{2.32}$$

where $r = 0, 1, 2, \ldots$

Example 2.11. *Dust chamber experiments.* Consider the problem of counting the density of dust particles in a chamber. An ultramicroscope is used to illuminate intermittently a small region of known volume, and the number of particles in it are counted or photographed. The number of particles in the chamber would normally be large, whereas the probability that one is in the small illuminated region is small. If therefore the probabilities for different particles are independent, and there is no electrostatic charge making the particles 'cluster', and if the particles are not too densely packed, the binomial distribution, and to a very good approximation, the Poisson distribution, will apply. One consequence of this is that

$$\Pr(\text{no particles in volume}) = e^{-\mu}$$

hence $$\mu = -\log_e \{\Pr(\text{no particles})\}$$

which gives us a very convenient practical method to estimate the particle density. This sort of argument applies to many physical and bacteriological experiments concerning suspensions of particles whose density is to be estimated. One classical experiment employing this was to counts of yeast cells per unit volume of a suspension, and the details of this were published by 'Student' (1907).   □ □ □

Example 2.12. *Congestion theory.* The Poisson distribution finds applications in what is termed congestion theory, and a typical example of this is to the theory of telephone systems. Briefly, if we consider that the probability that a subscriber makes a call in a small element of time δt is a small constant, say $\lambda\delta t$, then we expect the distribution of the number of calls made in a time t to be Poisson, with $\mu = \lambda t$. The application to telephone systems was developed by A. K. Erlang, and his work has been published by Brockmeyer *et al.* (1948). Another similar example of this type is accident proneness; see Greenwood and Yule (1920). ☐ ☐ ☐

(vi) Cumulative distribution function

We often use another method of representing a probability distribution. Instead of speaking of the probability that a random variable has a particular value, we speak of the probability that it is less than or equal to a given number,

$$\Pr(X \leqslant r) = \sum_{-\infty}^{r} \Pr(X = i).$$

This is represented graphically in Figure 2.3.

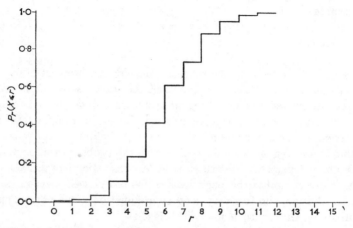

Figure 2.3 *Cumulative distribution function for the binomial distribution, $n = 20$, $\theta = 0.30$*

The cumulative distribution function, or c.d.f., is zero to the left of the range, and for discrete distributions it rises discontinuously to unity on the right of the range. This is the analogue for a probability distribution of the cumulative frequency histogram such as that in Figure 1.6.

Exercises 2.5

1. Calculate some of the terms of the binomial distribution for the following sets of parameters

n	5	10	50	100
θ	0·2	0·1	0·02	0·01

Compare these distributions with some of the early terms of the Poisson distribution for $\mu = 1$.

2. Compare the data of Example 1.2 with the Poisson distribution by the following procedure. Calculate the average of the results given in Example 1.2, and then calculate some of the terms of the Poisson distribution using this average as μ. Compare the distribution with the actual relative frequency table of the observed results. (Extensive tests on such data have shown good agreement with the Poisson distribution.)

3. Calculate the distribution of the number of digits between zeros in a series of truly random digits. Obtain a table of random digits and gather some empirical data for comparison.

4. Obtain a simple expression for the c.d.f. of the geometric distribution.

5.* Suppose that the probability that an event occurs in the interval $(t, t + \delta t)$ is $\lambda \, \delta t$, and denote by $P_x(t)$ the probability that there have been x events up to time t. Show that

$$P_x(t + \delta t) = P_x(t)(1 - \lambda \, \delta t) + P_{x-1}\lambda \, \delta t, \quad x = 1, 2, \ldots$$

and $\quad P_0(t + \delta t) = P_0(t)(1 - \lambda \, \delta t).$

Hence show that

$$P_x(t) = e^{-\lambda t} \frac{(\lambda t)^x}{x!}.$$

6.* Derive the Poisson distribution from the binomial by putting $n\theta = \mu$ in the formula for binomial probability, and then let $n \to \infty$. [See Mood and Graybill (1963), p. 71, or Parzen (1960), p. 105.]

2.6 Continuous random variables

(i) Generalities

In the previous section we dealt with discrete random variables, and now we turn to situations where observations are capable of falling anywhere within an interval, such as measurements of blood pressure, illustrated in Example 1.4. In practice such quantities are always recorded to a limited number of decimal places, so that, for example, the possible values of recorded blood pressures might be . . . 40·0, 40·1, 40·2, . . ., etc., but it is frequently easier and neater to work with the underlying continuous quantity.

Now a discrete probability distribution is obtained forming a relative frequency table of an infinite number of observations taken under constant conditions, and corresponds graphically to a diagram such as Figure 1.1, when made on a very large number of observations. The corresponding

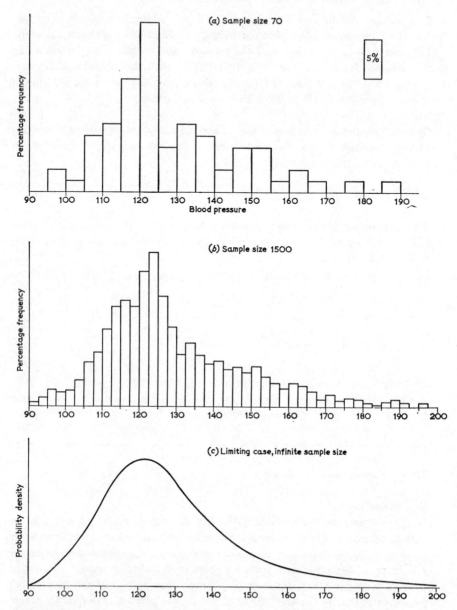

Figure 2.4 *Derivation of a probability density function*

diagram for the continuous case is to proceed to the limit with a histogram such as Figure 1.3, by increasing the number of observations indefinitely, and shrinking the grouping interval gradually to zero. How this limiting process might work out for Figure 1.3, is illustrated in Figure 2.4. For continuous random variables such as measurements of blood pressure, the limiting form of the histogram will be a smooth curve, and we call this a *probability density function*, often written p.d.f.

The meaning and properties of a p.d.f. follow from the fact that it is a limiting form of histogram.

(1) The total area under a p.d.f. is unity.
(2) The area under the p.d.f. between any two points a and b is the probability that the random variable lies between a and b.
(3) A p.d.f. is positive or zero, and is zero in any range where the random variable never falls.

Those in doubt should refer back to § 1.2, for the discussion on histograms. There is, however, one extra feature here. Suppose we have a continuous random variable, and we calculate the probability $\Pr(3\cdot99 < X < 4\cdot00)$, and then increase the number of decimal places and calculate $\Pr(3\cdot9999 < X < 4\cdot0000)$, then the probability is decreased. In the limit, for an indefinite number of decimal places, the calculated probability $\Pr(3\cdot99\ldots < X < 4\cdot00\ldots)$ is zero. This means that whereas we can speak of the probability that a discrete random variable is exactly $0, 1, 2, \ldots$, it is only physically meaningful to speak of the probability that a continuous random variable lies in an interval.

The cumulative distribution function for a continuous random variable is defined as for the discrete case. That is, it is the probability that a random variable X is less than or equal to some specified value x,

$$F(x) = \Pr(X \leqslant x),$$

and graphically, it is the area under the p.d.f up to the point x. In Figure 2.5, the area under the p.d.f. up to the point a, that is the area with vertical and horizontal hatching, is denoted $F(a)$, and in the cumulative distribution function $F(a)$ is plotted as the ordinate against the abscissa a. Similarly for the point b.

The cumulative distribution function is sometimes called simply the distribution function. For continuous random variables the d.f. is a function which increases gradually from zero to the left of the range, to unity to the right of the range. The p.d.f. $f(x)$ and the c.d.f. $F(x)$ of any random variable uniquely define each other; for continuous distributions we have

$$F(x) = \int_{-\infty}^{x} f(x)\, dx$$

$$f(x) = \frac{dF(x)}{dx}.$$

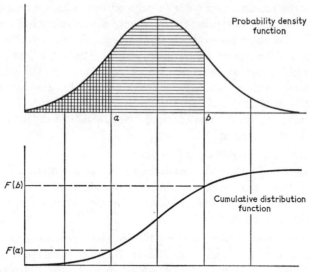

Figure 2.5 *The cumulative distribution function*

To obtain the probability $\Pr(a < x < b)$, we can use either the p.d.f. or the c.d.f. In terms of the p.d.f. we have

$$\Pr(a < x < b) = \int_a^b f(x)\,dx,$$

which is the area under the curve between a and b; see Figure 2.5. In terms of the c.d.f. it is

$$\Pr(a < x < b) = \int_{-\infty}^b f(x)\,dx - \int_{-\infty}^a f(x)\,dx$$
$$= F(b) - F(a)$$

which merely means the area under the p.d.f. up to b minus the area up to a. Examples of the calculation of probabilities using the d.f. are given below.

The meaning of the concepts p.d.f. and c.d.f. will become more clear in the following discussion of three important distributions.

(ii) The rectangular distribution

Suppose that the value of x always lies in the interval (α, β) and that the p.d.f. is constant throughout this interval. This means that if we consider intervals $(x, x + h)$, entirely within (α, β), then all such intervals have the same probability. The p.d.f. is

$$\begin{cases} 1/(\beta - \alpha) & \alpha < x < \beta \\ 0 & x < \alpha,\, x > \beta \end{cases}$$

and the c.d.f. is

$$\begin{cases} 0 & x < \alpha \\ (x - \alpha)/(\beta - \alpha) & \alpha \leqslant x \leqslant \beta \\ 1 & x \geqslant \beta \end{cases}$$

The value of the p.d.f. follows from the fact that the total area under the p.d.f. must be unity, and the c.d.f. is obtained by integrating the p.d.f. The p.d.f. is shown in Figure 2.6.

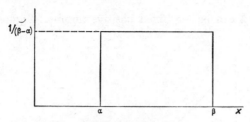

Figure 2.6 *The p.d.f. of the rectangular distribution*

The rectangular distribution is the simplest continuous distribution, but it does not arise very frequently in applications.

Example 2.13. *Random elements for computers.* In computer simulation experiments there is a need for random variables from various distributions. The usual method used to obtain these is to find a method for drawing approximate random deviates from the rectangular distribution on the range (0, 1) and then transform these so that they have the required distribution. Methods are described in Chapter 4 for drawing random deviates from the discrete rectangular distribution; the usual range is $(0, 2^N)$, where N is the number of binary digits which can be stored in each word of the computer store, and frequently N is about 40. By dividing these random deviates by 2^N we have a good approximation to random deviates from the rectangular distribution on the range (0, 1).   □ □ □

Example 2.14. *Rounding-off errors.* Suppose we are making a series of calculations, and we record the results, say to two decimal places, by rounding up or down to the nearest number. Thus we record, for example, 14·25748 . . ., as 14·26, and in doing so we make a *rounding-off error* of +0·00251 . . . If this is applied a large number of times in cases where the range of variation is large compared to the rounding-off error, then it is very plausible that the rounding-off error will have a rectangular distribution on the range (−0·005, +0·005). This fact can be proved mathematically under certain weak assumptions, and it is often useful in assessing the build-up of round-off errors in a long calculation involving many stages.   □ □ □

(iii) The exponential distribution
The exponential distribution is an important special distribution for random variables taking only positive values. The p.d.f. is

$$\begin{cases} \lambda\, e^{-\lambda x} & x > 0 \\ 0 & x < 0, \end{cases} \qquad (2.33)$$

and the c.d.f. is the integral of this, namely,

$$\begin{cases} 1 - e^{-\lambda x} & x > 0 \\ 0 & x \leqslant 0. \end{cases}$$

The parameter λ can be any fixed positive number. Figure 2.7 shows the p.d.f. for $\lambda = 1$.

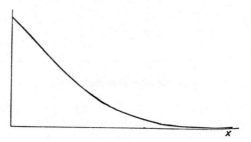

Figure 2.7 *The p.d.f. of the exponential distribution*

The most common way in which the exponential distribution arises in applications is in connection with series of point events such as accidents, radioactive emissions, breakdowns of a machine, etc., which occur completely randomly in time or space. Under certain conditions it can be proved that the distribution of the time interval between such events is exponential; see Exercise 2.6.4. It can never be proved, of course, that the exponential dis-

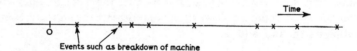

Figure 2.8 *Events occurring in a Poisson process*

tribution holds, say, for the breakdowns of a particular machine because we can never obtain an infinite amount of data. There are many applications, however, where the assumptions appear plausible, and where empirical evidence suggests agreement with the exponential distribution.

Example 2.15. The exponential distribution sometimes occurs in applications having little connection with a random series of events. For example, the duration of local telephone calls has a distribution which is very nearly exponential; so also in some applications do the survival times of animals injected with a fatal dose of a drug.　　　□ □ □

(iv) The normal distribution
This is in many ways the most important distribution for statistical work, although there is nothing peculiar about a distribution being non-normal.

A random variable is said to have a normal distribution if it takes values
in the range $(-\infty, \infty)$ and the p.d.f. is (2.34), which involves two parameters

$$\frac{1}{\sqrt{(2\pi)}\sigma} \exp\left\{-\frac{(x-\mu)^2}{2\sigma^2}\right\}, \tag{2.34}$$

μ and σ, which must satisfy $\sigma > 0$, $-\infty < \mu < \infty$.

For reasons which appear later μ is called the mean and σ is called the
standard deviation; see § 3.1. The c.d.f. of the normal distribution cannot be
expressed in terms of more elementary functions, but it has been extensively
tabulated and should be regarded as a known function. However, we can
check that the integral of the p.d.f. over the full range $(-\infty, \infty)$ is unity;
see Exercise 2.6.5.

The normal p.d.f. (2.34) is a bell-shaped curve which is symmetrical about
a mode at the value $x = \mu$. The parameter σ is a scale parameter. A plot of

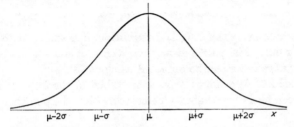

Figure 2.9 *The p.d.f. of the normal distribution*

the normal p.d.f. is shown in Figure 2.9. Approximately 95%of the total
area of the p.d.f. is enclosed in the interval $(\mu - 2\sigma, \mu + 2\sigma)$. The c.d.f. has
the shape given in Figure 2.5.

The normal distribution with parameters μ and σ is denoted $N(\mu, \sigma^2)$, and
the special case $\mu = 0$, $\sigma^2 = 1$, with a p.d.f.

$$\frac{1}{\sqrt{(2\pi)}} \exp\left(-\frac{x^2}{2}\right),$$

is called the *standard normal distribution* and denoted $N(0, 1)$. The c.d.f. of
the standard normal distribution is written

$$\Phi(x) = \int_{-\infty}^{x} \frac{1}{\sqrt{(2\pi)}} \exp\left(-\frac{x^2}{2}\right)dx. \tag{2.35}$$

Now we have also

$$\Phi(x) = 1 - \int_{x}^{\infty} \frac{1}{\sqrt{(2\pi)}} \exp\left(-\frac{x^2}{2}\right)dx \tag{2.36}$$

since the p.d.f. integrates to one. Also, since the p.d.f. of the $N(0, 1)$ distribution is symmetrical about zero, we have

$$\int_{x}^{\infty} \frac{1}{\sqrt{(2\pi)}} \exp\left(-\frac{x^2}{2}\right)dx = \int_{-\infty}^{-x} \frac{1}{\sqrt{(2\pi)}} \exp\left(-\frac{x^2}{2}\right)dx \tag{2.37}$$

$$= \Phi(-x).$$

Equation (2.37) merely states that the two shaded areas in Figure 2.10 are equal in area.

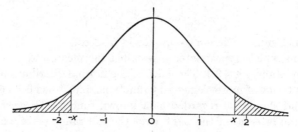

Figure 2.10 *The p.d.f. of the standard normal distribution*

By using (2.36) and (2.37) we have

$$\Phi(x) = 1 - \Phi(-x). \tag{2.38}$$

Therefore in tabulating the $N(0, 1)$ distribution, it is sufficient to tabulate it for positive x only. The p.d.f.s at any two values x and $-x$ are equal, while the c.d.f. for negative x can be obtained using (2.38).

Tables 1 and 2 of Appendix II give short tables of the $N(0, 1)$ distribution, and some examples of probability calculations based upon them are given below.

Suppose a random variable X has an $N(\mu, \sigma^2)$ distribution, and we want to calculate $\Pr(a < X < b)$. Then

$$\Pr(a < X < b) = \int_a^b \frac{1}{\sqrt{(2\pi)}\sigma} \exp\left\{-\frac{(x-\mu)^2}{2\sigma^2}\right\}dx$$

$$= \int_{(a-\mu)/\sigma}^{(b-\mu)/\sigma} \frac{1}{\sqrt{(2\pi)}} \exp\left(-\frac{z^2}{2}\right)dz$$

$$= \Phi\left(\frac{b-\mu}{\sigma}\right) - \Phi\left(\frac{a-\mu}{\sigma}\right) \tag{2.39}$$

Therefore all probability calculations based on normal distributions can be evaluated in terms of the standard normal distribution, and a tabulation of $N(0, 1)$ is sufficient.

The basic result is that if x is $N(\mu, \sigma^2)$, we use the transformation

$$y = (x - \mu)/\sigma,$$

and evaluate the probabilities treating y as distributed $N(0, 1)$.

Example 2.16. *Calculation of normal probabilities.* The reader should work carefully through the following examples of the calculation of normal probabilities, which can all be done using Appendix I table 1.

If x is distributed $N(0, 1)$, then;

(i) $\Pr(x < 1{\cdot}5) = 0{\cdot}9332$.

(ii) $\Pr(x > 1{\cdot}5) = 1 - \Pr(x < 1{\cdot}5) = 0{\cdot}0668$.

(iii) $\Pr(x < -1{\cdot}5) = \Pr(x > 1{\cdot}5) = 0{\cdot}0668$.

(iv) $\Pr(-1{\cdot}5 < x < 1{\cdot}5) = \Pr(x < 1{\cdot}5) - \Pr(x < -1{\cdot}5)$
$$= 0{\cdot}9332 - 0{\cdot}0668 = 0{\cdot}8664.$$

If y is distributed $N(4, 6^2)$ then

(v) $\Pr\{-5 < y < 13 \mid N(4, 6^2)\}$
$$= \Pr\left\{\left(\frac{-5 - 4}{6}\right) < x < \left(\frac{13 - 4}{6}\right) \middle| N(0, 1)\right\}$$
$$= \Pr\{-1{\cdot}5 < x < 1{\cdot}5 \mid N(0, 1)\} = 0{\cdot}8664$$
$$\text{by result } (iv). \quad \square\square\square$$

Reasons for importance of the normal distribution. In the sequel we shall frequently make the assumption that random variables are normally distributed. The reasons why the normal distribution holds such a central place in statistics are:

(a) Many practical distributions are fitted reasonably well by a normal distribution. Example 1.4, which is concerned with measurements of blood pressure, is one case where a normal distribution fits reasonably well, and this is true for measurements of heights of men, experimental errors, etc.

(b) An important result called the *central limit theorem* proves that the normal distribution is to be expected theoretically whenever the variation is produced by the addition of a large number of effects, non-predominant. It is plausible that these conditions hold quite often, such as in the cases quoted in (a) above.

(c) The central limit theorem also proves that many distributions such as the Poisson or binomial, tend to normality in the limit. The normal distribution is therefore important as an approximation to these distributions, for example, to the binomial when n is large and to the Poisson when the mean is large.

(d) Much of statistical theory takes on a very simple and elegant form when distributions are normal. Also, the results and techniques so obtained are often not too sensitive to the assumption of normality.

(e) When continuous random variables are not normally distributed, there is always a transformation of the variables which renders the distribution normal. Sometimes a very simple transformation suffices, such as $\log X$, $\sqrt{X}$, $1/X$, etc.

(v) *Other distributions*

Some common types of deviation from normality are shown in Figure 2.11. The most common cases are positive or negative skewness. Another case is that the curve is symmetrical, but either altogether too flat (platykurtosis), or too sharply peaked (leptokurtosis), for a normal distribution to fit.

One of the best ways to check that a particular set of data is fitted by a normal distribution is to plot the cumulative frequency distribution of the data on specially drawn paper – *normal probability paper* – which has the property that a straight line results when the distribution is normal. The

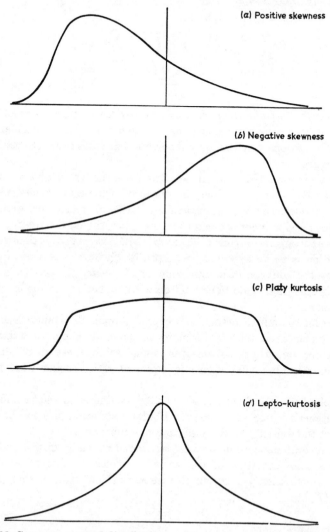

(*a*) Positive skewness

(*b*) Negative skewness

(*c*) Platy kurtosis

(*d*) Lepto-kurtosis

Figure 2.11 *Some non-normal distributions*

method of drawing normal probability paper is explained in Exercise 2.6.2, and in principle it is simply that the percentage frequency scale on a cumulative frequency diagram is transformed.

There are a vast number of probability distributions in common use. Pearson (see Kendall and Stuart, 1958, pp. 148–54) gave a whole system of distributions, and discussed methods of choosing one of these as most appropriate to a given set of data. However, for most applications our methods are not so critically dependent on the normality assumption as to matter.

In special cases some very peculiar distributions can arise. For example, consider the distribution of waiting time in a queue. With a probability p there will be no queue when a customer arrives, and the waiting time y is zero. If there is a queue, the waiting time may have, say, an exponential distribution. This can be summarized,

$$\begin{cases} \Pr(y = 0) = p, & \Pr(y > 0) = 1 - p \\ \text{p.d.f. when } y > 0, & (1 - p)\lambda e^{-\lambda y}. \end{cases}$$

This is a mixed distribution, being partly discrete and partly continuous. Cases much more difficult to treat mathematically could theoretically arise, and this point leads to severe problems in a rigorous treatment. For most practical purposes we can sidestep this issue by assuming our distributions to be mathematically well behaved.

Exercises 2.6

1. If x is distributed $N(4, 9)$, calculate the probabilities

(*i*) $\Pr(x < 0)$ (*iii*) $\Pr(-4 < x < 4)$

(*ii*) $\Pr(x > 9)$ (*iv*) $\Pr(-2 \cdot 5 < x < +1 \cdot 5)$.

2. *Normal probability paper.* By making the appropriate transformation of the ordinate when plotting the normal c.d.f., a straight line is obtained.

Represent the $2\frac{1}{2}\%$ to $97\frac{1}{2}\%$ range of the normal distribution by using graph paper of length $2l$ for the ordinate, and associate with x the probability

$$\Phi\left(\frac{1 \cdot 96x}{l}\right)$$

where x is measured from the mid-point of the axis of the ordinate.

On this paper, mark the 3, 4, 5, 6, 8, 10, 15, 20, . . ., 90, 92, 94, 95, 96, 97% points. Plot on your paper the normal distribution read off from tables. Also (on separate sheets), plot the data of Example 1.4, and the cumulative binomial distributions calculated in Exercise 2.5.1.

3.* Prove that the Poisson distribution tends to the normal as $\mu \to \infty$.

4.* Prove, as follows, that the time interval between events occurring in a Poisson process is exponentially distributed.

Definition. A *Poisson process* or completely random series of events is one for which events occur at random time points $T_1, T_2, \ldots$, satisfying the following conditions

(i) $\lim_{h \to 0} \Pr\{\text{an event occurs in the interval } (t, t+h)\}/h = \lambda$.

(ii) $\lim_{h \to 0} \Pr\{\text{more than one event in the interval } (t, t+h)\}/h = 0$.

(iii) The occurrence of events in $(t, t+h)$ is independent of the position and number of events before t.

From these definitions, the probability of an event in the time interval $(t, t+dt)$ is very nearly $\lambda \, dt$, independently of t and of other events.

Let the process start from time zero, and write

$$\Phi(t) = \Pr\{\text{no event in the interval } (0, t)\},$$

so that $\Phi(0) = 1$. By considering $\Phi(t+h)$, show that $\Phi(t)$ satisfies

$$\frac{d\Phi(t)}{dt} = -\lambda \, \Phi(t),$$

hence that

$$\Phi(t) = e^{-\lambda t}.$$

Hence, show that the c.d.f. of the time interval to the first event is the exponential distribution. Hint: see Exercise 2.5.5.

5.* Use the following method to prove that the integral of the normal p.d.f. over the full range is unity.

If

$$\int_{-\infty}^{\infty} \frac{1}{\sqrt{(2\pi)}} e^{-\frac{1}{2}x^2} \, dx = 1$$

then we must have

$$\left\{\int_{-\infty}^{\infty} \frac{1}{\sqrt{(2\pi)}} e^{-\frac{1}{2}x^2} \, dx\right\}\left\{\int_{-\infty}^{\infty} \frac{1}{\sqrt{(2\pi)}} e^{-\frac{1}{2}y^2} \, dy\right\} = 1.$$

In this last equation, change the variables to polar co-ordinates,

$$x = r \cos \theta$$
$$y = r \sin \theta$$

where $0 \leqslant r \leqslant \infty$, $0 \leqslant \theta \leqslant 2\pi$. The resulting integrals are easily calculated.

2.7 Several random variables

(i) Discrete distributions

We arrived at a discrete probability distribution, in the one-dimensional case, by supposing that an infinite number of observations were taken on an experiment, and the results plotted in terms of a relative frequency table. For example, in the radioactive emissions experiment, Example 1.2, we think of a table similar to Table 1.2 made on a basis of a very large number of trials. The generalization of this to the case of two or more dimensions is immediate. Consider the following example.

Example 2.17. First cousins of opposite sex are classified as A, very able; B, capable; C, intelligent but slow; D, dull, slow and mentally defective. The results on 1374 pairs of cousins are given in Table 2.3 below.

Table 2.3 *First cousins of opposite sex*

Female	Male				Totals
	A	B	C	D	
A	22	69	25	7	123
B	84	564	141	37	826
C	47	167	127	14	355
D	8	20	22	20	70
Totals	161	820	315	78	1374

□ □ □

The result from any pair of cousins, in Example 2.17, can be represented as a couplet, and we shall put the male classification first. Thus (A, B) denotes the result that the male cousin was Class A while the female cousin was class B. The probability of (A, B) is defined as the long-run relative frequency of this cell in a large number of observations taken on the same population. The actual relative frequency of (A, B) in Example 2.17 data is $84/1374 = 0.06$.

In general, a bivariate discrete random variable can be described as follows. We have two variables, x and y, both taking on a discrete set of values, which we shall take to be the set 0, 1, 2, . . ., in each case. The probabilities can be set out in a two-way array, as in Table 2.4. The probabilities $p_{0.}, p_{1.}$, etc., are the sums of all the probabilities in the respective rows, and $p_{.0}, p_{.1}$, etc., are the sums of the probabilities in the respective columns. The sum of all the p_{ij}'s must be unity, representing the fact that one and only one of the cells in this array occurs at any trial. An array such as Table 2.4 is called a *bivariate (discrete) probability distribution*.

Table 2.4 *A bivariate discrete probability distribution*

Values of y	Values of x					Totals
	0	1	2	3	...	
0	p_{00}	p_{01}	p_{02}	p_{03}	...	$p_{0.}$
1	p_{10}	p_{11}	p_{12}	p_{13}	...	$p_{1.}$
2	p_{20}	p_{21}	...	...		
.	.					
.	.					
.	.					
Totals	$p_{.0}$	$p_{.1}$	$p_{.2}$	...	...	1·00

The set of marginal probabilities for rows, $p_{0.}, p_{1.}, \ldots$, defines the *marginal probability distribution for rows*, ignoring the column classifications. It corresponds in Table 2.3 to the marginal distribution of the classification of female cousins, ignoring the classification of the male cousins. Similarly, we define the marginal distribution for columns, $p_{.1}, p_{.2}, \ldots$

Another important concept is that of a *conditional probability distribution*. For example, the conditional distribution of y given x is one, is

$$p_{01}/p_{.1}, \ p_{11}/p_{.1}, \ p_{21}/p_{.1}, \ \ldots$$

In terms of Example 2.17, this particular conditional distribution is the distribution of classification of all female cousins who have the male cousin in class B. Readers having difficulty with the concept of conditional probability distributions should refer back to the discussion of § 2.3.

A very special case arises when for all cells in Table 2.4, we have

$$p_{ij} = p_{i.}p_{.j}, \quad \text{all } i, j. \tag{2.40}$$

When this holds, the conditional distribution of y for given values of x, is identical for all x, and equal to the marginal distribution of y. Similarly (2.40) also implies that all the conditional distributions of x are identical. When condition (2.40) is satisfied we say that the random variables x and y are *statistically independent*. This is merely an application of the definition of statistical independence given in § 2.3. An application where statistical independence appears to hold is in Tables 1.8 and 1.9, which give a two-way table of successive emissions from a radioactive source. The cousins data given in Example 2.17 show very clearly that the intelligences of first cousins of the opposite sex are not statistically independent.

The generalization to three or more variables is immediate. For three random variables we can think of a three-dimensional array similar to Table 2.4, with the probabilities in all the cells adding to one. If the marginal probabilities satisfy

$$p_{ijk} = p_{i..}p_{.j.}p_{..k}, \quad \text{all } i, j, k, \tag{2.41}$$

then the classifications of i, j and k are said to be mutually independent.

(ii) Continuous distributions

For continuous distributions we proceed in a manner directly analogous to the discussion in § 2.6. For example, consider Table 1.10, which gives a two-way frequency table of the heights of fathers and daughters. Suppose in this example we took many more observations, and shrunk the grouping intervals. Then in the limit we have a surface which represents a two-dimensional p.d.f. The volume under the p.d.f. must add to unity. The discussion is exactly parallel to that given above for discrete random variables.

In general, for continuous random variables $x, y, z, \ldots$, the joint p.d.f. of these is a function $f(x, y, z, \ldots,)$, which is positive or zero, and which integrates to unity. If the joint p.d.f. factorizes into a product of terms in $x, y, z, \ldots$, separately,

$$f(x, y, z, \ldots,) = f_1(x) f_2(y) f_3(z) \cdots$$

then we say the random variables are (mutually) statistically independent.

For two random variables we have a joint p.d.f. $f(x, y)$, and

$$\int_{-\infty}^{\infty} \int_{-\infty}^{\infty} f(x, y) \, dx \, dy = 1.$$

The marginal distribution of y has a p.d.f.

$$\int_{-\infty}^{\infty} f(x, y) dx,$$

and the conditional distribution of y given that x has a value x', is

$$f(y \mid x') = f(x', y) \Big/ \left\{ \int_{-\infty}^{\infty} f(x', y) \, dy \right\}. \tag{2.42}$$

An important continuous bivariate distribution is the bivariate normal distribution, which has a p.d.f.

$$\frac{1}{2\pi\sigma_x\sigma_y\sqrt{(1-\rho^2)}} \exp\left[-\frac{1}{2(1-\rho^2)} \left\{ \left(\frac{x-\mu_x}{\sigma_x}\right)^2 - 2\rho\left(\frac{x-\mu_x}{\sigma_x}\right)\left(\frac{y-\mu_y}{\sigma_y}\right) \right. \right.$$
$$\left. \left. + \left(\frac{y-\mu_y}{\sigma_y}\right)^2 \right\} \right]. \tag{2.43}$$

The marginal distribution of x for this distribution is normal with mean μ_x and standard deviation σ_x; and similarly for the marginal distribution of y. The parameter ρ is called the correlation coefficient; see § 3.4.

The bivariate normal distribution is a suitable model for the distribution of heights of fathers and heights of daughters, discussed above.

Exercises 2.7

1. Variables x, y and z take on the values 0, 1, 2 only. Construct an example to show that it is possible for x and y to be statistically independent, for

any given value of z, while x and z, and y and z, are not statistically independent.

2. It is possible for pairwise independence to hold between x and y, x and z, and y and z, and yet for equation (2.41) not to be satisfied. Construct an example to show this.

3. Plot on the x–y plane the contours for which the bivariate normal p.d.f. is constant. What is the mode of the distribution?

4. Give examples from your own experience of random variables (a) which are statistically independent, and (b) which are not statistically independent.

CHAPTER THREE

Expectation and its Applications

3.1 Expectation

In Chapter 1 we explained that a given set of data is sometimes described by the full frequency table, or histogram, while in other cases we calculate one or two summarizing statistics which express the main aspects of the data in one or two numbers. One of these summarizing statistics is usually some kind of average, and another a measure of dispersion. In Chapter 2 we have described a probability distribution as a limiting form of relative frequency table, and in the same way, therefore, we can either work with the whole probability distribution, or else work with summarizing quantities. The analogous quantity to the sample mean is called the *expectation* in populations. In the development of statistical methods in the later part of this book, this concept is used repeatedly.

The sample is defined as

mean = sum of {observation × frequency}/no. of observations

= sum of {observation × relative frequency}   (3.1)

see § 1.3 and Appendix I,

so that in Example 1.2, we have, using Table 1.3,

mean = 1 × 0·010 + 2 × 0·020 + 3 × 0·065 + ... + 14 × 0·005

= 6·91.

When we are dealing with a probability distribution, and not a relative frequency table, the long-run relative frequencies become probabilities. This leads us to define expectation as

expectation = sum of {variable × probability}.   (3.2)

Expressing this mathematically, we say that the expectation of a random variable x is

$$E(x) = \sum x \Pr(x)$$ (3.3)

for discrete distributions, and

$$E(x) = \int_{-\infty}^{\infty} x f(x) \, dx$$ (3.4)

for continuous distributions, where $f(x)$ is p.d.f.

Example 3.1. If x takes the values 0 and 1 with equal probability $\frac{1}{2}$, then
$$E(x) = \tfrac{1}{2} \times 0 + \tfrac{1}{2} \times 1 = \tfrac{1}{2}.$$
This example shows that, when we say that the expected value is $\frac{1}{2}$, we do not mean that $\frac{1}{2}$ is a value likely to arise. Here $\frac{1}{2}$ is not a possible value of the random variable.   ☐ ☐ ☐

Example 3.2. Suppose x takes the values $0, 1, \ldots, 9$, with equal probabilities of 1/10, then
$$E(x) = \tfrac{1}{10} \times 0 + \tfrac{1}{10} \times 1 + \ldots + \tfrac{1}{10} \times 9 = 4 \cdot 5.$$
                                                                  ☐ ☐ ☐

Example 3.3 Suppose x is a continuous random variable with the p.d.f.
$$f(x) = \begin{cases} 2 - 2x, & 0 \leqslant x \leqslant 1 \\ 0 & \text{elsewhere.} \end{cases}$$

This is a triangular-shaped p.d.f. on the range (0, 1). The expectation is, by (3.4),
$$E(x) = \int x f(x)\, dx = \int_0^1 x(2 - 2x)\, dx$$
$$= [x^2 - 2x^3/3]_0^1 = 1/3. \qquad\qquad ☐ ☐ ☐$$
The expectation of a random variable taking various common distributions is as follows:

Binomial distribution. (Item (*iii*) of § 2.5)
$$E(x) = \sum_{x=0}^{n} x\, {}^nC_x\, \theta^x(1 - \theta)^{n-x} = n\theta. \tag{3.5}$$

Poisson distribution. (Item (*v*) of § 2.5)
$$E(x) = \sum_{x=0}^{\infty} x\, e^{-\mu}\, \mu^x/x! = \mu. \tag{3.6}$$

Rectangular distribution on (0, a). (Item (*ii*) of § 2.6)
$$E(x) = \int_0^a x \frac{1}{a}\, dx = \frac{1}{a}[x^2/2]_0^a = a/2. \tag{3.7}$$

Exponential distribution. (Item (*iii*) of § 2.6)
$$E(x) = \int_0^\infty x\lambda\, e^{-\lambda x}\, dx = 1/\lambda. \tag{3.8}$$

Normal distribution. (Item (*iv*) of § 2.6)
$$E(x) = \int_{-\infty}^{\infty} x \frac{1}{\sqrt{(2\pi)}\sigma} \exp\left\{-\frac{(x - \mu)^2}{2\sigma^2}\right\} dx = \mu. \tag{3.9}$$

The importance attached to the expectation of a random variable depends on the shape of the underlying distribution. This is demonstrated by the following example.

Example 3.4. If x denotes cloudiness measured on a scale from 0 to 1, then in some places the distribution of x is U-shaped; see Figure 3.1. In this case the value of x is near $\frac{1}{2}$, and will be in a region of low probability. This in no way conflicts with interpretation of $E(x)$ as a long-run average. □ □ □

The most important situations for the use of $E(x)$ are when (*a*) the underlying distribution is unimodal and nearly symmetrical, or (*b*) when $E(x)$ has a practical meaning, as for example when x is a yield and $E(x)$ is a long-run yield.

If the idea of expected value were restricted to the initial random variable

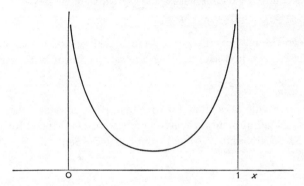

Figure 3.1 *A* U-*shaped p.d.f.*

of interest, it would be of very limited value. In the sequel we shall apply expectation to a variety of functions of random variables. That expectation can be generalized to cope with arbitrary functions of a random variable is seen by analogy with the equivalent consideration in samples. For example, suppose we wish to obtain the average value of x^2 in Example 1.2. By analogy with (3.1) this is

$$\text{average of } (x^2) = \text{sum of } \{x^2 \times \text{relative frequency}\}$$
$$= 1^2 \times 0{\cdot}010 + 2^2 \times 0{\cdot}020 + \ldots + 14^2 \times 0{\cdot}005$$
$$= 54{\cdot}45.$$

When we proceed to the limit, the relative frequencies become probabilities and we have

$$E(x^2) = \sum x^2 \Pr(x)$$

for the discrete case, and

$$E(x^2) = \int_{-\infty}^{\infty} x^2 f(x)\, dx$$

for the continuous case. [Below we shall frequently use the continuous case only, and leave the equivalent expression for discrete distributions to be understood.]

Example 3.5. For Example 3.2, the expectation of x^2 is

$$E(x^2) = \tfrac{1}{10}0^2 + \tfrac{1}{10}1^2 + \ldots + \tfrac{1}{10}9^2 = 28 \cdot 5. \qquad \square\ \square\ \square$$

Therefore in general we define the expectation of a function $g(x)$ of x as

$$E\{g(x)\} = \int_{-\infty}^{\infty} g(x) f(x)\, dx, \qquad (3.10)$$

where $f(x)$ is the p.d.f. of x. Examples of this are given below. When dealing with the expectation of functions of a random variable, E can be treated as an operator which obeys certain rules.

Let x be a random variable with a p.d.f. $f(x)$, then we have

*Rule E*1. $E(x + b) = E(x) + b$.

This rule corresponds to the proposition that if we add a constant, say 10, to every observation in a sample, then the sample mean is increased by this amount.

*Rule E*2. $E(ax) = a\, E(x)$.

In words, this rule says that, for example, if we measure observations in centimetres instead of inches, the average becomes multiplied by a constant. Rules E1 and E2 are readily proved from the definition.

From (3.10),

$$E(ax + b) = \int_{-\infty}^{\infty} (ax + b) f(x)\, dx$$

$$= a\int_{-\infty}^{\infty} x f(x)\, dx + b$$

$$E(ax + b) = a\, E(x) + b. \qquad (3.11)$$

*Rule E*3. If $g(x)$ and $h(x)$ are any two functions of a random variable x,

$$E\{g(x) + h(x)\} = E\{g(x)\} + E\{h(x)\}.$$

Example 3.6. Suppose we apply Rule E3 for $g(x) = x^2$, and $h(x) = x^4$, then

$$E(x^2 + x^4) = E(x^2) + E(x^4). \qquad \square\ \square\ \square$$

Rule E3 readily generalizes to the following result; the expectation of any sum of quantities is the sum of the expectations, provided all the expectations exist. In this connection notice the following points.

If x and y are two random variables with a joint p.d.f. $f(x, y)$, then by the expectation of x we mean

$$E(x) = \int_{-\infty}^{\infty} \int_{-\infty}^{\infty} x f(x, y)\, dy\, dx$$

$$= \int_{-\infty}^{\infty} x \left\{ \int_{-\infty}^{\infty} f(x, y) \, dy \right\} dx. \tag{3.12}$$

That is, the expectation of x is the expectation taken over the marginal distribution of x.

We can also define expectations over conditional distributions. The conditional distribution of x given $y = y'$ is, by (2.42),

$$f(x \mid y) = f(x, y) \Big/ \int_{-\infty}^{\infty} f(x, y) \, dx.$$

Hence we define

$$E(x \mid y = y') = \int_{-\infty}^{\infty} x f(x \mid y) \, dx$$

$$= \frac{\displaystyle\int_{-\infty}^{\infty} x f(x, y) \, dx}{\displaystyle\int_{-\infty}^{\infty} f(x, y) \, dx}. \tag{3.13}$$

Exercises 3.1

1. Prove the result given in Rule E3.

2. If x has a geometric distribution,
$$\Pr(x = r) = \theta(1 - \theta)^{r-1}, \quad r = 1, 2, \ldots,$$
find $E(x)$.

3. Let x and y be random variables with a joint p.d.f. $f(x, y)$. Prove that
$$E(x + y) = E(x) + E(y).$$

4. Random variables $x_1, x_2, \ldots, x_n$, not necessarily independent, all have expectation μ. Prove that the expectation of

$$\bar{x} = \sum_{1}^{n} x_i / n$$

is μ for all n.

5. Random variables x, y have a joint p.d.f. $k(x + y)$, for $0 \leqslant x \leqslant 1$, $0 \leqslant y \leqslant 1$, and where k is a constant. Determine the constant and draw the p.d.f. Also obtain the marginal distribution of x, the conditional distribution of x when $y = \frac{1}{2}$, and calculate

$$E(x), \quad E(x \mid y = \tfrac{1}{2}), \quad E(xy).$$

6. If x_1 and x_2 are independent random variables with p.d.f.s $f_1(x_1)$ and $f_2(x_2)$ respectively, show that

$$E(x_1x_2) = \int\int x_1 x_2 f_1(x_1) f_2(x_2) \, dx_1 \, dx_2 = \{E(x_1)\}\{E(x_2)\}.$$

[This does not hold for a general bivariate distribution, see § 3.4.]

3.2 Variance

Let a random variable x have a p.d.f. $f(x)$, and an expectation μ, then the quantity

$$V(x) = E(x - \mu)^2$$
$$= \int_{-\infty}^{\infty} (x - \mu)^2 f(x) \, dx \tag{3.14}$$

is called the (population) variance of x. This is the expected value of the square of the difference between the random variable x and the population mean μ; it is the corresponding population quantity to the sample variance, and is a measure of spread of a distribution. Notice that

$$V(x) = E(x - \mu)^2 = E(x^2 - 2\mu x + \mu^2)$$
$$= E(x^2) - \mu^2. \tag{3.15}$$
$$= \text{expected value of } x^2 - (\text{expected value of } x)^2.$$

Example 3.7. If x takes the values 0, 1, . . ., 9 with equal probabilities of 1/10, then from Examples 3.2 and 3.5, $E(x) = 4\cdot5$, $E(x^2) = 28\cdot5$, so that

$$V(x) = 28\cdot5 - (4\cdot5)^2 = 8\cdot25. \qquad \square\square\square$$

The (population) *standard deviation* is defined as

$$\text{standard deviation} = \sqrt{}(\text{variance}).$$

For Example 3.7, the standard deviation is $\sqrt{}(8\cdot25) = 2\cdot87$.

Example 3.8. *Poisson distribution.* Let X have a Poisson distribution with parameter μ, which is also the expectation, from equation (3.6), then

$$E(x^2) = \sum_0^\infty x^2 e^{-\mu} \mu^x / x!$$

$$= \sum_0^\infty (x - 1 + 1) e^{-\mu} \mu^x / (x - 1)!$$

$$= \mu^2 \sum_0^\infty e^{-\mu} \mu^{x-2} / (x - 2)! + \mu \sum e^{-\mu} \mu^{x-1} / (x - 1)!$$

$$= \mu^2 + \mu.$$

Therefore
$$V(x) = E(x^2) - E^2(x) = \mu^2 + \mu - \mu^2 = \mu.$$
Thus for the Poisson distribution, the variance is equal to the expectation.

□ □ □

The calculations of the variance for the other distributions mentioned in § 3.1 are set as exercises for the reader; see Exercise 3.2.1.

As for expectation, we often wish to calculate the variance of functions of the random variable x. We define the variance of $g(x)$ as
$$V\{g(x)\} = E\{g^2(x)\} - E^2\{g(x)\}.$$
The variance of simple functions is readily calculated, since the symbol V can also be regarded as an operator, which satisfies the following rules.

Rule V1. $V(x + b) = V(x)$.

This rule follows from the fact that if we add a constant to all of any set of observations, we do not alter the differences between the observations.

Rule V2. $V(ax) = a^2 V(x)$.

Since variance is a squared measure of dispersion, multiplying the scale of x by a has the effect of multiplying the variance by a^2. These rules follow immediately from the definition
$$\begin{aligned} V(ax + b) &= E\{ax + b - E(ax + b)\}^2 \\ &= E\{ax + b - a\,E(x) - b\}^2 \\ &= a^2\,E\{x - E(x)\}^2 = a^2\,V(x). \end{aligned}$$
Thus
$$V(ax + b) = a^2\,V(x). \tag{3.16}$$

Rule V3. If x_1, x_2 are independent observations with expectations μ_1 and μ_2 respectively,
$$V(x_1 + x_2) = V(x_1) + V(x_2). \tag{3.17}$$

To prove Rule V3, we put
$$y = x_1 + x_2$$
then
$$\begin{aligned} E(y^2) &= E(x_1^2 + 2x_1 x_2 + x_2^2) \\ &= E(x_1^2) + 2\mu_1 \mu_2 + E(x_2^2) \end{aligned} \tag{3.18}$$
provided x_1 and x_2 are independent, and
$$\{E(y)\}^2 = \{\mu_1 + \mu_2\}^2 = \mu_1^2 + 2\mu_1 \mu_2 + \mu_2^2. \tag{3.19}$$
By subtracting (3.18) from (3.19), we have (3.17).

Rule V3 readily extends to any set $x_1, \ldots, x_n$ of independent random variables, to give
$$V(x_1 + \ldots + x_n) = V(x_1) + \ldots + V(x_n). \tag{3.20}$$

Example 3.9. Suppose $x_1, \ldots, x_n$ are independent observations with the same expectation μ, and the same variance $V(x_i) = \sigma^2$, then

$$V(\bar{x}) = V\left(\frac{1}{n}\sum x_i\right)$$

$$= \frac{1}{n^2}V\left(\sum x_i\right) \quad \text{by Rule V2,}$$

$$= \frac{1}{n^2}\sum V(x_i) \quad \text{by Rule V3,}$$

$$= \frac{1}{n^2}n\sigma^2 = \sigma^2/n. \qquad \square\,\square\,\square$$

The following two examples give some important results.

Example 3.10. Independent random variables x_i for $i = 1, 2, \ldots, n$, have a known expectation μ and unknown variance σ^2. We can estimate σ^2 using

$$s'^2 = \frac{1}{n}\sum_{i=1}^{n}(x_i - \mu)^2.$$

Now
$$E(s'^2) = E\left\{\frac{1}{n}\sum(x_i - \mu)^2\right\}$$

$$= \frac{1}{n}E\left\{\sum(x_i - \mu)^2\right\} \quad \text{by Rule E2,}$$

$$= \frac{1}{n}\sum E(x_i - \mu)^2 \quad \text{by Rule E3.}$$

But $E(x_i - \mu)^2$ is defined as the variance σ^2, of the x_i. Hence

$$E(s'^2) = \frac{1}{n}\sum V(x_i) = \sigma^2. \qquad \square\,\square\,\square$$

Example 3.11. Suppose in Example 3.10 the expectation μ is also unknown then we usually estimate σ^2 using

$$s^2 = \frac{1}{n-1}\sum(x_i - \bar{x})^2.$$

To find $E(s^2)$ we proceed as follows.

$$\sum(x_i - \bar{x})^2 = \sum(x_i - \mu + \mu - \bar{x})^2$$

$$= \sum(x_i - \mu)^2 + 2(\mu - \bar{x})\sum(x_i - \mu) + n(\mu - \bar{x})^2$$

$$= \sum(x_i - \mu)^2 - n(\bar{x} - \mu)^2,$$

By rules E2 and E3,

$$E\left\{\sum (x_i - \bar{x})^2\right\} = E\left\{\sum (x_i - \mu)^2\right\} - n\, E(\bar{x} - \mu)^2,$$

and from Example 3.10,

$$E\left\{\sum (x_i - \mu)^2\right\} = n\sigma^2.$$

Also, by definition of the variance
$$E(\bar{x} - \mu)^2 = V(\bar{x})$$
$$= \sigma^2/n \quad \text{by Example 3.9.}$$
Therefore substituting back we have

$$E\left\{\sum (x_i - \bar{x})^2\right\} = n\sigma^2 - n\sigma^2/n = (n-1)\sigma^2,$$

or
$$E(s^2) = \sigma^2. \qquad \square\,\square\,\square$$

Exercises 3.2

1. The expectations for the binomial, rectangular, exponential and normal distributions are set out in § 3.1. Show that the variances are as follows.

Binomial: $\quad V(x) = n\theta(1 - \theta)$
Rectangular: $V(x) = a^2/12$
Exponential: $V(x) = 1/\lambda^2$
Normal: $\quad\;\; V(x) = \sigma^2$ (use integration by parts in this).

Also find $V(x)$ for the geometric distribution of Exercise 3.1.2.

2. For the distribution of Exercise 3.1.5, find
$$V(x), \quad V(x \mid y = \tfrac{1}{2}), \quad V(xy).$$

3. Let x_i, $i = 1, 2, \ldots, n$, be independent, with
$$\Pr(x_i = 1) = \theta, \quad \Pr(x_i = 0) = 1 - \theta.$$
Show that $V(x_i) = \theta(1 - \theta)$, and hence use (3.20) to obtain the variance of the binomial distribution.

3.3 Higher moments

If some data are known to have been sampled randomly from a normal population, then the sample mean and variance do in fact summarize all the information contained in the data. More usually it may not be known that the population is normal, or perhaps it may be known to be exponential, etc.; in such situations some measures may be needed to represent further shape characteristics of a distribution, such as skewness. We shall deal here with population measures, although analogous definitions hold for sample data.

Now the expectation and variance of a random variable are $E(x) = \mu$, say, and $E(x - \mu)^2$. An obvious generalization is to consider $E(x - \mu)^3$, called the third moment, and we notice that this is zero in symmetrical distributions, and tends to be non-zero for skew distributions. The third moment will be positive for distributions shaped like Figure 2.11(a) and negative for those shaped like Figure 2.11(b) respectively.

The third moment cannot be used as it stands as a measure of skewness, for it can be made arbitrarily larger or smaller by altering the scale, and presumably we would want a given shape to have the same value for a measure of skewness, whatever its scale. We therefore consider

$$\gamma_1 = \frac{E(x - \mu)^3}{\{E(x - \mu)^2\}^{3/2}} \qquad (3.21)$$

as a measure of skewness.

Example 3.12. Poisson distribution with expectation μ.
$$E(r - \mu)^3 = \mu$$
$$\gamma = \mu/\mu^{3/2} = \mu^{-1/2}.$$

Thus as $\mu \to \infty$, $\gamma_1 \to 0$, and the Poisson distribution for a large mean is nearly symmetrical.   □ □ □

Another coefficient which measures the shape of a distribution is the coefficient of *kurtosis*, usually defined

$$\gamma_2 = \frac{E(x - \mu)^4}{\{E(x - \mu)^2\}^2} - 3. \qquad (3.22)$$

By defining kurtosis in this way, $\gamma_2 = 0$ for the normal distribution; see Exercise 3.3.1. Kurtosis measures the amount of probability distant from the mean μ; for distributions which are symmetrical and unimodal, this is a measure of flatness or peakedness of the distribution; see Figure 2.11(c) and (d).

So far we have dealt with moments about the mean, denoted

$$\mu_r = E(x - \mu)^r.$$

If moments are taken about the origin instead of about the expectation μ, we have moments about the origin, denoted μ'_r

$$\mu'_r = E(x^r).$$

When calculating the moments $\mu_2, \mu_3, \ldots$, it is useful to have the relations between moments about the mean, and moments about the origin: these relations can be obtained as follows.

$$\mu_r = E\{x - \mu'_1\}^r = E(x^r) - r\,E(x^{r-1})\mu'_1 + \ldots$$
$$= \mu'_r - r\mu'_{r-1}\mu'_1 + \frac{r(r-1)}{1\cdot2}\mu'_{r-2}\mu'^2_1 + \ldots$$

Thus we have

$$\left.\begin{array}{l} \mu_2 = \mu_2' - \mu_1'^2 \\ \mu_3 = \mu_3' - 3\mu_2'\mu_1' + 2\mu_1'^3 \\ \mu_4 = \mu_4' - 4\mu_3'\mu_1' + 6\mu_2'\mu^2 - 3\mu_1'^4 \end{array}\right\} \qquad (3.23)$$

Also, for moments about the origin in terms of moments about the mean, and the first moment μ_1',

$$\mu_r' = E(x - \mu_1' + \mu_1')^r = E(x - \mu_1')^r + r\,E(x - \mu_1')^{r-1}\mu_1' + \dots$$

Thus we have,

$$\left.\begin{array}{l} \mu_2' = \mu_2 + \mu_1'^2 \\ \mu_3' = \mu_3 + 3\mu_2\mu_1' + 4\mu_1'^3 \\ \mu_4' = \mu_4 + 4\mu_3\mu_1' + 6\mu_2\mu_1'^2 + 5\mu_1'^4 \end{array}\right\} \qquad (3.24)$$

Sample moments are defined in a way exactly analogous to the population moments defined above. Thus

$$m_r = \frac{1}{n}\sum (x_i - \bar{x})^r f_i \qquad (3.25)$$

where $\bar{x}$ is the sample mean, f_i the frequency of x_i, and $\sum f_i = n$ is the total number of observations.

The methods of summarizing information in a large number of observations by means of the location and dispersion statistics, third and fourth moments, etc., will in general only make sense for unimodal distributions. For some illustrations involving skewness and kurtosis, see Pearson and Please (1975).

Exercises 3.3

1. Find γ_1 and γ_2 for the binomial, rectangular, exponential and normal distributions.

2. Show that if independent random variables x_i, $i = 1, 2, \dots, n$, have $E(x) = \mu$, $E(x - \mu)^2 = \sigma^2$, and $E(x - \mu)^4 = \mu_4$, then

$$E(s'^4) = \{\mu_4 + (n - 1)\sigma^4\}/n$$

and hence that

$$V(s'^2) = (\mu_4 - \sigma^4)/n,$$

where

$$s'^2 = \frac{1}{n}\sum_{i=1}^{n} (x_i - \mu)^2.$$

Find $V(s'^2)$ if the x_i are normally distributed.

3.4 Dependence and covariance

Suppose x_1, x_2 are random variables having a joint p.d.f. $f(x_1, x_2)$, then by analogy with (3.10) and (3.13), the expectation of any function $g(x_1, x_2)$ is defined

$$E\{g(x_1, x_2)\} = \int \int g(x_1, x_2) f(x_1, x_2) \, dx_1 \, dx_2. \qquad (3.26)$$

In particular,

$$E(x_1 x_2) = \int \int x_1 x_2 f(x_1, x_2) \, dx_1 \, dx_2.$$

Now by analogy with the result of Exercise 3.1.6, it would be tempting to conclude that

$$E(x_1 x_2) = E(x_1) \, E(x_2), \qquad (3.27)$$

but this is in general false. The reason is that if we take arbitrary sets of numbers, the arithmetic average of a product is *not* in general the product of the separate averages. That is, the equation

$$\frac{1}{n}(x_1 y_1 + \ldots + x_n y_n) = \frac{1}{n}(x_1 + \ldots + x_n) \times \frac{1}{n}(y_1 + \ldots + y_n)$$

is in general false. This is to be contrasted with

$$\frac{1}{n}\{(x_1 + y_1) + \ldots + (x_n + y_n)\} = \frac{1}{n}(x_1 + \ldots + x_n)$$
$$+ \frac{1}{n}(y_1 + \ldots + y_n),$$

which is always true.

The equation (3.27) is, however, true if x_1 and x_2 are independent. Under independence we have from § 2.7,

$$f(x_1, x_2) = f_1(x_1) f_2(x_2)$$

so that

$$E(x_1 x_2) = \int \int x_1 x_2 f_1(x_1) f_2(x_2) \, dx_1 \, dx_2$$
$$= \int x_1 f_1(x_1) \, dx_1 \int x_2 f_2(x_2) \, dx_2$$
$$= E(x_1) \, E(x_2).$$

Example 3.13. Suppose x_1 and x_2 take on the values 0 and 1 with the following probabilities:

x_2	x_1		Marginal probability
	0	1	
0	0·10	0·40	0·50
1	0·40	0·10	0·50
Marginal probability	0·50	0·50	

The (marginal) expectations of x_1 and x_2 are each equal to $\frac{1}{2}$. Thus
$$E(x_1)\,E(x_2) = 0 \cdot 25.$$
On the other hand
$$E(x_1 x_2) = (0 \times 0)(0 \cdot 10) + (0 \times 1)(0 \cdot 40) + (1 \times 0)(0 \cdot 40)$$
$$+ (1 \times 1)(0 \cdot 10)$$
$$= 0 \cdot 10$$
which is not equal to $E(x_1) \cdot E(x_2)$. □ □ □

This property leads to a possible measure of statistical dependence. We define the *covariance* of x_1 and x_2 as
$$C(x_1, x_2) = E[\{x_1 - E(x_1)\}\{x_2 - E(x_2)\}]$$
$$= E(x_1 x_2) - E(x_1) \cdot E(x_2). \tag{3.28}$$

If the covariance between two variables is zero, we say that they are *uncorrelated*. However, this does *not* imply that the variables are statistically independent. Although variables which are independent are also uncorrelated, the reverse statement is not true, as may be easily seen from the following example

Example 3.14. Suppose the joint distribution of two random variables x_1 and x_2 is as given in the following table:

x_2	x_1			*Marginal probability*
	0	1	2	
0	0	0·25	0	0·25
1	0·25	0	0·25	0·50
2	0	0·25	0	0·25
Marginal probability	0·25	0·50	0·25	

The variables x_1 and x_2 are not independent, since the conditional distribution, say of x_1 given x_2, depends on the value of x_2. The reader will readily check that
$$E(x_1) = E(x_2) = 1, \quad E(x_1) \cdot E(x_2) = 1$$
$$E(x_1 x_2) = 0 \cdot 25 \times 2 + 0 \cdot 25 \times 2 = 1$$
$$C(x_1 x_2) = 1 - 1 = 0.$$

The variables x_1 and x_2 have zero covariance, even though they are dependent.

In spite of this last result, we shall see that the covariance often provides a useful measure of dependence. For a coefficient measuring dependence, we would like it to be independent of the scales in which x_1 and x_2 were measured. The *correlation coefficient* is therefore defined as

$$\rho(x_1, x_2) = \frac{C(x_1, x_2)}{\sqrt{\{V(x_1)\ V(x_2)\}}}. \tag{3.29}$$

The main properties of the correlation coefficient are as follows:

(*i*) The correlation coefficient always lies between -1 and $+1$, and takes the extreme values if and only if x_1 and x_2 are exactly linearly related. Proofs of these results are outlined in the exercises.

(*ii*) If x_1 and x_2 are independent, then $\rho = 0$ but the converse of this is not true.

The calculation of the correlation coefficient for a continuous distribution is set in the exercises.

Another use of covariance is in obtaining the variance of the sum of random variables. Let x_1 and x_2 be any two random variables, then

$$\begin{aligned}
V(x_1 + x_2) &= E\{(x_1 + x_2) - E(x_1 + x_2)\}^2 \\
&= E[\{x_1 - E(x_1)\} + \{x_2 - E(x_2)\}]^2 \\
&= E\{x_1 - E(x_1)\}^2 + E\{x_2 - E(x_2)\}^2 \\
&\quad + 2E[\{x_1 - E(x_1)\}\{x_2 - E(x_2)\}] \\
V(x_1 + x_2) &= V(x_1) + V(x_2) + 2C(x_1, x_2).
\end{aligned} \tag{3.30}$$

An important extension of this result is given in the exercises.

Now the definition (3.26) was given in general terms, and we could proceed to discuss the higher cross-moments, such as $E(x_1 x_2^3)$, $E(x_1^2 x_2^2)$, etc. However, the covariance is by far the most important function of this type, and only very rarely are the higher cross-moments needed.

Exercises 3.4

1. Random variables x_1 and x_2 take on values 0, 1, 2 with probabilities set out below. Find $C(x_1, x_2)$.

x_2	x_1		
	0	1	2
0	0·16	0·10	0·06
1	0·10	0·16	0·10
2	0·06	0·10	0·16

2. Find $C(x, y)$ for the distribution of Exercise 3.1.5, and hence find the correlation coefficient.

3. Show that if x_i are any random variables, and a_i are constants, for $i = 1, 2, \ldots, n$, then

$$V\left(\sum_1^n a_i x_i\right) = \sum_1^n a_i^2 V(x_i) + 2 \sum \sum_{i < j} a_i a_j \, C(x_i, x_j).$$

4. Random variables $x_1, \ldots, x_n$ have equal expectations μ and equal variances σ^2. The covariances are

$$C(x_i, x_{i+1}) = \rho \sigma^2, \quad i = 1, 2, \ldots, (n-1),$$

and all other covariances are zero. [This form of dependence between observations is a particular case of *serial correlation*.] Show that

$$V\left(\sum x_i\right) = \sigma^2 \{n + 2(n-1)\rho\},$$

and hence that

$$V(\bar{x}) = \sigma^2 \left\{ 1 + 2\rho\left(1 - \frac{1}{n}\right) \right\} \Big/ n.$$

Discuss the difference between this result, and the result of Example 3.9. Suggest a practical application in which serial correlation would hold to a reasonable approximation.

5. Let y and x be random variables (not necessarily independent), with variances of 1 and a^2 respectively. By considering $V\{y + (x/a)\}$ and $V\{y - (x/a)\}$, show that $|\rho| \leqslant |$. Also show that if $\rho = 1$, there is an exact relationship

$$x = ay + b. \quad a > 0,$$

and if $\rho = -1$, there is an exact relationship

$$x = -ay + b.$$

6. Calculate the correlation coefficient between x and y when the joint distribution of x and y is

$$f(x, y) = 2, \quad x + y \leqslant 1, \quad 0 \leqslant x, \quad 0 \leqslant y,$$
$$= 0, \text{ otherwise.}$$

3.5 Normal models

We have explained that instead of specifying a distribution in detail, we sometimes give only the expectation and variance, When our observations are normally distributed, these two quantities completely specify the distribution, since they define the parameters μ and σ^2 of (2.34). Therefore, frequently we begin an analysis of a set of data by assuming a model for the

expectation and variance of the observations. The model assumed is frequently chosen after a cursory analysis of the data, and we must be careful
throughout the analysis to include checks of the model. Consider the following
example.

Example 3.15. The data given in Table 3.1 below are extracted from the
results of an experiment by Gohn and Lysaght, which concerned the effect of
surface preparation on Rockwell hardness readings of metal blocks. In effect,
the hardness readings are measurements of the depth of impression made by a
diamond brale under a given load. The machine on which the measurements
are taken has to be reset each time a new series of observations is taken, and
'standard blocks' of known hardness are used for this. It is vital to determine what effect, if any, the surface preparation of these test blocks has.

Table 3.1 *The effect of surface preparation of metal blocks on Rockwell hardness
readings*

Laboratory 1	2	3
Original surface (denoted O)		
45·6	46·1	46·3
45·9	46·1	45·6
45·8	45·8	46·2
45·7	45·5	46·2
45·5	45·9	46·1
45·4	45·9	46·4
45·4	45·8	46·1
45·3	45·8	46·4
Mechanically polished surface (denoted M)		
46·9	46·4	46·3
46·9	46·6	46·3
46·7	46·2	45·9
46·9	45·9	46·8
47·0	46·2	46·5
46·8	46·3	46·0
46·9	46·4	46·4
46·5	46·2	46·3
Electropolished surface (denoted E)		
47·0	46·7	45·6
46·6	46·2	46·1
47·0	46·6	46·2
46·9	47·0	46·1
47·1	47·2	45·7
47·0	47·0	46·7
46·8	46·1	46·4
46·4	46·6	46·6

A circular test block was split into three segments, and one segment was electropolished, another mechanically polished, and the third left with the original ground surface. Table 3.1 records eight repeat measurements taken by each of three different laboratories on each segment; this data is part of a very much larger body of data, but it will be sufficient for our discussion. The aim of the experiment is to discover the effect of surface preparation on the observations.

Let us start to build up a model for this data as follows. For any given laboratory, different measurements on the same segment will vary due to measurement errors, variations in hardness over the segment, etc. Let the expectations of the measurements of, say, laboratory 1 on segments O, M and E be denoted μ_{1O}, μ_{1M} and μ_{1E}, respectively. Similarly, we can denote the variances by σ_{1O}^2, σ_{1M}^2, σ_{1E}^2, etc. We expect that the distributions of the results are approximately normal with these expectations and variances, but we can test for this; see below.

The next step is to examine the sample means and variances in the data, and these values are given in Table 3.2. This table can be regarded either as a table of summarizing statistics describing the data, or else, anticipating results of the next chapter, we can regard these as estimates of the 'true' μ's and σ^2's defined above.

Table 3.2 *Sample means and variances for Table* 3.1 *data*
(Variance entries are 100s²)

	Laboratory 1	2	3	Average over Labs
	Means			
O	45·57	45·86	46·16	45·86
M	46·82	46·27	46·33	46·47
E	46·85	46·68	46·18	46·57
	Variances			
O	4·50	3·70	6·55	4·92
M	2·50	4·21	7·39	4·70
E	5·71	15·07	15·36	12·04

The sample means are presented graphically in Figure 3.2. We see that the order of the three laboratories is different for each segment. For both laboratories 1 and 2, mechanical polishing and electropolishing both lead to increased hardness, but laboratory 3 shows almost identical means for the surface treatments. When interpreting such a set of means, we must be aware of the possibility of wild observations. For the experiment under discussion, such wild shots could easily arise for various reasons. For example, some observations may be taken too close to the edge of the test piece, or perhaps a series of observations had to be stopped for some reason, and the machine would then have had to be reset before continuing. A few wild observations can sometimes completely upset an otherwise consistent picture. There are methods available for examining data for possible outliers, see Tukey (1962); these suspect observations can then be examined, to see whether there was something unusual about the conditions under which they were taken, etc. In the data given in Table 3.1, there are two low values in the observations on segment E by Laboratory 3, namely 45·6 and 45·7, all other values being

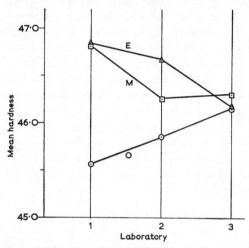

Figure 3.2 *Plot of the means of Table 3.1 data*

over 46.1. However, such gaps occur elsewhere in this data, and in any case the overall picture is not greatly simplified by omitting them. Great care should be taken before rejecting such observations merely on internal statistical evidence.

The nine sample means can be reduced further by assuming some further structure. We could consider each mean μ_{1O}, μ_{1M}, etc., as a combination of laboratory effect and a surface finish effect, such as

$$\left.\begin{aligned}
\mu_{1O} &= \mu_1 + \mu_O \\
\mu_{1M} &= \mu_1 + \mu_M \\
\mu_{1E} &= \mu_1 + \mu_E \\
\mu_{2O} &= \mu_2 + \mu_O,
\end{aligned}\right\} \text{ etc.} \tag{3.31}$$

If this picture holds, the means can be summarized merely in terms of the six quantities μ_1, μ_2, μ_3, μ_O, μ_M, μ_E. However, this structure implies that apart from random variation, the effect of surface preparation is the same in all three laboratories. If the data fitted model (3.31) exactly, it would look something like Figure 3.3, with three sets of parallel lines. The difference between the results for laboratories 1 and 2 on any segment would be simply

$(\mu_2 - \mu_1)$, and the difference between the results on segments M and O for any laboratory would be $(\mu_M - \mu_O)$. In fact, Figure 3.2 contrasts so greatly with Figure 3.3 that it is most unlikely that an additive model holds. We say that the effects of laboratories and surface preparation are *not additive* or alternatively we say that there is an *interaction* between these effects. Now since the observed sample means are subject to random fluctuations, it is possible for the underlying model to be additive, but a sizeable interaction appears due to these random errors. We must find a method of testing whether an apparent interaction is real or whether it is due to error. Other

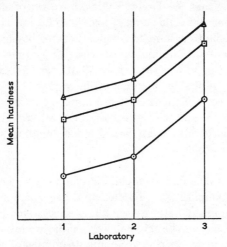

Figure 3.3 *Completely additive effects*

questions arise, such as whether the analysis is being done in the best scale but we shall not proceed with these points here. A more detailed discussion of models and interactions is given in Chapters 9 and 12.

We can analyse the sample variances in a similar way. The over-riding features of the variances of Table 3.2 are the large variances for laboratory 2 and 3 results on the electropolished segment. Even for laboratory 1 the electropolished segment has the largest variance. Unfortunately sample variances are more difficult to estimate than sample means, and they are more sensitive to the presence of wild observations. The data of Table 3.1 alone are not sufficient to be able to see much of the underlying pattern. We shall therefore leave this discussion at this point.

If we wish to check whether the distribution of the observations is normal, two lines of attack are immediately apparent. Firstly we could represent the eight observations in each sample as

$$\text{(observation} - \text{sample mean)}/\sqrt{\text{(sample variance)}} \qquad (3.32)$$

and then plot the cumulative distribution of these for the whole set of data, on normal probability paper. [The transformation (3.32) should reduce all observations to quantities having the same mean and variance. If this were not done we would have to keep each set of eight observations separate.] A second possibility is to calculate averaged values of skewness and kurtosis from the nine sets of observations. These problems are set as exercises for the reader. □ □ □

This example has been introduced at this point to illustrate the way in which we build up a *statistical model* for a set of data. In Example 3.15 we begin by assuming that the data of Table 3.1 are sets of independent random variables from nine different populations. As indicated above, we can check that it is reasonable to assume the populations to be normal, possibly each with different expectations and variances. We can now start to simplify further, by trying a structure such as (3.31) for the expectations, and a similar structure, probably in terms of $\log \sigma^2$, for the variances. A simple model provides a convenient and concise summary of the data, facilitates the comparison of the results with other sets of data, etc. Example 3.16 below illustrates further the idea of statistical model building, with reference to a more complicated problem.

Example 3.16. For a second example, we refer back to the sampling of baled wool problem, Example 1.10. A fundamental point here is that there are two stages in the sampling. Firstly we randomly select seven bales from the very large number of bales in the shipment. Secondly, we take four cores of wool from each bale; there are, of course, an infinite number of possible cores which could be taken from each bale. We are sampling from at least two populations therefore, and the set-up is summarized in Table 3.3. The quantity

Table 3.3 *Sampling operations for Example 3.2*

	First stage	*Second stage*
Population	All bales in shipment	All possible cores in a bale
Individual	A single bale	A single core
Sample	The 7 bales selected	The 4 cores selected

measured was the percentage clean content of the wool. Denote the actual percentage clean content of a bale by m_i, where the suffix i runs over all the bales in the shipment. We shall approximate to the population of all m_i in the shipment by a normal population with expectation μ and variance τ^2. The quantity μ is therefore the average (over all bales) of the actual percentage clean content in each bale, and τ^2 is a measure of the dispersion of the values of m_i among all the bales.

The second sampling operation is from each of the selected bales, and the percentage clean content will vary from one part of the bale to another. Therefore each selected bale represents a population from which we have sampled four cores. Let us suppose that these populations are also normal, with expectations m_i and variance σ_i^2, $i = 1, 2, \ldots$ The value m_i for the average of all possible core measurements, must of course be equal to the true percentage clean content of the bale. As a first approximation we might assume that all the σ_i^2's are equal, say, to σ^2, provided we incorporate some checks on the reasonableness of this assumption into the analysis. If this model is a good approximation, the whole set-up is described in terms of the three parameters μ, τ^2, and σ^2.

Therefore our statistical model is as follows. We denote the observation x_{ij}, where i runs over the seven bales and j runs over the four cores. Then we put

$$x_{ij} = \mu + m_i + \varepsilon_{ij}, \quad i = 1, 2, \ldots 7,$$
$$j = 1, 2, \ldots 4, \quad (3.33)$$

where μ is a constant, and m_i and ε_{ij} are independent normally distributed quantities with expectations zero, and variances τ^2 and σ^2 respectively.

Now clearly σ^2 can be estimated from the sample variances between cores within bales, and an average of these seven sample variances would seem appropriate. However, the between bale sample variance contains a component of variation due to σ^2, as well as that due to τ^2. We are still not ready for a full discussion of this problem, and it is considered further in Exercise 3.5.2 and Chapter 12.   □ □ □

In Example 3.16 our statistical model is (3.33), involving three parameters, μ, τ^2 and σ^2, but whereas in Example 3.15 it was fairly clear how to obtain estimates of parameters in the model, it is more difficult in Example 3.16. Problems connected with the estimation of unknown parameters in a statistical model are discussed in the next chapter. The important point for the reader to grasp from Examples 3.15 and 3.16 is the concept of a statistical model.

Exercises 3.5

1. Discuss possible statistical models for the bioassay experiment, Example 1.8, and the tobacco moisture experiment, Example 1.9.

2. Continue the discussion of Example 3.16 as follows. A combined within bales estimate of σ^2 is

$$s_1^2 = \frac{1}{21} \sum_{i=1}^{7} \sum_{j=1}^{4} (x_{ij} - \bar{x}_{i.})^2$$

where
$$\bar{x}_{i.} = \frac{1}{4}\sum_{j=1}^{4} x_{ij}.$$

Find $E(s_1^2)$.

The sample variance of the bale means is
$$s_2^2 = \frac{1}{6}\sum(\bar{x}_{i.} - \bar{x}_{..})^2$$

where
$$\bar{x}_{..} = \frac{1}{28}\sum_{i=1}^{7}\sum_{j=1}^{4} x_{ij}.$$

Use the model (3.33) to show that

$$6s_2^2 = \sum_{i=1}^{7}(m_i - \bar{m}_.)^2 + \sum(\bar{\varepsilon}_{i.} - \bar{\varepsilon}_{..})^2 + 2\sum(m_i - m_.)(\bar{\varepsilon}_{i.} - \bar{\varepsilon}_{..}).$$

Now since we have two sampling operations, we can take expectations over each separately, over the ε's, E_ε, and over the m's, E_m. Show that
$$E_m E_\varepsilon(s^2) = (\tau^2 + \sigma^2/4).$$

Hence suggest a method of obtaining estimates of σ^2 and τ^2. Does it matter whether s_1^2 and s_2^2 are independent or not?

3. Make the transformation (3.32) on each of the nine sets of observations in Table 3.1, and then

(*i*) Plot the cumulative distribution of all 81 transformed observations on normal probability paper. How many restrictions are there on these 81 transformed observations, and what effect do you think they will have on the normal plot?

(*ii*) Calculate averaged values of skewness and kurtosis for the whole set of data.

CHAPTER FOUR

Sampling Distributions and Statistical Inference

4.1 Statistical inference*

We have seen that the science of statistics is concerned with the interpretation of data in which random variation is present. There are two main aims in the statistical analysis, depending on the purpose for which the data were collected: one purpose is to reach a decision, such as which of a number of varieties of a crop gives a higher yield; another purpose is to summarize the information contained in the data. In either case we are faced with the problem of arguing from the particular to the general; that is, we attempt to argue – say from the particular set of crop yields observed – to a statement of what will generally happen if the experiment is repeated under similar circumstances. The theory of methods of reasoning from the particular to the general with data involving random variation, is called statistical inference. This chapter sets out the basic methods of inference, and a clear understanding of the principles set out here is essential to all that follows.

The first step is to set up a statistical model in which the data are assumed to correspond to random variables. The precise population sampled is unknown of course, otherwise there would be no problem of inference, but it is often possible to assume, for example, that the population is a normal population, or a binomial population, with unknown parameters, and in this case the problem of inference is a problem of making statements about possible values of the unknown parameters.

Example 4.1. In the manufacture of some measuring equipment, the meters are adjusted to make them work synchronously with a standard meter. After adjustment, a sample of 9 meters is tested using precision instruments. The constant of a standard meter is 1·000, and the deviations from this of the 9 meters tested are as follows:

$$-0{\cdot}009 \quad -0{\cdot}002 \quad +0{\cdot}005 \quad -0{\cdot}008 \quad +0{\cdot}011$$
$$-0{\cdot}007 \quad -0{\cdot}014 \quad -0{\cdot}006 \quad -0{\cdot}015$$

* Readers who find the first section difficult should work through the rest of the chapter and then re-read it.

The standard deviation of these observations is known from past tests of this type to be 0·0080. The question is whether there is any evidence that the constants of the adjusted meters deviate systematically from unity. If there is such evidence, an estimate of the average amount of the deviation is required in order to try and correct for it.   ☐ ☐ ☐

Example 4.1 illustrates the points made above. We assume a model that the population of results is $N(\theta, \sigma^2)$. If $\theta = 0$, $\sigma^2 = 1$, and we work to two places of decimals, this population can be thought of as being composed of a large number of cards each having a number, and the proportion having, say, 1·35 on them is the area of the normal curve between 1·345 and 1·355. If θ is not zero, and if σ^2 is not unity, the population is the same, but each number is multipled by σ, and θ is added. Our model is then, that the observations are a set of 9 independent drawings from this population, with some unknown value of θ, and $\sigma = 0·008$.

Two types of problem which arise with Example 4.1 can be written in the following form:

(i) Do a particular set of data give any evidence that θ, the population mean of the deviations, is different from zero?

(ii) If there is evidence that θ is not zero, an estimate of θ is required to decide on the amount of adjustment necessary. Before acting on this estimate, we would want to know how much it could be in error, and so we might also ask the question 'what is the best estimate of θ from the data, and between what limits is θ likely to lie?'

We shall proceed immediately to discuss an answer to question (i), and later on in the chapter discuss an answer to question (ii).

If θ is exactly zero, then our assumptions state that the observations are a random sample of 9 from a population $N(0, (0·008)^2)$. We can now consider a *reference set* defined by the collection of all samples of 9 which could be drawn from this particular normal population, each sample occurring with its appropriate frequency; we then consider whether it is reasonable to assume that the observed sample is randomly chosen from this reference set. Deviations from this population are expected to be in the average of the observed sample differing from zero, and therefore we consider the distribution of averages among the samples in the reference set. We can now judge whether the data of Example 4.1 give any evidence of a systematic deviation of the constants from unity, by comparing the mean of the observed sample with the reference set. Suppose the distribution of means among samples of the reference set is as given in Figure 4.1. That is, the relative frequency with which sample means occur in any specified interval is governed by the p.d.f. shown. In the figure, the observed sample is so far out in the tails of this distribution, that sample means as different or more different from zero as this would occur only very rarely *if* the data are sampled from a $N(0, (0·008)^2)$

population. In these circumstances we would believe rather that the meter constants deviate systematically from unity, so that although the population may be normal, the mean of the deviations is unlikely to be zero. One measure of the strength of evidence against the mean being zero is the probability of a sample mean as different or more different from zero, in the reference set.

This illustrates the main principle behind all methods of inference – the observed sample is compared to a reference set. In the discussion above we

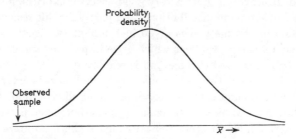

Figure 4.1 *Observed sample mean in relation to distribution of means in reference set*

have used the distribution of means among the samples in the reference set, and this is called the *sampling distribution* of means of 9 observations from a $N(0, (0 \cdot 008)^2)$ population. This sampling distribution is the distribution representing the means of all possible samples of nine from a $N(0, (0 \cdot 008)^2)$ population, with each possible sample represented with its appropriate frequency. Very often we shall refer to such distributions as sampling distributions, leaving the detailed specification of the distribution to be understood. In other problems we might use the sampling distribution of the variance or the range, that is, the distribution of these statistics in the reference set, and very much more complicated statistics arise. We can give the following general definition of a sampling distribution.

Definition. Suppose $x_1, \ldots, x_n$ are independent random variables from a distribution $F(x)$, and that we have a statistic $h_n(x_1, \ldots, x_n)$ of the sample values, then the distribution

$$\Pr\{h_n < h \mid F(x)\} \tag{4.1}$$

is called the *sampling distribution* of h_n. The *standard deviation* of the sampling distribution is called the standard error of h_n.

Thus the distribution specified by (4.1) would be generated empirically by drawing successive samples of n independent random variables from $F(x)$, computing $h_n(x_1, \ldots, x_n)$ in each case, and forming the distribution of the h_n's. This distribution will usually – but not always – depend on $F(x)$.

Most statisticians are agreed that nearly all problems of statistical inference are solved by considering sampling distributions. In a few cases several different reference sets have been proposed to answer the same problem, but a discussion of this is outside the scope of this volume. All our inferences will be based on sampling distributions.

There are three ways of examining sampling distributions:

(*i*) We could artificially sample from the specified population, and generate the distribution empirically. This is a very laborious method, and has only really become feasible on a large scale with the availability of large electronic computers, but when other methods fail, this method must be used. However, it gives a very good basis for explaining the meaning and uses of sampling distributions, and partly for this reason, a detailed discussion of the method is given in the next two sections. There are many complicated problems which arise in practice which defy mathematical analysis, and this method is employed.

(*ii*) We could approach this as a problem in probability theory, and attempt to derive mathematically the exact sampling distribution. If the exact distribution is too difficult to examine, sometimes asymptotic properties can be derived. No theoretical derivations of this kind are given in this volume, and the reader is referred to books on probability theory such as Cramér (1946) and Lindley (1964).

(*iii*) Frequently, the expectation and variance of a sampling distribution can be derived exactly or approximately, using the methods outlined in Chapter 3. If, further, the sampling distribution is known to be normal in the limit, by the central limit theorem, see § 2.6(*iv*), then the expectation and variance may be all we need to work out approximate methods of inference. Expectation and variance methods will be used frequently in this volume, since they provide a simple and often essential method of understanding statistical techniques.

Before proceeding to outline the basic methods of inference, we discuss in the next section how to draw random samples from a population artificially, and in § 4.3 a sampling experiment is described which illustrates method (*i*) above of examining sampling distributions.

Exercises 4.1

1. Write a short note on any problem calling for statistical analysis that you have met in science or everyday life.

Set up a statistical model, and consider the problems of inference which might arise.

2. Consider in detail Examples 3.15, 3.16, 1.8 and 1.9. For each one, discuss possible statistical models, and list the questions of inference you would wish to answer. Specify these questions as clearly as you can.

4.2 Pseudo random deviates

Random numbers
Nearly all methods of sampling start by obtaining random variables from the discrete rectangular distribution, which was described in § 2.5(*ii*). By use of a transformation, these deviates can be transformed into random variables from any other distribution. For example, random digits are observations on the distribution

$$\Pr(X = 0) = \Pr(X = 1) = \ldots = \Pr(X = 9) = 1/10,$$

and Kendall and Babington-Smith (1951) published a table of 100,000 random digits. Kendall and Babington-Smith (1938, 1939) produced their 'random digits' by use of a machine which involved a rapidly rotating disc with numbered sectors, illuminated by a flashing light. The series of digits produced by this or any other method have to be carefully tested before use, and the tests Kendall and Babington-Smith used are described below.

With the advent of electronic computers, new methods were required, capable of generating random digits rapidly inside the computer. A survey and discussion of some possible methods is given by Page (1959). One of the best methods to date is the multiplicative congruential method, and this method illustrates the kind of procedure used. It uses the equation

$$u_{r+1} = ku_r + c(\text{Mod } M)$$

where k, c and M are suitable constants, and the sequence of random numbers is u_r, starting at some fixed value u_1. The 'Mod M' in this equation means that we subtract as many multiples of M as possible, and take the *remainder* as our answer. The calculations outlined in Table 4.1 illustrate the method.

Table 4.1 *Illustration of multiplicative congruential method of generating random numbers*

Equation

$u_{r+1} = 29\ u_r(\text{Mod } 100)$ $u_1 = 2$

u_r	$29u_r$	u_{r+1}
2	58	58
58	1682	82
82	2378	78
78	2262	62
62	1798	98
98	2842	42

The reader should work through Exercise 4.2.1 and convince himself that the method seems to work well. Two features stand out:
(*i*) Depending on the constants k, c, M and u_1, the series generated is either all even numbers, all odd numbers, or alternating even and odd. However,

if we use a large enough number of decimal places the last digit can be ignored.

(*ii*) The method must generate a closed cycle of numbers, or possibly degenerate, producing the same number for all r above some value. This is clear, since the method could be continued to produce an endless sequence, but for instance, in the example in Table 4.1, there are only 100 different pairs of digits. Therefore the sequence must either repeat itself at some point, and be a closed cycle from then onwards, or else it degenerates. (For a degenerate series, try $u_1 = 25$ in Table 4.1.) If the constants are well chosen, enormous closed cycles can result, and the sequences then behave very much like random variables from the discrete rectangular distribution. Some suitable sets of constants are given in Exercise 4.2.1. For some theorems relating to the choice of the constants, see Hammersley and Handscomb (1964). If M is chosen to be large, and the results are divided by M, the output from this method approximates to a continuous rectangular distribution on the range (0, 1).

As mentioned above there are a number of other reasonably satisfactory methods of generating pseudo random numbers, besides the multiplicative congruential method. Whether any particular method is satisfactory or not depends to some extent on the use made of the numbers. The multiplicative congruential method has been found unsatisfactory from some points of view; see Peach (1961).

Before proceeding, an entirely different source of random series should be mentioned. The method is based on the emissions from a radioactive source with a very long half-life, and in § 2.6(*iii*) and Exercise 2.5.5 we stated that exponential or Poisson random variables can be obtained directly from such a source. This method is the only one which might properly be called random, and must be used in any situation such as the (British) Premium Bond prize draw, in which no deterministic sequence can be permitted; see Thompson 1959) for an analysis of this method. For most purposes, however, there is a great advantage in having sequences which can be reproduced exactly, to facilitate the checking of calculations and computer programmes.

'Tests' of randomness

Suppose we have a series of digits, which we hope are random, produced by one of the methods listed above. The next step is to examine the series for departures from randomness. A thorough discussion of various tests is beyond the scope of this volume, but the four tests used by Kendall and Babbington-Smith are as follows.

Gap test. The numbers of digits between successive zeros were counted, and a frequency distribution calculated. The theoretical distribution which applies in a truly random series is easily calculated, see Exercise 2.5.3 (cf. Example 2.4), and this can be compared with the observed distribution.

The remaining three tests were done on blocks of 1000 numbers, the series being artificially broken up for the purpose.

Frequency test. The frequencies of occurrence of each digit 0, 1, ... , 9, were calculated in each block, and compared with the theoretical expectation of 100.

Serial test. Similarly, the frequencies of occurrence of all pairs of digits, 00, 01, ... , 99, were tested.

Poker test. Each block of 1000 digits was broken up into 250 sets of four. Each set of four was classified as type (*a*) all four digits the same, (*b*) three digits the same and one different one, (*c*) two pairs, (*d*) one pair, or (*e*) all different digits. The frequency of occurrence of these types in a block of 1000 was calculated and compared with expectation.

A precise method of comparing observed and theoretical frequencies for these tests is discussed in Chapter 8, but in any case there is one very serious difficulty with 'tests of randomness'. If we break a series up into blocks, then in a truly random series, occasional blocks will occur which show glaring discrepancies with the theoretical frequencies, for some test. Indeed, if such glaring discrepancies do not occur with their appropriate frequency, the series cannot be random! The usual solution to this is to mark blocks which fail any test, and only use them in conjunction with other blocks. The whole series would not be rejected unless rather more blocks fail a given test than would be expected from chance fluctuations.

Random variables from discrete distributions

Let us suppose we have obtained a series of pseudo random numbers, and that these have been converted by division into pseudo random variables from the rectangular distribution on the range (0, 1). For example, we might have the results shown in Table 4.2.

Table 4.2 *Some typical pseudo random rectangular deviates*

0·3874	0·3195	0·3451	0·7304	0·9502
0·1773	0·4512	0·4340	0·2812	0·8408
0·9363	0·6269	0·1218	0·7263	0·1968

Suppose we want to generate a series of observations from a binomial population, with $n = 10$, $p = 0.20$.

The individual and cumulative probabilities are set out in Table 4.3. We now take the results from Table 4.2, and proceed as follows:

For values between   0·0000 and 0·1073   take $r = 0$
                      0·1074   0·3757 $r = 1$
                      0·3758   0·6777 $r = 2$
                                                    etc.

Table 4.3 *Individual and cumulative binomial probabilities for* $n = 10$, $p = 0.20$

	r 0	1	2	3	4
Prob.	0·1074	0·2684	0·3020	0·2013	0·0881
Cum. prob.	0·1074	0·3758	0·6778	0·8791	0·9672

	r 5	6	7	8
Prob.	0·0264	0·0055	0·0008	0·0001
Cum. prob.	0·9936	0·9991	0·9999	1·0000

The limiting numbers for choosing the values of r are determined by the cumulative probability distribution. If the original observations are rectangular on the range $(0, 1)$, then to four places of decimals, each of the numbers 0000, 0001, . . . , 9999, has the probability of $1/1000$. Now to four places of decimals there are 0000 to 1073, or 1074 numbers of this set for which we take $r = 0$, and 1074 to 3757, or 2684 numbers for which we take $r = 1$, etc. This procedure therefore generates the numbers 0, 1, 2, . . . , with the probabilities given in Table 4.3. Table 4.2 leads to the sequence

$$2, 1, 1, 3, 4, 1, 2, 2, 1, 3, 4, 2, 1, 3, 1.$$

This procedure can be followed for any discrete distribution – all we need is a table similar to Table 4.3 giving the cumulative probability distribution, and a set of random rectangular deviates.

Random variables from continuous distributions
The method outlined above also works for continuous distributions, provided we use a limited number of decimal places. Suppose we want to generate random normal deviates, which are random variables from the standard normal distribution, and suppose one decimal place is required. We start by considering the cumulative distribution, and Table 4.4 shows some values

Table 4.4 *Extracts from the standard normal c.d.f.*

X	0·00	0·05	0·10	0·15	0·20	0·25
c.d.f.	0·5000	0·5199	0·5398	0·5596	0·5793	0·5987

If we are using a large number of decimal places, any rectangular deviate between 0·5000 and 0·5198 inclusive has an associated value of X less than 0·05, and greater or equal to zero. Also any rectangular deviate between 0·5199 and 0·5595 inclusive, has an associated value of X in the range

$0.05 \leqslant X < 0.15$. When these are rounded up to one decimal place they are represented by 0·0 and 0·1 respectively. Working in this way we see that the table of values to transform rectangular deviates to random normal deviates with one decimal place will be as set out in Table 4.5. The reader should work through Exercise 4.2.2 at this point.

Table 4.5 *Table for obtaining random normal deviates**

	X 0	0·1	0·2	0·3	0·4	0·5
from	0·5000	0·5199	0·5596	0·5987	0·6368	0·6736
to	0·5198	0·5595	0·5986	0·6367	0·6735	0·7087

	X 0·6	0·7	0·8	0·9	1·0	
from	0·7088	0·7422	0·7734	0·8023	0·8289	
to	0·7421	0·7733	0·8022	0·8288	0·8530	etc.

* For $p < 0.5000$, find x for $(1 - p)$ and make it negative.

If this procedure is used in an electronic computer, we would have to store a very large table indeed to get many decimal places, and looking up the table inside the computer may involve a great deal of computer time, even if interpolation is used. For these reasons mathematical ways are sought to transform rectangular deviations into deviates of the required distribution. Suitable transformations to do this are given by Hastings (1955) and Wetherill (1965) for the standard normal distribution. As a very simple example of the basic method employed the reader should demonstrate that by taking logarithms of rectangular deviates, exponential deviates are obtained. See Exercise 4.2.3.

For small sampling experiments, not using a computer, tables of random deviates from the common continuous distributions are available. These are particularly valuable for class exercises.

Exercises 4.2

1. Use the multiplicative congruential method of generating random numbers, with one of the following sets of numbers,

$$k: \quad 11 \quad 51 \quad 201 \quad 20{,}001$$
$$c: \quad 3 \quad 1 \quad 1 \quad 1$$
$$M: \quad 100 \quad 1000 \quad 10{,}000 \quad 1{,}000{,}000$$

and $u_1 = 1$ in each case.

By dividing by M, produce a series of approximate rectangular (0, 1) deviates. Devise and carry out some suitable checks of randomness on your data.

2. Use the data from Exercise 4.2.1 above, or else random digits from a table, to produce random normal deviates from the distribution $N(0, 1)$, rounded up to one decimal place. Generate 20 such deviates, and then calculate the sample mean and variance, which should be close to zero and one respectively

3. Use the data from Exercise 4.2.1 above, or else random digits, and form pseudo rectangular deviates on the range $(0, 1)$. Take 30 such numbers and calculate the modulus of their logarithms. These variables should have an exponential distribution, with parameter $\lambda = 1$. Compare the empirical and theoretical cumulative distributions.

4.3 A sampling experiment

We are now in a position to return to the discussion of § 4.1, and to generate empirically a sampling distribution. The discussion of Example 4.1 led to the problem of examining the sampling distribution of means of 9 observations from a normal distribution. The sample mean was used because we thought that the expectation θ of the observed data was not zero, and that the sample mean would reveal this. However, we could have used the sample median instead, or a number of other quantities. The question is: which statistic is best, and why? The sampling experiment we discuss next helps us to answer this. We shall generate the sampling distribution of a number of statistics at the same time, and study the relationships and differences between them.

Suppose we have available a series of random normal deviates with variance one and expectation *five*. An expectation of five is used to reduce the number of negative values, and make the calculations easier; see below. The random deviates are presented in groups of *nine*, and for each group $x_1, \ldots, x_9$, we calculate:

(*i*) The sample mean, $\bar{x} = \dfrac{1}{9} \sum_1^9 x_i.$

(*ii*) The (sample) standard deviation,

$$s = \sqrt{\left\{ \frac{1}{8} \sum_1^9 (x_i - \bar{x})^2 \right\}}.$$

(*iii*) The sample median, $\dot{x}$.

(*iv*) The sample range, w.

(*v*) The mean of the seven observations remaining after the largest and the smallest have been omitted in each group. We denote this $'\bar{x}'$ and refer to it as the truncated mean.

Some specimen calculations are shown in Table 4.6, and the reader should work through these and check that he can repeat them. In calculating the

Table 4.6 *Specimen calculations for sampling experiment*
Data: 5 groups of 9

Group										Total	Uncorrected sum of squares (USS)
1	4·33	3·41	6·57	5·35	6·85	6·28	4·14	3·90	6·72	47·55	266·0333
2	5·97	6·82	3·37	5·36	3·19	5·33	5·57	5·88	4·17	45·66	243·8130
3	5·28	5·91	5·57	3·92	5·13	6·04	5·33	5·20	3·79	46·17	241·8093
4	3·39	6·39	5·57	5·62	3·38	5·04	5·12	4·63	4·47	43·61	219·3917
5	5·86	3·76	3·17	5·23	3·89	4·13	2·27	4·94	5·83	39·08	181·6134

Calculations

	Group 1	2	3	4	5
Total	47·55	45·66	46·17	43·61	39·08
Mean	5·28	5·07	5·13	4·84	4·34
USS	266·0333	243·8130	241·8093	219·3917	181·6134
Corr.	251·2225	231·6484	236·8521	211·3147	169·6940
CSS	14·8108	12·1646	4·9572	8·0770	11·9194
Variance, s^2	1·8514	1·5206	0·6197	1·0096	1·4899
St. deviation, s	1·36	1·23	0·79	1·00	1·22
Median, $\dot{x}$	5·35	5·36	5·28	5·04	4·13
Range	3·44	3·63	2·25	3·01	3·59
Truncated mean, $'\bar{x}'$	5·33	5·09	5·19	4·83	4·42

sample mean and variance on a hand calculating machine, the sum and the sum of squares of the numbers can be calculated at the same time provided there are no negative numbers; see Appendix I on 'Desk Computation'. Making the population mean five instead of zero does not alter the range or the variance of any sample, and makes both easier to calculate; the other quantities calculated are changed merely by the addition of five. The reader should work through these calculations with at least ten groups of nine using random normal deviates from Wold (1954). However, the best value from the exercise can be had in a class, if several students are given different sets of random deviates, the answers duplicated and issued, and then analysed along the following lines. At least 100 groups of nine are needed.

Firstly, suitable grouping intervals should be chosen for the statistics, frequency tables calculated, and histograms plotted. Frequency tables for statistics (*i*), (*ii*) and (*iii*) above are given in Table 4.7, and histograms of (*i*) and (*iii*) shown in Figure 4.2. Figure 4.2(*a*) gives the histogram of the sample means, and it indicates how much sample means of nine observations vary about the population mean. If a very large number of groups of nine deviates were sampled, and the grouping interval of the histogram shrunk, the histogram would approach to the smooth curve representing the sampling distri-

Table 4.7 *Results of sampling experiment. Frequency tables of the mean, median and standard deviation in 200 samples of nine from a* $N(5, 1)$ *population*

	(a) mean	(b) median	(c) standard deviation	
<4·10	1	1	<0·45	0
>4·10	0	2	>0·45	5
>4·20	0	1	0·55	12
4·30	6	9	0·65	21
4·40	5	6	0·75	11
4·50	8	14	0·80	12
4·60	19	8	0·85	26
4·70	19	15	0·90	16
4·80	22	17	0·95	13
4·90	17	17	1·00	18
5·00	18	22	1·05	10
5·10	23	10	1·10	10
5·20	17	21	1·15	6
5·30	19	16	1·20	7
5·40	12	12	1·25	9
5·50	7	13	1·30	11
5·60	4	4	1·35	7
5·70	2	2	1·45	3
5·80	1	4	1·55	2
5·90	0	2	1·65	1
>6·00	0	4	1·75	0
Total	200	200	Total	200

bution of the mean. The interpretation of this smooth curve is that in repeated sampling from the same population, the limiting relative frequency of sample means of nine deviates lying between any two values, say between 4·5 and 5·0, is the area under the sampling distribution between these values. This interpretation is clear from the way in which the histogram is generated.

If the data of Table 4.7(*a*) are plotted on normal probability paper, see Exercise 4.3.1, a very nearly linear graph is obtained, with a slope representing a standard deviation of about 1/3. It is clear from expectation and variance arguments, see Example 3.9 and Exercise 3.1.4, that

$$E(\bar{x}) = \theta \quad V(\bar{x}) = \sigma^2/n$$

where θ and σ^2 are the population mean and variance, in this case five and unity respectively, and n is the number of observations, which is nine in this case. Further, it is proved in Exercise 4.3.4 that the sampling distribution of the mean is exactly a normal distribution, if the original population is normal.

It is clear that because the population is symmetrical the sampling distribution of the median must be a symmetrical distribution centred on the

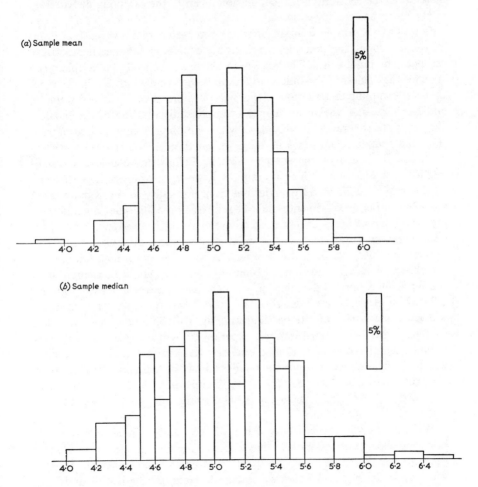

Figure 4.2 *Histograms of sample mean and median*

population mean. However, a normal probability plot of the data of Table 4.7(*b*) confirms that the distribution is not exactly normal, although it can be proved that it is asymptotically normal for large sample sizes; see Cramér (1946). In comparing the mean and median, it is obvious from Table 4.7 and Figure 4.2 that the sampling distribution of the median is much more spread than the sampling distribution of the mean. In fact, the variances of Table

4.7(a) and 4.7(b) frequency distributions are about 0·12 and 0·19 respectively.*
It can be proved that the ratio of the variance of the median to the variance
of the mean is $\pi/2 = 1·57$ in large samples. Thus although both the mean
and the median have a sampling distribution with almost the same symmetrical shape centred on the population mean θ, the sampling distribution
of the mean has the lower spread.

If we have a sample from any normal population with expectation θ unknown, we could, for example, use the sample mean or the sample median to
estimate the unknown θ. The reasoning above shows that the sample mean
is on average closer to the unknown θ than is the sample median, in the sense
that the sampling distribution of the mean is more closely grouped round θ,
and has a smaller variance, than the sampling distribution of the median.
We can state this rather loosely, that when sampling from a normal population, the sample mean gives us more information about the unknown expectation θ than does the sample median. Before this conclusion can be
applied to a situation such as Example 4.1, we must examine how this conclusion might be affected by deviations from normality in the population. A
series of sampling experiments could be run to answer this question, and there
are two different kinds of deviation from normality to examine. Firstly, the
data might be drawn from one population which has skewness or kurtosis,
or both. Secondly, most of the data may be very nearly normal, but some of
the observations are wild shots or blunders and these might be thought of as
coming from a normal distribution with the same mean, but with a grossly
inflated variance. The reader should consider this question, and think out
what the qualitative effects on the sampling distributions of the mean and
median might be. A few sampling experiments will help considerably providing extreme conditions are chosen: see Exercise 4.3.2. A further point
which is left as an exercise for the reader is how the sampling distribution of
the truncated mean, statistic (v) above, compares with those for the mean
and median, and further how this statistic might be affected by non-normality of the population.

We have as yet covered only a few of the points which arise from the
sampling experiment detailed above. Some further questions are left as
exercises for the reader, but one question of great importance will be dealt
with explicitly. Figure 4.3 shows a scatter diagram of $\bar{x}$ against s^2, and Table
4.8 shows frequency distributions of s for $\bar{x} > 5·0$ and $\bar{x} < 5·0$. There is no
evidence that the sampling distribution of s^2 is in any way affected by the
value of $\bar{x}$ from the same sample. Exercise 4.3.5 shows that when sampling
from a normal population, $\bar{x}$ and s^2 are uncorrelated, and further it can be
shown that $\bar{x}$ and s^2 are statistically independent; see Cramér (1946) and
Exercise 5.4.2. [See § 3.4 for a brief discussion of statistical independence.]
The reader will find it illuminating to compare Figure 4.3 with the plots of $\bar{x}$

* The extreme observations were counted individually.

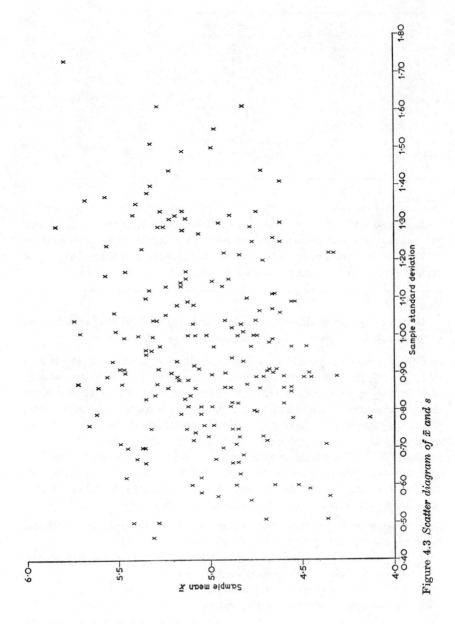

Figure 4.3 *Scatter diagram of* $\bar{x}$ *and* s

Table 4.8 *Frequency distribution of the sample standard deviation for two ranges of the sample mean*

S	$\bar{x} < 5 \cdot 0$	$\bar{x} > 5 \cdot 0$
$<0 \cdot 65$	10	7
$<0 \cdot 80$	16	16
$<0 \cdot 90$	22	16
$<1 \cdot 00$	13	16
$<1 \cdot 10$	13	15
$<1 \cdot 20$	7	9
$<1 \cdot 30$	8	8
$<1 \cdot 45$	6	12
$<1 \cdot 65$	2	3
$<1 \cdot 85$	0	1
Total	97	103

versus $\dot{x}$ and s versus w, see Exercise 4.3.3(*ii*) and (*iii*), since both the latter plots reveal dependence between the relevant statistics. The independence of $\bar{x}$ and s^2 in normal samples is an important feature in the development of certain statistical methods in subsequent chapters.

Exercises 4.3

1. Plot the frequency distributions for cases (*a*) and (*b*) of Table 4.7, on normal probability paper. What conclusions do you draw?

2. Generate random variables as follows. Take a random digit, and if it is 0, 1, . . . , 7, generate a random deviate from $N(0, 1)$, but if the random digit is 8 or 9, generate a random deviate from $N(0, 36)$. [This procedure simulates the situation in which 20% of data is made up of wild observations, which are thought of as having a standard deviation 6 times that ordinarily obtained.] Do this independently, 25 times, and group the results into 5 groups of 5 observations. For each group, obtain the mean and the median, and then calculate the sample variances of the means, and of the medians. Which statistic is more seriously affected by wild observations? Design a set of sampling trials to answer this question more effectively.

3. This question is for those who have sets of results from the empirical sampling experiment described in this section.
   (*i*) Calculate and plot the frequency distribution for '$\bar{x}$'. Also calculate the mean and variance of the distribution.
   (*ii*) Plot the scatter diagram of $\bar{x}$ versus $\dot{x}$.
   (*iii*) Plot the scatter diagram of s versus w.
   (*iv*) Plot a histogram of s, and of log s. Calculate the mean and variance of the distribution of s.
Discuss the results you obtain.

4.* If x is $N(\mu_1, \sigma_1^2)$ and y is $N(\mu_2, \sigma_2^2)$, the p.d.f. of $z = x + y$ is

$$\int_{-\infty}^{\infty} \frac{1}{\sqrt{(2\pi)}\sigma_1} \exp\left\{-\frac{(x - \mu_1)^2}{2\sigma_1^2}\right\} \cdot \frac{1}{\sqrt{(2\pi)}\sigma_2} \exp\left\{-\frac{(z - x - \mu_2)^2}{2\sigma_2^2}\right\} dx$$

Carry out this integration and find the distribution of z. Hence find the distribution of $\bar{x}$, when x_i are independent $N(\mu, \sigma^2)$.

5. Assume that x_i, for $i = 1, 2, \ldots, n$, are independent $N(0, \sigma^2)$. Put

$$\bar{x} = \frac{1}{n}\sum_1^n x_i, \quad s^2 = \frac{1}{(n-1)}\left\{\sum_1^n x_i^2 - \frac{1}{n}\left(\sum_1^n x_i\right)^2\right\}$$

and show that $C(x_i, s^2) = E(x_i s^2) = 0$ and hence that $C(\bar{x}, s^2) = 0$. Show also that $C(\bar{x}, s^2)$ remains zero if x_i are $N(\theta, \sigma^2)$ instead of $N(0, \sigma^2)$.

6. Explain the terms sampling distribution and standard error in relation to the statistics $\bar{x}$ and s^2. (See the definition (4.1).)

4.4 Estimation

Point estimation
Let us now refer back to Example 4.1, the measuring equipment example, and summarize our position with regard to it. The observations made were deviations of the meter constants from 1·000, and these deviations are assumed to be normally distributed with a known variance σ^2, but unknown mean θ. We therefore have the problem of estimating the unknown value of this parameter from the set of 9 results observed.

Definition. An *estimator* $\tilde{\theta}$ of θ is defined as any function of sample values which is calculated in order to be close to the true value of some unknown parameter θ.

Some possible estimators of θ in Example 4.1 are the sample mean, the sample median, the truncated mean, and there are many others. A choice between these alternative estimators is based upon the sampling distributions of these statistics. The comparison between the mean and the median, discussed in the previous section, is particularly clear cut.

If the observations x_i are sampled from a $N(\theta, \sigma^2)$ population, then from Example 3.9 and Exercise 3.1.4,

$$E(\bar{x}) = \theta \quad \text{and} \quad V(\bar{x}) = \sigma^2/n.$$

Since the population is symmetrical it is clear that $E(\dot{x}) = \theta$, and it can be shown that

$$V(\dot{x}) \simeq \pi\sigma^2/2n$$

for all θ. The sampling distribution of the median is known to be asymptotically normal, so that the sampling distributions of $\bar{x}$ and x are nearly identical in shape, and they are both symmetrical about θ, but the sampling distribution of $\bar{x}$ has the smaller spread. Clearly, if the assumptions of normality and independence are valid, $\bar{x}$ is a better estimator of θ than $\dot{x}$.

Many comparisons between estimators are not so clear cut as the comparison between $\bar{x}$ and $\dot{x}$, and in the last analysis are a matter of judgment based on sampling distribution properties. For example, the discussion in the previous section shows that a comparison between the sample mean and the truncated mean would be more difficult. A decision as to which to use will depend on whether we think any outlying observations represent wild shots or alternatively that they indicate non-normality of the population; the decision may also depend on other factors. Indeed, a thorough discussion of estimation would be very difficult, and out of place here. The reader should concentrate on the basic point, that the relative merits of estimators are judged by their sampling distributions, and that comparisons are sometimes, but not always, clear cut.

One property of estimators which is frequently referred to is *unbiasedness*.

Definition. An *unbiased estimator* of θ is one for which the expectation over the sampling distribution is equal to θ. That is,

$$E(\tilde{\theta}) = \theta,$$

for all values of θ.

We nearly always want estimators to be approximately unbiased, but there is no reason why we should insist on *exact* unbiasedness. It is more important that estimators have a compact sampling distribution.

For another example of comparisons between estimators, consider the estimation of the standard deviation σ of a normal distribution of unknown mean θ. The usual estimator of σ is the square root of the sample variance,

$$s = \sqrt{\left\{ \frac{1}{(n-1)} \sum_{i=1}^{n} (x_i - \bar{x})^2 \right\}}$$

where $x_1, x_2, \ldots, x_n$, are the observations. Another estimator can be based on the range

$$w_n = x_{(n)} - x_{(1)},$$

where $x_{(n)}$ is the largest observation and $x_{(1)}$ the smallest. It can be shown that for a normal population,

$$E(w_n) = \alpha_n \sigma$$

and $$V(w_n) = \beta_n \sigma^2$$

where α_n and β_n are functions of n only, and are tabulated for example, in Biometrika Tables, Pearson and Hartley (1966). Therefore (w_n/α_n) will be an unbiased estimator of σ, with a variance $\sigma^2(\beta_n^2/\alpha_n^2)$; see § 5.2.

In order to decide whether to use (w_n/α_n) or s to estimate σ, we shall com-

pare the spread of the sampling distributions of these statistics as measured
by their variances. Table 4.9 below shows some values of the ratio

$$\frac{\text{variance (range estimate)}}{\text{variance } \{\sqrt{\text{(sample variance)}}\}}$$

For $n = 2$ the estimators are identical, but for $n \geqslant 3$, the sample variance
estimate of σ has the sampling distribution with the smaller spread, as measured by the variance. However, for $n \leqslant 10$, the range estimate has a
reasonably high efficiency, and since it is so easy to calculate, it may be preferred on the grounds of simplicity in small sample sizes. Thus on practical
grounds we may on occasions prefer an estimator which is clearly not the
best from a sampling distribution viewpoint.

Table 4.9 *Ratio of variances of unbiased estimates of the standard deviation*
Variance (range estimate)/variance (sample st. dev. estimate)

n	2	3	4	5	6	7	8	9	10	15	20
Ratio	1·00	1·01	1·02	1·05	1·08	1·10	1·12	1·15	1·17	1·30	1·43

Interval estimates
In many applications it is important to be able to state limits within which
we expect a parameter to lie, and this question arose in our discussion of the
testing of measuring equipment example, Example 4.1.

Suppose we have observations $x_1, x_2, \ldots, x_n$, which are taken from a
normal distribution with known variance σ^2, but unknown mean θ, and we
wish to obtain an interval estimate for θ. We know that the sampling distribution of $\bar{x}_n$, the sample mean, is $N(\theta, \sigma^2/n)$, so that we know the shape of
this distribution but not its location. The situation is represented in Figure
4.4, where A or B are possible locations for the distribution. Suppose the

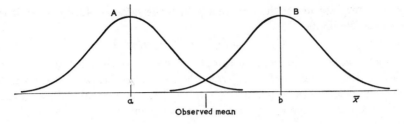

Figure 4.4 *Interval estimates for a normal mean*

observed sample mean is indicated by the line, then if the sampling distribution were located at A, or to the left of A, an event has occurred which we
know happens very rarely. On this basis it seems unlikely that θ is to the left
of some rather vaguely defined point 'a', and similarly it is unlikely to be to
the right of b. This intuitive approach can be formalized as follows.

Since the sampling distribution of $\bar{x}_n$ is $N(\theta, \sigma^2/n)$, then

$$\Pr\{\theta - \lambda_{\alpha/2}\sigma/\sqrt{n} < \bar{x}_n < \theta + \lambda_{\alpha/2}\sigma/\sqrt{n}\} = 1 - \alpha \qquad (4.2)$$

where
$$\alpha/2 = 1 - \Phi(\lambda_{\frac{1}{2}\alpha}),$$

so that $\lambda_{0 \cdot 01} = 2 \cdot 326$, $\lambda_{0 \cdot 025} = 1 \cdot 96$, etc. This statement means that if a series of independent samples of size n are taken from the same normal population – as in the sampling experiment – we obtain a series of sample means, and in the long run, 95% of these sample means will be within $1 \cdot 96$ standard errors of the true value of θ. This simply states the meaning of a sampling distribution, and the reader could check empirically how many sample means came within $1 \cdot 96$, $2 \cdot 326$, etc., standard errors of $5 \cdot 0$ in the sampling experiment. Now the inequalities in (4.2) can be restated, for

$$\theta - \lambda_{\alpha/2}\sigma/\sqrt{n} < \bar{x}_n \quad \text{is equivalent to} \quad \bar{x}_n + \lambda_{\frac{1}{2}\alpha}\sigma/\sqrt{n} > \theta$$

so that (4.2) can be written

$$\Pr\{\bar{x}_n - \lambda_{\alpha/2}\sigma/\sqrt{n} < \theta < \bar{x}_n + \lambda_{\alpha/2}\sigma/\sqrt{n}\} = 1 - \alpha. \qquad (4.3)$$

This restatement of (4.2) is equivalent to the intuitive argument given above, that θ is unlikely to be outside (a, b) in Figure 4.4.

In equation (4.3) the two parts of the inequalities either side of θ involve quantities which are known, once $\bar{x}_n$ is observed, and it looks like a probability statement about inequalities on the unknown value θ. However, the unknown value of θ is a perfectly definite quantity even though it is unknown, and θ does not have a probability distribution. Suppose, for example, that a chemist performs a number of repeat weighings on a chemical which does not deteriorate or evaporate. The chemist's weighings will differ slightly, and a confidence interval could be constructed for the true weight of the chemical. This does not imply that the true weight of the chemical has a probability distribution, for this true weight is a precise value, even though it is unknown.

The interpretation of the intervals (4.3) is the same as (4.2); that is, if we calculate the limits

$$(\bar{x}_n - \lambda_{\frac{1}{2}\alpha}\sigma/\sqrt{n}, \ \bar{x}_n + \lambda_{\frac{1}{2}\alpha}\sigma/\sqrt{n}) \qquad (4.4)$$

for a large series of independent samples of size n, we shall obtain a series of different limits, and in the long run $100(1 - \alpha)\%$ of these will contain θ. Therefore when we put a particular number $\bar{x}_n$ in the interval (4.4), we do not know whether or not this particular interval contains θ; but we do know that $100(1 - \alpha)\%$ of intervals constructed in this way would in fact contain θ.

The intervals (4.4) for a normal mean can be restated,

$$\text{sample mean} \pm \lambda_{\alpha/2} \text{ (standard error of mean)}, \qquad (4.5)$$

which holds for the mean of any normally distributed quantity with known variance. The coefficients $\lambda_{\alpha/2}$ must be obtained from standard normal tables.

It is important to recognize the basic method used in this derivation of confidence intervals. A probability statement is written down which says that a sample statistic falls in an interval with a given probability, where the interval depends on the unknown value of θ. A value of the sample statistic is then observed, and the statement rewritten so as to look like a probability statement about θ lying in an interval. For example, the method could be used in connection with the median instead of the mean to obtain confidence intervals for the mean of a normal distribution. If we assume the sampling distribution of the median to be approximately $N(\theta, \pi\sigma^2/2n)$, then approximately,

$$\Pr\left\{\theta - \lambda_{\alpha/2}\sigma\sqrt{\left(\frac{\pi}{2n}\right)} < \tilde{x} < \theta + \lambda_{\alpha/2}\sigma\sqrt{\left(\frac{\pi}{2n}\right)}\right\} = 1 - \alpha.$$

By inverting the inequalities we obtain

$$\left\{\tilde{x} - \lambda_{\alpha/2}\sigma\sqrt{\left(\frac{\pi}{2n}\right)}, \tilde{x} + \lambda_{\alpha/2}\sigma\sqrt{\left(\frac{\pi}{2n}\right)}\right\} \tag{4.6}$$

for confidence intervals for θ.

We notice that the intervals (4.6) are wider than (4.4) for the same value of n. In fact, a large number of different sets of confidence intervals could be calculated for a normal mean, and there is need for a theory to compare them and choose an optimum set. However, this point is more of theoretical than practical interest, and we shall not discuss it.

One-sided confidence intervals

If we look again at the derivation of the confidence intervals (4.4), we see that it is actually a combination of two one-sided statements. We can write

$$\Pr\{\theta - \lambda_{\alpha/2}\sigma/\sqrt{n} < \bar{x}\} = 1 - \alpha/2$$

instead of (4.2), and hence obtain

$$\Pr\{\theta < \bar{x} + \lambda_{\alpha/2}\sigma/\sqrt{n}\} = 1 - \alpha/2$$

parallel to (4.3). The quantity $(\bar{x} + \lambda_{\frac{1}{2}\alpha}\sigma/\sqrt{n})$ is called an upper confidence limit for θ, with confidence $100(1 - \alpha/2)\%$. In the same way we can obtain $(\bar{x} - \lambda_{\alpha/2}\sigma/\sqrt{n})$ *as a lower confidence limit* for θ. The correct way to interpret the interval (4.4) is to regard it simply as a combination of the upper confidence limit $(\bar{x} + \lambda_{\frac{1}{2}\alpha}\sigma/\sqrt{n})$ and the lower confidence limit $(\bar{x} - \lambda_{\frac{1}{2}\alpha}\sigma/\sqrt{n})$, each of these limits having a confidence level $100(1 - \frac{1}{2}\alpha)\%$. Whether one- or two-sided statements are required in any application is determined by the problem. Consider the following example.

Example 4.2. A chromatographic method is employed in order to determine the percentage impurity contained in a dye used in foodstuffs. The error variance of an estimate is known to be 0·8. Three independent determinations give an average of 4·2%. Calculate a 95% upper confidence bound, and a 90% confidence interval for the true percentage impurity, assuming that each estimate is normally distributed.

The standard error of the average 4·2 is $\sqrt{(0\cdot8/3)}$, so that a 95% upper confidence bound is

$$4\cdot2 + 1\cdot64\sqrt{(0\cdot8/3)} = 5\cdot05,$$

since 1·64 is the upper 95% point of the standard normal distribution. A lower confidence bound at the 95% level is

$$4\cdot2 - 1\cdot64\sqrt{(0\cdot8/3)} = 3\cdot35,$$

and on putting these together, (3·35, 5·05) is a 90% confidence interval for the true percentage impurity. □ □ □

Example 4.3. The mean of the nine observations in Example 4.1 is −0·0050, and the standard error $0\cdot008/\sqrt{9} \simeq 0\cdot0027$. Therefore a 99% confidence interval for the deviations of the meter constants from unity is

$$-0\cdot0050 \pm 2\cdot58.0\cdot0027 = -0\cdot0120, +0\cdot0020$$

since 2·58 is the 99·5% point of the standard normal distribution. □ □ □

In Example 4.2, it is important to have an upper confidence bound on the percentage impurity, since some countries have legal limits on the maximum permitted percentage. It would not matter much how *low* the percentage was – it only matters that it should not be *above* a set limit. In Example 4.3, there is no obvious reason for wanting to ensure the meter constants do not deviate too much in one way or the other – simply that they should not deviate too much in either direction from unity. For these reasons, a one-sided bound is appropriate in Example 4.2, and a two-sided bound in Example 4.3. The choice between the two is always determined by the circumstances of the problem, even though it may not be clear in every case.

Example 4.4. How many meters must be tested in Example 4.1, to reduce the width of 99% confidence intervals for the meter constants to 0·005?

The width of 99% confidence intervals is

$$2 \times 2\cdot58 \times \text{standard error} = 2 \times 2\cdot58 \times \frac{0\cdot008}{\sqrt{n}}$$

Therefore n must satisfy

$$\frac{2 \times 2\cdot58 \times 0\cdot008}{\sqrt{n}} = 0\cdot005$$

$$n = 68\cdot16.$$

Therefore 69 meters must be tested. □ □ □

Example 4.5. Suppose that in Example 4.1, sixteen meters of a new batch are tested, and the standard deviation of the observations is the same as before, 0·008. If the average of the sixteen deviations of the meter constants from unity is $+0.012$, find a 95% confidence interval for the difference of meter constants in the two batches.

We shall have to go through this problem a little more slowly than the others. We assume that the observations on the two batches of meters come from normal populations with means μ_1 and μ_2 respectively, for the first and second batch, and with a common variance $(0.008)^2$. If we denote the sample means by $\bar{x}_1$ and $\bar{x}_2$, on nine and sixteen observations respectively, then

$$V(\bar{x}_1) = \frac{(0.008)^2}{9}, \quad \text{and} \quad V(\bar{x}_2) = \frac{(0.008)^2}{16}.$$

Further, $(\bar{x}_1 - \bar{x}_2)$ is a normally distributed variable with unknown mean $(\mu_1 - \mu_2)$, and known variance,

$$V(\bar{x}_1 - \bar{x}_2) = \frac{(0.008)^2}{9} + \frac{(0.008)^2}{16} = \frac{(0.008)^2 25}{9.16}.$$

Therefore the standard error of $(\bar{x}_1 - \bar{x}_2)$ is

$$\sqrt{V(\bar{x}_1 - \bar{x}_2)} = 0.003.$$

We can now use formula (4.5) to obtain 95% confidence intervals for $(\mu_1 - \mu_2)$ as

$$(\bar{x}_1 - \bar{x}_2) \pm 1.96\{\text{standard error of } (\bar{x}_1 - \bar{x}_2)\}$$

or

$$-0.017 \pm 1.96(0.003) = -0.023, -0.011. \qquad \square\square\square$$

Confidence intervals, like any other statistical tool, must be used with a certain amount of judgment, and difficulties in interpretation can occur. For example, consider the bioassay of wood preservatives experiment, Example 1.8. If we consider a high retention level, then negative weight losses sometimes occur, because of experimental error, although the mean weight loss cannot be negative. Thus if we obtain confidence intervals for the percentage weight loss at high retention levels, it would be possible to obtain, for example, $(-0.9, +0.2)$. Here we have a confidence interval which straddles zero, for a quantity which cannot be negative.

Exercises 4.4

1. In an experiment concerning the percentage yield of sugar beet, the beet are grown on square plots. The standard deviation of the percentages from each plot is expected to be 2.50. What will be the standard error of the mean sugar percentage of ten plots? How many plots are needed to make the width of 95% confidence intervals for the mean percentage one unit or less?

2. Suppose there are available n independent observations $x_1, \ldots, x_n$ from a normal population with unknown mean θ but known variance σ^2. Consider each of the following as estimators of θ:

(i) the sample mean, $\theta_1 = \bar{x}$,

(ii) $\theta_2 = k\bar{x}$, for $0 < k < 1$,

(iii) the trivial estimator which ignores the observations and makes a guess θ_3 at θ.

(a) Show that only θ_1 is an unbiased estimator.

(b) One basis for comparing sampling distributions is the mean squared error, $E(\hat{\theta} - \theta)^2$. Evaluate the M.S.E. for all three estimators.

(c) Hence show that the estimator which minimizes the M.S.E. for a given θ, is to take $\theta_3 = \theta$. Is this a helpful result?

4.5 Significance tests

Consider the following example.

Example 4.6. A production process gives components whose strengths are normally distributed with mean 40 lb and standard deviation 1·21 lb. The process is modified and 12 components are selected at random giving strengths (in lb)

$$39\text{·}8 \quad 40\text{·}3 \quad 43\text{·}1 \quad 39\text{·}6 \quad 41\text{·}0 \quad 39\text{·}9$$
$$42\text{·}1 \quad 40\text{·}7 \quad 41\text{·}6 \quad 42\text{·}1 \quad 40\text{·}8 \quad 42\text{·}5.$$

If the modified process gives on average stronger components than the unmodified process, it will be preferred for general use (the standard deviation can be assumed unchanged). Do the data give any evidence that the modified process is stronger?   □ □ □

The example given above raises a question which can be put in the following form: we have a hypothesis about the distributional form of some observations, and we want to know whether the observations are consistent with this hypothesis. In Example 4.6 we can put as our hypothesis that modification of the process has no effect, and inquire if the observations are consistent with this being true. In the measuring equipment example, Example 4.1, a similar problem arises, and here we wish to know whether the measured constants of nine meters can be regarded as coming from a population with mean unity.

Consider a simple case, and suppose we have sixteen observations from a distribution known to be normal with unit variance, and the sample mean $\bar{x}_{16} = 0\text{·}68$. The standard error of $\bar{x}_{16}$ is $\sigma/\sqrt{n} = 1/4$. Suppose we have a hypothesis that the population mean is zero, then if this is true the observed sample mean is $0\text{·}68/(1/4) = 2\text{·}72$ standard errors from the mean. From normal tables we see that this result would be well in the tails of the distribution – the probability of a result greater than 0·68 is about 0·003. Therefore although the hypothesis cannot be ruled as impossible, it appears very unlikely, and the true population mean is probably greater than zero. On the other hand if the sample mean were close to zero, say 0·068, the hypothesis that the population mean is zero is supported, but it is not thereby proved. The reason for this is that any of a range of hypothesis values close to the observed sample mean would also be 'supported'. However, this discussion does suggest a general method by which hypotheses can be tested, which we formalize as follows.

Formal development
We shall discuss briefly the formal development under a series of headings, using Example 4.6 as an illustration, and assuming normality.

NULL HYPOTHESIS (denoted H_0). We have a hypothesis relating to the distribution of some observations $x_1, x_2, \ldots, x_n$, for example, that the distribution is normal with mean μ_0 and variance σ^2. The null hypothesis must be precise, that is, statements such as 'the mean is not zero', are inadmissible.

ALTERNATIVE HYPOTHESIS (denoted H_1). If we wish to test whether the null hypothesis is consistent with certain data, some alternative hypotheses are implied, although our knowledge of them is usually very vague. Thus we might consider alternatives, 'the mean is greater than zero, with known population variance σ^2'.

TEST STATISTIC. From the data we calculate some statistic which indicates departures from the null hypothesis. Such a statistic should have a distribution strongly dependent on the departures from the null hypothesis which are of interest (for example, location), and should preferably not be too dependent on other assumptions or possible deviations.

The null hypothesis should be such that if it is true, the distribution of the test statistic is completely specified. For our example, consider as a test statistic

$$Z = (\bar{x} - \mu_0)\sqrt{n}/\sigma. \tag{4.7}$$

If H_0 states that x is distributed $N(\mu_0, \sigma^2)$, then Z is distributed $N(0, 1)$. If $E(x) \neq \mu_0$, the distribution of Z is still normal, but $E(z) \neq 0$.

SIGNIFICANCE LEVEL. The statistical significance level of an observed result is the probability of as extreme or more extreme values of the test statistic occurring by chance alone, under the null hypothesis. This is the probability

$$\Pr\{z \geqslant Z \mid z \text{ is } N(0, 1)\}, \tag{4.8}$$

where Z is the observed value of the test statistic, and z denotes a random variable having the distribution of the test statistic under the null hypothesis, which is $N(0, 1)$ here.

If the significance level is small enough this is taken to be evidence against the null hypothesis being true. A rough general rule which seems to work well in practice for the interpretation of statistical significance is,

s. sig. at 1%: almost conclusive evidence against the null hypothesis.
s. sig. at 5%: reasonable evidence against the null hypothesis.
not s. sig. at 5%: no evidence against the null hypothesis.

These are not hard and fast boundaries, and further guidance is given below.

Example 4.7. For Example 4.6 data we have $\bar{x} = 41 \cdot 125$, $\mu_0 = 40$, so that

$$Z = \frac{(41 \cdot 125 - 40)\sqrt{12}}{1 \cdot 21} = 3 \cdot 22.$$

The significance level of this is 0·0006. Thus the data of Example 4.6 give very strong evidence (significant at the 0·06% level) that the modification increases the average strength of the process. □ □ □

Interpretation of significance tests

There are a number of important points to notice about the interpretation of the results of significance tests.

(i) Even if a particular alternative H_1 is specified, H_0 and H_1 are not treated symmetrically. Only when the significance level against H_0 being true is small, is this hypothesis rejected.

(ii) No final proof of the truth or otherwise of H_0 or H_1 can be established, although results can show H_0 very unlikely. Suppose we are testing for changes in the mean of normal distribution, then a non-significant result does not imply that the mean has not changed. We can only say, 'The results are compatible with H_0. There may be a change in the mean, but our results do not justify claiming one.'

(iii) If we apply a test, say for a mean of a normal distribution, and use this repeatedly when H_0 is true, then in 1% of cases we shall find differences significant at 1%. If we accept this as sufficient evidence that a real difference exists, then in 1% of such cases we are wrong.

(iv) The level of significance at which we begin to suspect that a real difference exists must be fixed using general knowledge of the experiment on hand. A result s. sig. at 5% when a difference is strongly expected from prior arguments may be good evidence; on the other hand a result s. sig. at 5% when previous experiments suggest no difference may quite likely be a chance effect. Sometimes the correct conclusion to draw is that further experimentation is required.

(v) We must distinguish very clearly between technical significance and statistical significance. With small samples and large σ, quite large differences may not be statistically significant, although these differences may be real and of great practical importance (and vice versa).

In a well-designed experiment the number of observations is arranged
to give just enough accuracy to detect the smallest difference considered
of practical importance.

(*vi*) The conclusion of an experiment is *not* summed up in a significance
level such as 3·293486 . . . %! Conclusions should always have a practical meaning, in terms of the experiment. Significance tests are only a
guide to the interpretation of data, and should be used as such.

One-sided and two-sided tests

We have discussed the formulation of significance tests in terms of the production process example, Example 4.6, in which interest lay in departures
from the null hypothesis in one direction only. The aim of the experiment is
to decide whether the modified process gives components with a higher
average strength, and if the modification does not increase the average
strength there is no interest in it. A similar situation arises if we wish to compare a standard medical treatment with a new treatment. If the new treatment does not show itself to be better than the standard, the standard would
be used. Therefore although in these examples departures from the null
hypothesis in both directions are possible, we are only interested in detecting
departures in one direction. The significance test appropriate to these situations is said to be one-sided.

The measuring equipment example, Example 4.1, is typical of a rather
different situation which is in fact by far the most common. Here the average
of the adjusted meters may be above or below 1·000, and in either case we
want to detect such a difference. If therefore we calculate the statistic Z,
(4.7), large deviations of Z from zero, positive or negative, must be regarded
as significant. The significance level is therefore the probability

$$\Pr\left\{|z| > |Z| \,|\, z \text{ is } N(0, 1)\right\}. \tag{4.9}$$

The two-sided test for a normal mean is summarized next for convenience.

Example 4.8. *Two-sided significance test for Example* 4.1. The null hypothesis
states that the distribution of the nine differences reported in Example 4.1
is normal, mean zero and variance $(0\cdot008^2)$. From Example 4.3 we see that
the mean of the nine observations is $-0\cdot0050$, and the test statistic (4.7) is
therefore,

$$\frac{-0\cdot0050\sqrt{9}}{0\cdot008} = -1\cdot88.$$

The probability of a result greater than 1·88 in the standard normal distribution is 3·0%, and this is also the probability of a result less than $-1\cdot88$.
Therefore significance levels are:

one-sided   3·0%,

two-sided   6·0%.

If we were carrying out a one-sided significance test, we would regard the level 3·0% as reasonable evidence against the null hypothesis. However, on a two-sided test, the level of 6·0% is merely high enough to be indicative of a possible deviation from the null hypothesis. If a deviation of the meter constants from unity of the order of $-0·0050$ is considered important, then further observations must be taken to establish this.   □ □ □

In any practical example of this type, it would be usual to work into the conclusions a confidence statement of the type given in Example 4.3.

Relationship of significance tests and confidence intervals
Finally it is important to notice the connection between confidence intervals and significance tests. A confidence statement at a level $(1 - \alpha)$ for a normal mean, is the set of all μ which could be used as a significance test, with a given set of observed results, and not yield a significance level higher than α. For example, with reference to the calculations of Example 4.3 based on Example 4.1, a two-sided test of the null hypotheses

$$\mu = 0·0020 \quad \text{or} \quad \mu = -0·0120$$

with Example 4.1 data would yield a significance level of exactly 1%.

Exercises 4.5

1. If $x_1, x_2, \ldots, x_n$ is a random sample from $N(\mu_1, \sigma_1)$, and $y_1, y_2, \ldots, y_m$, is a random sample from $N(\mu_2, \sigma_2)$, where σ_1 and σ_2 are known, set up a test of the null hypothesis $\mu_1 = \mu_2$ against one-sided alternatives $\mu_1 > \mu_2$. Show how to obtain lower confidence bounds for $\mu_1 - \mu_2$.

2. Suppose observations are normally distributed with $\sigma = 3$ known, and we wish to carry out a two-sided test of the null hypothesis $\mu = 0$. If $\mu = 0$, the probability that a random sample leads to a test result which is regarded as significant at the 5% level is by definition 5%. If $\mu \neq 0$, then this probability is not 5%. Calculate the probability that a random sample of observations from $N(\mu_1, 3^2)$ leads to a significant test result (at the 5% level) for $\mu_1 = 0·5$, 1·0, 1·5, 2·0. The probability that a random sample leads to a significant result for some given value of μ is called the *power* of the test.

3. For a one-sided test of a normal mean, with the null hypothesis $\mu = 0$ and $\sigma = 1$ known, calculate the sample size n which is such that the probability that a random sample from $N(0·5, 1)$ leads to a test result which is regarded as significant (at the 5% level) is at least 99%, i.e. calculate the lowest n to satisfy

$$\Pr\left\{\frac{(\bar{x} - \mu)\sqrt{n}}{\sigma} > w_{5\%} \,\middle|\, \begin{array}{l} x_i \text{ is } N(0·5, 1) \\ w \text{ is } N(0, 1) \end{array}\right\} \geqslant 0·99.$$

CHAPTER FIVE

Single Sample Problems

5.1 Introduction

In this chapter, we consider various problems of inference which arise when we are presented with a single set of data, as in the measuring equipment problem, Example 4.1. Following the reasoning given in § 2.6(*iv*) and § 3.5, it is frequently possible to assume that our data are independently and normally distributed with some (unknown) expectation and variance, μ and σ^2. All the methods of this chapter are based on these assumptions. A simple check of normality can be carried out by plotting the data on normal probability paper, but this technique is unlikely to be fruitful with less than 20 or 30 observations. Shapiro and Wilk (1965) have devised a more elaborate test for normality which achieves quite high power even in small sample sizes, see the reference for details. Simple checks of statistical independence can also be made, but these depend upon the form of dependence thought to be present, and will be discussed later. The vital question here is not whether the assumptions are valid, but whether possible deviations from them are important. This problem is considered in § 5.7.

Once we have made the assumptions of normality and independence, all questions of inference reduce to questions concerning μ and σ^2. Consider the following situation.

Example 5.1. A production process gives components whose strengths are approximately normally distributed with mean 40 lb, and standard deviation 1·15 lb. A modification is made to the process which cannot reduce the mean strength of the components, but it is not certain that it will lead to any appreciable increase in mean strength. The modification may also alter the standard deviation of the strength measurements. The strengths of nine components selected from the modified process are

$$39\cdot6, \quad 40\cdot2, \quad 40\cdot9, \quad 40\cdot9, \quad 41\cdot4, \quad 39\cdot8, \quad 39\cdot4, \quad 43\cdot6, \quad 41\cdot8.$$

The following questions may arise in this situation.

(*i*) What are the best point estimates of μ and σ^2 from the data?

(ii) In § 4.4 we gave a method for obtaining interval estimates for μ when σ^2 is known. More commonly σ^2 is not known, and we may require interval estimates either for μ by itself, or for σ^2 by itself, or for μ and σ^2 simultaneously.

(iii) We may wish to test the hypothesis that the expectation μ is some particular value, say $H_0 : \mu = \mu_0$. In Example 5.1 it would be relevant to test $H_0 : \mu = 40$.

(iv) Similarly, we may wish to test the hypothesis that the variance is some particular value, say $H_0 : \sigma^2 = \sigma_0^2$. In Example 5.1 we would test $H_0 : \sigma^2 = (1 \cdot 15)^2$.

Subsequent sections of this chapter give answers to these problems. Although our discussion is relevant to the single sample problem, the techniques given are widely used, and of great importance.   □ □ □

Exercises 5.1

1. One possible departure from independence is that successive observations are correlated. Suggest a graphical technique of examining for the presence of this type of dependence, and try your technique out on data generated by the multiplicative congruential method of generating pseudo random deviates (§ 4.2).

2. Obtain a table of random rectangular deviates on the range (0, 1) and let x_i denote any entry in the table. Generate a sequence of variables y,

$$y_1 = x_1 + \ldots + x_m$$
$$y_2 = x_{m+1} + \ldots + x_{2m}$$

etc. Generate 30 y-values for $m = 2, 6$, and plot the cumulative distributions on normal probability paper.

5.2 Point estimates of μ and σ^2

We have already discussed problems relating to point estimates of μ and σ^2 in § 1.3, § 4.3 and § 4.4, and in this section we merely collect together results already noted.

Firstly, if our raw data is not normally distributed, some simple transformation may be necessary before proceeding further. Some suitable transformations are described in Figure 1.9 and the surrounding text.

Secondly, our raw data may contain outliers. For example, in Example 5.1, the observation $43 \cdot 6$ is some distance from the other observations. If possible, the actual conditions underlying such observations should be checked, by consulting the laboratory notebooks, etc., to see if there is some reason to justify rejecting the observations. Great caution should be taken before arbitrarily rejecting observations, and it is often best to carry out our statistical procedures both with and without the outliers. Practical data probably contains a considerable number of outliers, and some technique is required to aid spotting them; such techniques are described in § 10.7.

Let us assume that these general points have been dealt with, and that our observations are $x_1, \ldots, x_n$, then in general we use the sample mean

$$\bar{x} = \sum_{i=1}^{n} x_i/n$$

as an estimate for μ. On occasions we may prefer to use truncated or Winsorized means, see § 1.3, but in the majority of situations we simply use $\bar{x}$.

There are three main possibilities for estimators of σ or σ^2.

(i) In § 4.4 the range estimator was discussed. The procedure is simply to use $w_n \cdot \gamma_n$ as an estimate of σ, where w_n is the range of the sample, and γ_n is the function tabulated in Appendix II, table 4. Thus for Example 5.1 we have,

$$\text{range} = 43 \cdot 6 - 39 \cdot 4 = 4 \cdot 2$$

$$\gamma_9 = 0 \cdot 3367$$

$$\text{range estimate of } \sigma = 4 \cdot 2 \times 0 \cdot 337 = 1 \cdot 42.$$

If we have more than about 9 observations it is better to break up the data arbitrarily into subsets of size 6 or 7, and then average the ranges of each subset before multiplying by γ_n. To illustrate this latter method, we will use the Table 1.14 data on baled wool; here the data are naturally broken up into 7 sets of 4, and the quantity being estimated is σ^2 of (3.33), the variance of core sampling within bales. The seven ranges are

$$10 \cdot 53, \ 1 \cdot 70, \ 4 \cdot 65, \ 5 \cdot 18, \ 2 \cdot 73, \ 1 \cdot 92, \ 1 \cdot 18$$

and the average of these is $3 \cdot 98$. The range estimate of σ is therefore

$$3 \cdot 98 \times 0 \cdot 486 \simeq 1 \cdot 93.$$

(ii) If the population mean μ is known, we can use

$$s'^2 = \sum_{i=1}^{n} (x_i - \mu)^2/n$$

as an estimate of σ^2. This is discussed in Example 3.10 and Exercise 3.3.2. Since this estimate is very sensitive to the value of μ assumed, it is used only very rarely.

(iii) The usual estimate of σ^2 is the sample variance,

$$s^2 = \sum_{i=1}^{n} (x_i - \bar{x})^2/(n-1).$$

This is discussed in § 1.3, equation (1.1), and § 4.4.

All of these estimates of σ^2 are very sensitive to the presence of wild observations.

Exercises 5.2

1. Use Example 4.1 data and obtain (a) the range estimate of σ, (b) the estimate s'^2 of σ^2 assuming the deviations have mean zero and (c) the sample variance, s^2.

2. Let x_i, $i = 1, 2, \ldots, n$, be independent random variables with $E(x_i) = \mu$, $V(x_i) = \sigma^2$, and third and fourth moments about the mean μ_3, μ_4. Show that the coefficients of skewness and kurtosis of the mean $\bar{x}$ are

$$\gamma_1(\bar{x}) = \left(\frac{\mu_3}{\sigma^3}\right)\Big/\sqrt{n}, \quad \gamma_2(\bar{x}) = \left\{\frac{\mu_4}{\sigma^4} - 3\right\}\Big/n.$$

3. Let x_i be the random variables defined in question 5.2.2 above. Show that the sample variance

$$s^2 = \sum_{i=1}^{n} (x_i - \bar{x})^2/(n-1)$$

can be written

$$s^2 = \left\{\left(1 - \frac{1}{n}\right)\sum_{i=1}^{n}(x_i - \mu)^2 - \frac{2}{n}\sum_{i>j}(x_i - \mu)(x_j - \mu)\right\}\Big/(n-1).$$

Hence show that

$$E(s^4) = \frac{\mu_4}{n} + \frac{(n^2 - 2n + 3)}{n(n-1)}\sigma^4$$

and that

$$V(s^2) = \frac{\mu_4}{n} - \frac{(n-3)}{n(n-1)}\sigma^4.$$

Also give $V(s^2)$ in normal samples.

4. Let x_i, $i = 1, 2, \ldots, n$, be random variables with $E(x_i) = \mu$, $V(x_i) = \sigma^2$, and $C(x_i, x_{i+1}) = \rho\sigma^2$, $i = 1, 2, \ldots, (n-1)$, and all other covariances are zero. Show that

$$E(s^2) = \sigma^2\left(1 - \frac{2\rho}{n}\right).$$

Compare this result with the result of Exercise 3.4.4, and discuss the relevance of both results to the problem of making point estimates of μ and σ^2 when serial correlation may be present.

5.3 Interval estimates for μ (σ² unknown)

The basic method of obtaining confidence intervals was outlined in § 4.4. The starting point of the argument is that

$$(\bar{x} - \mu)\sqrt{n}/\sigma, \tag{5.1}$$

has a distribution which is known exactly and if σ^2 is known we can make the probability statement (4.2), and then invert it to obtain confidence intervals. When σ^2 is unknown, the argument cannot be carried through, since there are then two unknowns in the statement (4.2).

Consider the statistic

$$t = \frac{(\bar{x} - \mu)\sqrt{n}}{s}, \tag{5.2}$$

where s is the sample variance. It is easy to see that the sampling distribution of t does not depend on σ^2. If we multiply all the observations $x_1, \ldots, x_n$ by a, and put $y_1 = ax_1$, $y_2 = ax_2, \ldots$, then the t-statistic for the y's is

$$t_y = \frac{(\bar{y} - a\mu)\sqrt{n}}{\sqrt{\left\{ \dfrac{\sum (y_i - \bar{y})^2}{n - 1} \right\}}} = \frac{a(\bar{x} - \mu)\sqrt{n},}{as}$$

which is the same as for the x's. But the variance of the x's is σ^2, and of the y's is $a^2\sigma^2$, and the distribution of the y's is still normal, and therefore the distribution of t cannot depend on the value of σ^2, since altering σ^2 leaves t unchanged. The t-distribution must however depend on the distribution of s. The distribution of s is discussed in the next section, and depends on a quantity called the *degrees of freedom, ν*.

The sampling distribution of t is in fact

$$\left(1 + \frac{t^2}{\nu}\right)^{-(\nu+1)/2} \Gamma\left(\frac{\nu + 1}{2}\right) \Big/ \left\{ \sqrt{(\pi\nu)} \Gamma\left(\frac{\nu}{2}\right) \right\} \tag{5.3}$$

where $\nu = n - 1$ is the degrees of freedom of the estimate s^2 of σ^2. This is a symmetrical distribution and it tends to the normal distribution as a limit when ν is large; see Figure 5.1 and Table 5.1. The line for $\nu = \infty$ in Table 5.1

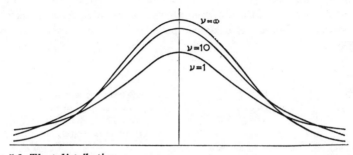

Figure 5.1 *The t-distribution*

Table 5.1 *Percentiles of the t-distribution*

		50%	60%	70%	80%	90%	95%	97·5%	99·0%	99·9%
$\nu =$	8	0	0·262	0·546	0·889	1·397	1·860	2·306	2·896	4·501
	10	0	0·260	0·542	0·879	1·372	1·812	2·228	2·764	4·144
	20	0	0·257	0·533	0·860	1·325	1·725	2·086	2·528	3·552
	100	0	0·254	0·526	0·845	1·290	1·660	1·984	2·365	3·174
	∞	0	0·253	0·524	0·842	1·282	1·645	1·960	2·326	3·090

is obtained from the standard normal distribution. A more detailed table of *t*-values is given in Appendix II, table 3.

Thus if in Example 5.1 the observations are in fact normally distributed with an expectation of 40 and any σ^2, and if we carried out repeated sampling of sets of nine observations and calculated

$$(\bar{x} - 40)\sqrt{9}/s$$

on each occasion, we would generate the *t*-distribution with $\nu = 8$. We would find that in the long run, 98% of our *t* values would lie between $\pm 2\cdot896$, etc. We shall denote the $100(1-\alpha)\%$ point of the *t*-distribution for ν degrees of freedom by $t_\nu(\alpha)$, so that for example, $t_8(0\cdot01) = 2\cdot896$.

Therefore if μ is the true unknown expectation of the *x*'s, we can write

$$\Pr\left\{-t_\nu\left(\frac{\alpha}{2}\right) < \frac{(\bar{x} - \mu)\sqrt{n}}{s} < t_\nu\left(\frac{\alpha}{2}\right)\right\} = 1 - \alpha, \tag{5.4}$$

parallel to (4.2). We can now express this as an inequality on μ,

$$\Pr\left\{\bar{x} - t_\nu\left(\frac{\alpha}{2}\right)s/\sqrt{n} < \mu < \bar{x} + t_\nu\left(\frac{\alpha}{2}\right)s/\sqrt{n}\right\} = 1 - \alpha$$

and $100(1 - \alpha)\%$ confidence intervals for μ are

$$\bar{x} \pm t_\nu\left(\frac{\alpha}{2}\right)s/\sqrt{n}. \tag{5.5}$$

That is, in comparison with (4.4), when σ^2 is unknown, we use s instead of σ, and the percentile of the *t*-distribution instead of the normal distribution.

Example 5.2. Suppose that in Example 5.1 we require 95% confidence intervals for the mean of the modified process. The calculations are as follows; see Appendix I, 'Notes on desk computation', for an explanation of the calculations.

Total	367·60	
Mean	40·84	
Unc. SS	15,028·38	Degrees of freedom
Corr.	15,014·42	$= \nu = n - 1 = 8.$
CSS	13·96	From Table 5.1 we have
Est. var	1·745	$t_8(0\cdot025) = 2\cdot306.$
Est. var ($\bar{x}$)	0·1939	
St. error ($\bar{x}$)	0·4403	

Hence $s/\sqrt{n}$ is 0·4403, and s is estimated on 8 d.f. Therefore 95% confidence intervals are

$$40\cdot84 \pm 2\cdot306 \times 0\cdot4403 = (39\cdot82, 41\cdot86).$$ □ □ □

The set of calculations outlined in Example 5.2 to obtain the standard error of $\bar{x}$ is very important, and this form of calculation is required frequently. The reader should make himself familiar with the method, and would be well advised to set the results out in the form shown.

The mathematics of the derivation of the t-distribution (5.3) is given in many books on probability theory, for example Cramér (1946), or Mood and Graybill (1963). The moments of t are as follows.

$$\left.\begin{array}{ll} E(t) = 0 & V(t) = \nu/(\nu - 2) \\ \gamma_1(t) = 0 & \gamma_2(t) = (18\nu - 42)/(\nu - 3)(\nu - 5) \end{array}\right\}. \qquad (5.6)$$

In fact, the t-distribution is so close to normal that a normal distribution fitted with the exact expectation and variance gives very good results for 5% points down to four degrees of freedom, and reasonably good results for 1% points down to about 25 degrees of freedom; see Exercise 5.3.2.

Exercises 5.3

1. Give (a) 95% confidence intervals, (b) a lower 99% confidence bound for μ for Example 4.1.

2. If t was exactly normally distributed with expectation zero and variance $\nu/(\nu - 2)$, the 97·5% point of the distribution would be

$$\lambda\sqrt{\{\nu/(\nu - 2)\}}$$

where

$$\int_{-\infty}^{\lambda} \frac{1}{\sqrt{(2\pi)}} e^{-\frac{1}{2}u^2} du = 0\cdot975.$$

Calculate these approximate percentiles for $\nu = 5, 10, 15, 20, 30$, and compare your results with exact values obtained from tables.

3. Obtain an approximate expectation and variance of t as follows. Put

$$s = (\sigma^2 + s^2 - \sigma^2)^{\frac{1}{2}} = \sigma\left(1 + \frac{s^2 - \sigma^2}{\sigma^2}\right)^{\frac{1}{2}}.$$

Hence obtain

$$t = \frac{\sqrt{n}\bar{x}}{\sigma}\left\{1 - \frac{1}{2}\left(\frac{s^2 - \sigma^2}{\sigma^2}\right) + \frac{3}{8}\left(\frac{s^2 - \sigma^2}{\sigma^2}\right)^2 - \cdots\right\}$$

$$t^2 = \frac{n\bar{x}^2}{\sigma^2}\left\{1 - \left(\frac{s^2 - \sigma^2}{\sigma^2}\right) + \left(\frac{s^2 - \sigma^2}{\sigma^2}\right)^2 - \cdots\right\}.$$

Take expectations to obtain

$$E(t) = \frac{\sqrt{n}\mu}{\sigma}\left(1 + \frac{3}{4(n-1)} - \cdots\right)$$

$$V(t) = 1 + \frac{2}{n-1} + \cdots$$

You may assume that $\bar{x}$ and s^2 are statistically independent; this is justified empirically in § 4.3, and proved in Exercise 5.4.2.

---

4. The wording of Example 5.1 indicates that a lower confidence bound might be more appropriate than the interval given in Example 5.2. Calculate a lower 99% confidence bound for this example and comment on your result, in view of the statement that the modification cannot decrease μ.

5.4 Interval estimates for σ^2

In order to obtain confidence intervals for σ^2 we shall need the distribution of estimates, and for s'^2 and s^2, these are related to the χ^2-distribution.

Definition. If z_i, $i = 1, 2, \ldots$, are independent standard normal variables the distribution of

$$y = z_1^2 + \ldots + z_\nu^2$$

is called the χ^2- distribution with ν degrees of freedom.

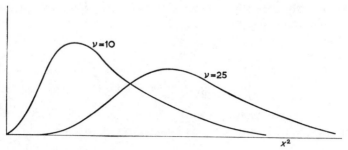

Figure 5.2 *The χ^2 distribution*

Table 5.2 *Percentiles of the χ^2-distribution*

		1%	5%	10%	90%	95%	99%
$\nu =$	8	1·65	2·73	3·49	13·36	15·51	20·09
	10	2·56	3·94	4·87	15·99	18·31	23·31
	20	8·26	10·85	12·44	28·41	31·41	37·57
	100	70·06	77·93	82·36	118·50	124·34	135·81

We could generate the χ^2-distribution on, say, 9 degrees of freedom, by taking sets of nine independent random normal deviates and summing their squares; if this is repeated many times and a histogram made, the histogram will tend to the smooth curve shown in Figure 5.2. Some percentiles of the χ^2-distribution are given in Table 5.2, and Appendix II, Table 5.

The expectation and variance of χ^2 are easily obtained from the definition; we have

$$E(y) = E(z_1^2) + \ldots + E(z_\nu^2) = \nu. \tag{5.7}$$

Also
$$E(y^2) = E(z_1^4) + \ldots + E(z_\nu^4) + 2E(z_1^2 z_2^2) + \ldots,$$
$$= 3\nu + \nu(\nu - 1),$$

hence
$$V(y) = 2\nu. \tag{5.8}$$

Now consider the estimate

$$s'^2 = \sum_{i=1}^{n} (x_i - \mu)^2/n.$$

Rewrite this
$$\frac{ns'^2}{\sigma^2} = \sum_{i=1}^{n} (x_i - \mu)^2/\sigma^2$$

$$= \sum_{i=1}^{n} z_i^2 \tag{5.9}$$

where $z_i = (x_i - \mu)/\sigma$, and has a standard normal distribution if x_i is $N(\mu, \sigma^2)$. The distribution of (5.9) is χ^2 on n degrees of freedom, and therefore this is also the distribution of (ns'^2/σ^2).

Let the pth percentile of the χ^2-distribution with ν degrees of freedom be denoted $\chi_\nu^2(p)$. That is, $\chi_\nu^2(p)$ is the number such that the area of the p.d.f. up to this point is p. Then since (ns'^2/σ^2) has a χ_n^2-distribution, we have

$$\Pr\left\{\chi_n^2\left(\frac{\alpha}{2}\right) < \frac{ns'^2}{\sigma^2} < \chi_n^2\left(1 - \frac{\alpha}{2}\right)\right\} = 1 - \alpha. \tag{5.10}$$

For example, if $n = 10$, $\sigma^2 = 1$, then from Table 5.2 the statement

$$\Pr\{3 \cdot 940 < 10s'^2 < 18 \cdot 307\} = 0 \cdot 90$$

means that in repeated sampling of 10 observations from a normal population with variance unity, 90% of values of s'^2 would lie between 0·3940 and 1·8307.

We now proceed in a manner parallel to the argument in § 4.4 and § 5.3, and invert (5.10) to make it look like a probability statement on σ^2,

$$\Pr\left\{\frac{ns'^2}{\chi_n^2\left(1 - \frac{\alpha}{2}\right)} < \sigma^2 < \frac{ns'^2}{\chi_n^2\left(\frac{\alpha}{2}\right)}\right\} = 1 - \alpha \tag{5.11}$$

Confidence intervals for σ^2 are therefore

$$\left(\frac{ns'^2}{\chi_n^2\left(1 - \frac{\alpha}{2}\right)}, \frac{ns'^2}{\chi_n^2\left(\frac{\alpha}{2}\right)}\right). \tag{5.12}$$

However, the most common estimate of σ^2 is s^2, not s'^2. From Example 3.11 and Exercise 5.2.3 we have

$$E(s^2) = \sigma^2, \quad V(s^2) = \frac{2\sigma^4}{(n-1)} \tag{5.13}$$

which compares with

$$E(s'^2) = \sigma^2, \quad V(s'^2) = \frac{2\sigma^4}{n},$$

which are proved in Example 3.10 and Exercise 3.3.2. Therefore, from (5.13)

$$E\left\{\frac{(n-1)s^2}{\sigma^2}\right\} = (n-1), \quad V\left\{\frac{(n-1)s^2}{\sigma^2}\right\} = 2(n-1), \tag{5.14}$$

which are the expectation and variance respectively of a χ^2 distribution on $(n-1)$ degrees of freedom. This indicates a result we might expect, that $\{(n-1)s^2/\sigma^2\}$ has a χ^2 distribution on $(n-1)$ degrees of freedom. A partial proof of this result is given in Exercise 5.4.2.

It therefore follows that $100(1-\alpha)\%$ confidence intervals for σ^2 based on s^2 are

$$\left\{\frac{(n-1)s^2}{\chi^2_{(n-1)}\left(1 - \frac{\alpha}{2}\right)}, \frac{(n-1)s^2}{\chi^2_{(n-1)}\left(\frac{\alpha}{2}\right)}\right\}. \tag{5.15}$$

Example 5.3. A 95% confidence interval for the variance of the modified process in Example 5.1 is as follows.

From Example 5.2 calculations,

$$s^2 = 1 \cdot 745$$

on 8 d.f. From tables of χ^2,

$$\chi_8^2(2 \cdot 5\%) = 2 \cdot 18$$

$$\chi_8^2(97 \cdot 5\%) = 17 \cdot 53.$$

Therefore 99% confidence intervals are

$$\left(\frac{8 \times 1 \cdot 745}{17 \cdot 53}, \frac{8 \times 1 \cdot 745}{2 \cdot 18}\right) = \left(\frac{13 \cdot 96}{17 \cdot 53}, \frac{13 \cdot 96}{2 \cdot 18}\right)$$

$$= (0 \cdot 796, 6 \cdot 404).$$

In calculating confidence intervals for σ^2, the reader should notice that $(n-1)s^2$ is merely the corrected sum of squares.   □ □ □

Exercises 5.4

1. Obtain (a) 95% confidence intervals, (b) a lower 99% confidence bound for σ^2 for Example 4.1.

2.* Let $x_1, \ldots, x_n$ be independent $N(\mu, \sigma^2)$ variables. Consider the following transformation

$$y_1 = (x_1 + \ldots + x_n)/\sqrt{n}$$
$$y_2 = (x_1 - x_2)/\sqrt{2}$$
$$y_3 = (x_1 + x_2 - 2x_3)/\sqrt{(3 \times 2)}$$

$$\begin{array}{c} \cdot \\ \cdot \\ \cdot \end{array}$$

$$y_n = \{x_1 + \ldots + x_{n-1} - (n-1)x_n\}/\sqrt{\{n(n-1)\}}.$$

Prove by induction that

$$y_1^2 + \ldots + y_n^2 = x_1^2 + \ldots + x_n^2.$$

Prove that

$$C(y_i, y_j) = 0$$

for $i \neq j$, and that

$$V(y_i) = \sigma^2$$
$$E(y_1) = \sqrt{n}\,\mu,$$
$$E(y_i) = 0,$$

for $i \geqslant 2$. We now rely on a result that if variables are normally distributed and uncorrelated, then they are independent, so that the y_i are independent. Express $\{(n-1)s^2/\sigma^2\}$ in terms of the y_i's, and hence prove that this quantity is distributed as χ^2 on $(n-1)$ degrees of freedom.

(Also it follows that since the y_i are independent, $\bar{x}$ and s^2 are independent. A simpler proof of the independence of $\bar{x}$ and s^2 is given in Feller (1966), pp. 84, 85.)

5.5 Significance test for a mean

We now consider question (iii) of Example 5.1, and give a method for testing the hypothesis $H_0 : \mu = \mu_0$, when the variance σ^2 is unknown. All the necessary theory has in fact been given in § 5.3. We need a test statistic which has a distribution strongly dependent on the true value μ of the mean, and independent of σ^2. Such a statistic is provided by

$$t = \frac{(\bar{x} - \mu_0)\sqrt{n}}{s} \tag{5.16}$$

and if x_i are independently and normally distributed with expectation μ_0, then t has a t-distribution (5.3) with $(n-1)$ degrees of freedom.

If the null hypothesis is not true, and $\mu \neq \mu_0$, then (5.16) no longer has a distribution symmetrical about zero, and from Exercise 5.3.3 it is clear that the expectation is approximately

$$E(t) = \sqrt{n}(\mu - \mu_0)\left(1 + \frac{3}{4(n-1)} + \dots\right)\Big/ \sigma. \qquad (5.17)$$

Thus the statistic (5.16) should provide a satisfactory test. In fact it can be shown that in a certain sense, (5.16) is the best statistic which can be used for testing a normal mean.

Example 5.4. To test the hypothesis $H_0 : \mu = 40$, in Example 5.1, we can use the calculations given in Example 5.2. We have

$$s/\sqrt{n} = 0\cdot4403$$
$$\bar{x} = 40\cdot84$$
$$\bar{x} - 40 = 0\cdot84$$
$$t = \frac{0\cdot84}{0\cdot4403} = 1\cdot91 \text{ on 8 d.f.}$$

We now refer to tables of percentage points of t, but before doing so, we must decide whether we have a one-sided or a two-sided test. We are told in Example 5.1 that the modification cannot reduce the mean strength, but it is not certain that it will increase it appreciably. In this example therefore, we have a one-sided test. The one-sided 5% point for t on 8 d.f. is 1·86, and therefore the observed value of 1.91 is just significant at this level.

□ □ □

Two points about this significance test are especially important.
(*i*) The conclusions from analysing a set of data should **never** be stated simply as 'significant' or 'not significant'. A full statement of conclusions should always be given, and a suitable statement for Example 5.1 is given after Example 5.5 in the next section.
(*ii*) A non-significant result does **not** prove that the null hypothesis is true. For example, if a two-sided test of $H_0 : \mu = \mu_0$ is carried out for Example 5.1, then it is clear from Example 5.2 calculations that a test of any μ_0 within the range (39·82, 41·86) would yield a result not significant at the 5% level on a two-sided test.

Significance tests are a useful tool in analysis of data, and *not* an end in themselves. The main conclusions from Example 5.4 are that, if an increase in the mean strength of 0·84 lb is considered important, further observations are required to establish more definitely that such an increase has taken place.

Note on one-sided and two-sided tests
The effect of using a one-sided rather than a two-sided test is to halve the

significance level of an observed result. Thus in Example 5.4 the observed value of t would be assigned a significance level of 10% on a two-sided test, and this would not be regarded as giving any grounds to dispute the null hypothesis; on a one-sided test the same result is just significant at the 5% level and we do doubt the validity of the null hypothesis.

Now let us suppose that the result of Example 5.4 had yielded a t-value of -4.0; this value of t is beyond the one-sided $\frac{1}{2}$% level, so that it is highly significant in the wrong direction, and would be highly significant even on a two-sided test. What would we conclude in such a case? In Example 5.1 we had stated that the new process could not reduce the mean strength of the components. With $t = -4.0$, we must either conclude, (i) that the mean strength has not reduced, but a chance fluctuation of a most unlikely size has occurred, or (ii) that our understanding of the mechanism of the process proved inadequate, and that it is after all possible for the mean strength to be reduced by the new process, or alternatively (iii) that our statistical model is wrong, and our estimate of the standard error much too small. If we make conclusion (ii), we are not taking our one-sided test seriously, and it would seem to be inadvisable to use it. If the observed t-value is high enough to be regarded as significant even on a two-sided test, the question of the validity of one-sided tests is unimportant, but when results such as that in Example 5.4 occur we must take care when drawing conclusions.

One situation in which a one-sided test can be safely used is the case mentioned in § 4.5 of comparing a standard medical treatment with a new treatment. The result is regarded as significant only if the new treatment is better than the standard; if the new treatment is worse than the standard treatment, even very much worse, there is no interest whatever in it.

Therefore a critical question to consider before using a one-sided test is what you would conclude if the result was highly significant in the wrong direction. Further examples of one- and two-sided tests occur in subsequent chapters of the book.

Exercises 5.5

1. Carry out a significance test for the mean for Example 4.1, regarding the population variance as unknown.

5.6 Significance test for a variance

In this section we consider question (iv) of Example 5.1, and give a test for the hypothesis $H_0 : \sigma^2 = \sigma_0^2$, which is valid regardless of the value of the mean μ.

If our original data are independently and normally distributed with a variance σ_0^2, then the results of § 5.4 show that if s^2 is the sample variance on ν degrees of freedom, then

$$\chi^2 = (\nu s^2/\sigma_0^2) = \frac{\text{corrected sum of squares}}{\sigma_0^2} \qquad (5.18)$$

has a χ^2-distribution on ν degrees of freedom. If the variance of the observations is $\sigma^2 \neq \sigma_0^2$, then

$$E(\chi^2) = \nu\sigma^2/\sigma_0^2$$

and the statistic χ^2 will therefore tend to show up differences of the variance from σ_0^2

Example 5.5. In Example 5.1 we are told that the unmodified process had a variance of $(1\cdot15)^2$. We would therefore be interested in testing whether the data indicate that any change has taken place in the variance. This indicates that we need a two-sided test of $H_0 : \sigma^2 = (1\cdot15)^2$. (The variance may have increased or decreased, and both directions are of interest.)

From the calculations of Example 5.2, we see that the statistic (5.18) is

$$\chi^2 = 13\cdot96/(1\cdot15)^2 = 10\cdot56$$

on 8 degrees of freedom. From χ^2 tables we see that the one-sided 10% level – the two-sided 20% level – is $13\cdot36$, and our observed sample is not even significant at this level. We therefore conclude that the data of Example 5.1 are consistent with the variance of the modified process being the same as the variance of the unmodified process.   □ □ □

We are now in a position to make a summary of conclusion for our analysis of Example 5.1 data.

The 9 components of the modified process which were measured have a mean strength of $40\cdot84$ lb with an estimated standard error of $0\cdot44$ lb, and the sample variance of the 9 components is $1\cdot75$. Confidence intervals at the 95% level are

for the mean $(39\cdot82, 41\cdot86)$

and for the variance $(0\cdot796, 6\cdot404)$.

By taking the square root of the confidence bounds for the variance, we obtain 95% confidence intervals for the standard deviation as $(0\cdot892, 2\cdot531)$. There is no evidence that modification of the process has changed the variance of the components, but a significance test for the mean yielded a borderline result (just above 5%). If an increase in mean strength of $0\cdot84$ is considered important, a further sample must be selected and measured, in order to decide more certainly whether such an increase has taken place.

One of the observations $(43\cdot6)$ is more than one standard deviation greater than any other, and this contributes considerably to the impression that the mean strength has increased, the mean of the 8 remaining components being $40\cdot50$ lb. However, there appears to be no basis for rejecting this observation.

A full conclusion of this type must be included with every analysis the reader makes.

Exercises 5.6

1. Complete your analysis of Example 4.1 data by testing whether there is any evidence that the variance has changed from its previously known value of $(0 \cdot 008)^2$. Then write up a full conclusion of your analyses on this data.

5.7 Departures from assumptions

The methods given in this chapter for obtaining confidence intervals and significance tests for μ and σ^2 have assumed that the original observations are (a) normally distributed, and (b) independent. We must therefore investigate the effects of lack of normality and of lack of independence on our methods. One of the best discussions of this available is given in Chapter 10 of Scheffé (1959); for other useful references see Bartlett (1935), Geary (1936), Gayen (1949, 1950), Ghurye (1949), Pearson and Please (1975) and Tiku (1963). In this section we give a brief summary of the main points.

(a) Inferences about μ based on the t-statistic

It follows from the central limit theorem, that for nearly all possible distributions of x_i the sample mean is asymptotically normally distributed. This effect is seen in Exercise 5.2.2, from which we have that if γ_1 and γ_2 are the coefficients of skewness and kurtosis of the x_i, then the coefficients $\gamma_1(\bar{x})$, $\gamma_2(\bar{x})$ of the mean are

$$\gamma_1(\bar{x}) = \gamma_1/\sqrt{n}$$
$$\gamma_2(\bar{x}) = \gamma_2/n.$$

Therefore whatever γ_1 and γ_2, $\gamma_1(\bar{x})$ and $\gamma_2(\bar{x})$ become small for large n. Furthermore, if $V(x_i) = \sigma^2$, then $V(\bar{x}) = \sigma^2/n$ regardless of γ_1 and γ_2, and we also see that $E(s^2) = \sigma^2$ regardless of γ_1 and γ_2.

Now the t-statistic is

$$t = (\bar{x} - \mu)\sqrt{n}/s \tag{5.19}$$

and it is fairly clear from the statements made above that if n is large enough, deviations from normality will have a negligible effect on the distribution of t and hence also on the probabilities of error based on this distribution. For further evidence on this point see Gayen (1949), Tiku (1963) and Scheffé (1959). This still leaves open the possibility that if there are deviations from normality, a different test would be more sensitive.

The effect of lack of independence on the t-statistic is more serious.

Suppose we have serial correlation, as defined in Exercise 3.4.4, then the results of this exercise show that the variance of the sample mean is

$$V(\bar{x}) = \sigma^2\left\{1 + 2\rho\left(1 - \frac{1}{n}\right)\right\}/n.$$

Further, it is easy to show that

$$E(s^2) = \sigma^2(1 - 2\rho/n).$$

Therefore, when serial correlation is present, the denominator of the t-statistic is *not* an estimate of the standard error of the numerator. Even where $E(x_i) = \mu$, the distribution of (5.19) is asymptotically $N(0, 1 + 2\rho)$ instead of $N(0, 1)$. This leads to drastic errors in the significance levels, even for quite small values of ρ; see Exercise 5.7.1. For a more general type of dependence, even more serious effects may be expected.

(b) *Inferences about σ^2*

Interval estimates and significance tests regarding σ^2 are based on the fact that, if the observations are independent and normal, the quantities

$$\{ns'^2/\sigma^2\} \quad \text{and} \quad \{(n-1)s^2/\sigma^2\} \tag{5.20}$$

have χ^2-distributions on n and $(n-1)$ degrees of freedom respectively. It is easy to see that the distributions of the quantities (5.20) are seriously affected if there is appreciable kurtosis, γ_2, of the observations, for although the expectations of (5.20) are still correct the variances are

$$V\left\{\frac{ns'^2}{\sigma^2}\right\} = 2n + \gamma_2 n$$

and

$$V\left\{\frac{(n-1)s^2}{\sigma^2}\right\} = 2(n-1) + \gamma_2(n-1)^2/n,$$

from Exercises 3.3.2 and 5.2.3. Even when n is large, these variances are still $(1 + \frac{1}{2}\gamma_2)$ times their values when γ_2 is zero. Scheffé (1959) shows that for $\gamma_2 = \pm 0.5$, the true confidence coefficient associated with a nominal 95% confidence interval is about 3% above or below the nominal value, and larger values of $|\gamma_2|$ lead to even more serious errors; see Exercise 5.7.2.

By similar methods, it can be shown that the effect of serial correlation on inferences about σ^2 is also serious.

(c) *Conclusions*

Lack of independence can have serious effects on inferences about both μ and σ^2. Fortunately, in many experiments, each observation is made on a physically different unit, and independence can be safely assumed. However, the remarks given above indicate that it is worth while going to some trouble, when planning an experiment, to ensure independence.

Lack of normality does not seriously affect inferences about μ, but does, unfortunately, have serious effects on inferences about σ^2.

Exercises 5.7

1. Suppose x_i for $i = 1, 2, \ldots, n$, are normally distributed random variables with properties given in Exercise 5.2.4. Argue that the t-statistic (5.19) is asymptotically $N(0, 1 + 2\rho)$, and hence find the true probability that t exceeds the 95% point of the t-distribution (for large n), for $\rho = \pm 0\cdot 1, \pm 0\cdot 2, \pm 0\cdot 3$. (See Scheffé, 1959, p. 339.)

2. Assume the conditions of Exercise 5.2.2, and use the results of Exercise 5.2.3 to find the expectation and variance of

$$\sqrt{(n-1)}(s^2 - \sigma^2)/\sigma^2.$$

Hence, assuming that the distribution of s^2 is asymptotically normal, find the probability that, for large n, the statistic (5.18) exceeds the 95% point of the χ^2-distribution, when $\gamma_2 = \pm 0\cdot 5, \pm 1, \pm 1\cdot 5$. (See Scheffé, 1959, p. 337).

CHAPTER SIX

Two Sample Problems

6.1 Introduction
Consider the following experiment.

Example 6.1. In order to assess the effect on the extensibility of yarn of washing, 20 lengths are taken and 10 selected at random to serve as controls. All are measured for breaking extension, but one observation is missing because of a defect in the measuring apparatus. The results are as follows:

Group	% Breaking extension								
Washed	13	16	8	12	14	11	13	14	9	
Control	15	15	10	18	17	13	9	12	12	13

□ □ □

In Example 6.1 we have two samples of data presented, and in the present instance we may safely assume them to be independent samples. Let us assume that the observations are normally distributed, with distributions $N(\mu_w, \sigma_w^2)$ and $N(\mu_c, \sigma_c^2)$ for the washed and control group respectively. If we require point and interval estimates for the μ's and σ^2's, then those can be done separately for each sample, by the methods of Chapter 5. However, the main object of this particular experiment lies in comparisons between μ_w and μ_c, and between σ_w^2 and σ_c^2. Some lengths of yarn had been subjected to a *treatment* (washing), and others had not; differences between the two groups of results can be ascribed to a *treatment effect*, if the lengths of yarn are otherwise treated identically in the experiment.

We are therefore interested in deriving confidence intervals or significance tests about $(\mu_w - \mu_c)$, and about σ_w^2/σ_c^2. Methods for these problems are given in the following sections.

Exercises 6.1
1. A sample of n_1 observations from a normal population with known variance σ^2 gave a mean $\bar{x}_1$, and a sample of n_2 observations from another normal population with the same (known) variance σ^2 gave a mean $\bar{x}_2$.

In order to test the null hypothesis that the two population means are equal, the following procedure is suggested. Obtain 95% confidence limits for the first population, l_1 and u_1, and similarly obtain 95% confidence limits l_2 and u_2 for the second population. If the two confidence intervals do not overlap then the sample means differ significantly at the 5% level of significance.

Explain why this reasoning is wrong. (Based on B.Sc. Gen. London, 1959)

6.2 The comparison of two independent sample means

Confidence intervals and significance tests about $(\mu_w - \mu_c)$ can be constructed very simply by methods given in § 5.3 and § 5.5, provided we assume that the two populations have equal variance.

Let the two groups be written

	Observations	Population distribution
Group I	$x_1, \ldots, x_n$	$N(\mu_x, \sigma^2)$
Group II	$y_1, \ldots, y_m$	$N(\mu_y, \sigma^2)$,

and also write

$$\bar{x} = \sum_{i=1}^{n} x_i/n, \quad \bar{y} = \sum_{j=1}^{m} y_j/m$$

$$s_x^2 = \sum_{i=1}^{n} (x_i - \bar{x})^2/(n-1), \quad s_y^2 = \sum_{j=1}^{m} (y_j - \bar{y})^2/(m-1).$$

Then an unbiased estimate of $(\mu_x - \mu_y)$ is $(\bar{x} - \bar{y})$, and this estimate has a variance

$$V(\bar{x} - \bar{y}) = \frac{\sigma^2}{n} + \frac{\sigma^2}{m} = \sigma^2(m + n)/(mn). \tag{6.1}$$

Information on σ^2 is provided by both s_x^2 and s_y^2, and these estimates must be combined in some way. The best combined estimate of σ^2 is

$$s^2 = \{(n-1)s_x^2 + (m-1)s_y^2\}/(m+n-2); \tag{6.2}$$

one reason for using this combination is given in Exercise 6.2.3. The distribution of (6.2) is proportional to χ^2 on $(m + n - 2)$ degrees of freedom. An estimate of the standard error of $(\bar{x} - \bar{y})$ is therefore

$$s\sqrt{\{(m+n)/mn\}}.$$

The reader will now see by comparison with § 5.3, that

$$t = \frac{\{(\bar{x} - \bar{y}) - (\mu_x - \mu_y)\}}{s\sqrt{\{(m+n)/mn\}}} \tag{6.3}$$

has a t-distribution on $(m + n - 2)$ degrees of freedom.

Confidence intervals for $(\mu_x - \mu_y)$ are therefore

$$\bar{x} - \bar{y} \pm t_\nu\!\left(\frac{\alpha}{2}\right) s\sqrt{\{(m+n)/(mn)\}} \tag{6.4}$$

at the level $100(1 - \alpha)\%$, where $\nu = m + n - 2$, the combined degrees of freedom.

A significance test of the hypothesis $H_0 : \mu_x = \mu_y$ is obtained by calculating

$$t = \frac{(\bar{x} - \bar{y})}{s\sqrt{\{(m+n)/mn\}}} \tag{6.5}$$

and referring to the tables of the t-distribution on $(m + n - 2)$ degrees of freedom.

Example 6.2. The calculations for Example 6.1 are as follows:

	Washed	Control
Total	110	134
No. of obs.	9	10
Mean	12·22	13·40
Unc. SS	1396	1870
Corr.	1344·44	1795·60
CSS	51·56	74·40
Pooled CSS	125·96	(on 17 d.f.)
Pooled estimate of variance	7·409	
Pooled estimate of variance of difference of means	$7·409(\tfrac{1}{10} + \tfrac{1}{9}) = 1·564$	
Standard error	1·251	

The difference of the group means is $13·40 - 12·22 = 1·18$, and 95% confidence intervals for the true difference are

$$1·18 \pm 2·11 \times 1·25 = -1·46, \, 3·82$$

where 2·11 is the two-sided 5% value for t on 17 degrees of freedom.

A significance test that washing has no effect on the breaking extension of yarn is given by referring

$$t = \frac{1·18}{1·251} = 0·9$$

to the t-tables for 17 degrees of freedom. This is clearly nowhere near significance.

Conclusion for Example 6.1

A significance test showed that the data give no evidence that washing affects extensibility of yarn. The average decrease in extensibility of the washed group relative to the controls was 1·18%, with a standard error of

1·25%. The 95% confidence intervals for the decrease in extensibility due to washing ranged from 3·8% to a possible increase of 1·46%.   □ □ □

This method for confidence intervals and significance tests for the difference of two means assumes
(*i*)   that the variances of the two groups are equal,
(*ii*)   that the observations are independent,
(*iii*) that the observations are normally distributed.
Some comments on the assumptions are given in § 6.6.

Exercises 6.2
1. An experiment to discover the movement of antibiotics in a certain variety of broad bean plants was made by treating 10 cut shoots and 10 rooted plants for 18 hours with a solution containing 200 μg/ml of chloramphenicol. After treatment, the roots of the rooted plants and the base of the cut shoots that had been immersed in the solution were discarded, and the concentration of chloramphenicol present in the remainder of the treated plants was determined by an assay. The results are given below. Carry out a two-sample *t*-test to examine whether there is any evidence of a difference in the (population) concentration of chloramphenicol between cut shoots and rooted plants, and also estimate 99% confidence intervals for the difference.

	Concentration of chloramphenicol ($\mu g/g$ fresh weight)								
Cut shoots	55	61	57	60	52	65	48	58	68	63
Rooted plants	53	50	43	46	35	48	39	44	56	51

(Based on London, Anc. Stats., 1956)

2. Assume that the 12 sets of data given in Table 1.7 for the quality control problem, Example 1.5, are all normally distributed with homogeneous variance σ^2 but with possibly different means. Combine the corrected sum of squares within samples to obtain a combined estimate of σ^2. Then use your estimate of σ to place warning lines on the $\bar{x}$-chart, Figure 1.10, such that if production is stable with a mean of 2·00, only 1% of sample means should be outside these lines. This last calculation will involve use of the *t*-distribution on 48 degrees of freedom. Usually, the normal distribution percentage points are used in such calculations, as the extra accuracy obtained by use of the *t*-distribution is not warranted in this application.

Would you have any reservations about the assumption of homogeneous variance for Table 1.7 data?

3.* Suppose that s_x^2 and s_y^2 are two estimates of σ^2 based on ν_x and ν_y degrees of freedom, when the observations are independently and normally distributed. Find the variance of

$$as_x^2 + (1 - a)s_y^2, \quad 0 < a < 1,$$

and find the value of a for which this variance is a minimum.

4.* In a two sample t-test, n_1 observations are drawn from a population $N(0, \sigma_1^2)$ and n_2 from a population $N(0, \sigma_2^2)$. Demonstrate that the distribution of the t-statistic (6.5) is approximately $N\left(0, \dfrac{1 + RU}{R + U}\right)$ when n_1 and n_2 are large, where

$$R = n_1/n_2, \quad U = \sigma_2^2/\sigma_1^2.$$

Hence calculate the probability that the 95% point of the t-distribution is exceeded for $R = 1, 2, 5$ and $U = \frac{1}{5}, \frac{1}{2}, 1, 2, 5$. Hence discuss the possible effect of unequal variances on the methods of this section; see Scheffé (1959, p. 340).

6.3 The comparison of two independent sample variances

It frequently arises that we wish to compare two independent estimates of variance, and test the null hypothesis that the variances are equal.

Example 6.3. In a problem similar to Example 6.1 in which the observations are normally distributed, two sample variances were $s_1^2 = 6 \cdot 32$ on 8 d.f. and $s_2^2 = 2 \cdot 34$ on 16 d.f. Are these compatible with there being a common population variance? □ □ □

Suppose s_1^2 and s_2^2 are independent estimates of σ_1^2 and σ_2^2 on ν_1 and ν_2 degrees of freedom respectively. Then it can be shown that the distribution of

$$F = \frac{s_1^2/\sigma_1^2}{s_2^2/\sigma_2^2} \tag{6.6}$$

depends only on ν_1 and ν_2; it is called the F-distribution on (ν_1, ν_2) d.f. The expectation and variance of F are obtained approximately in Exercise 6.3.4; the exact values are

$$E(F) = \nu_2/(\nu_2 - 2) \qquad\qquad \nu_2 > 2 \tag{6.7}$$

$$V(F) = \frac{2\nu_2^2(\nu_1 + \nu_2 - 2)}{\nu_1(\nu_2 - 2)^2(\nu_2 - 4)} \qquad \nu_2 > 4. \tag{6.8}$$

The F-distribution is markedly skew, so that this expectation and variance cannot be used in a normal approximation; this contrasts with properties of the t-distribution discussed in § 5.3. One property of the F-distribution is that the inverse of ratio (6.6)

$$\frac{s_2^2/\sigma_2^2}{s_1^2/\sigma_1^2}$$

must have an F-distribution on $(\nu_2,\ \nu_1)$ d.f. Therefore the $100\alpha\%$ point on $(\nu^1,\ \nu_2)$ d.f. must be the reciprocal of the $100(1-\alpha)\%$ point on $(\nu_2,\ \nu_1)$ d.f.,

$$F_\alpha(\nu_1,\ \nu_2) = 1/F_{(1-\alpha)}(\nu_2,\ \nu_1). \tag{6.9}$$

This means that we need not tabulate both ends of the F-distribution, and it is usual to tabulate upper percentage points only. A short table of percentage points is given in Appendix II, table 6.

We can now construct confidence intervals for the ratio σ_2^2/σ_1^2 by the method used in § 4.4, etc. The F-distribution is such that

$$\Pr\left\{F_{\alpha/2}(\nu_1,\ \nu_2) < \frac{s_1^2/\sigma_1^2}{s_2^2/\sigma_2^2} < F_{(1-\alpha/2)}(\nu_1,\ \nu_2)\right\} = 1 - \alpha.$$

We rewrite the inequalities to obtain confidence limits for σ_2^2/σ_1^2

$$\frac{s_2^2\,F_{\alpha/2}(\nu_1,\ \nu_2)}{s_1^2},\quad \frac{s_2^2\,F_{1-\alpha/2}(\nu_1,\nu_2)}{s_1^2}.$$

Since only upper percentage points of F are tabulated, we use (6.9) to obtain

$$\frac{s_2^2}{s_1^2\,F_{1-\alpha/2}(\nu_2,\ \nu_1)},\quad \frac{s_2^2\,F_{1-\alpha/2}(\nu_1,\ \nu_2)}{s_1^2}. \tag{6.10}$$

Example 6.4. For Example 6.3, 90% confidence intervals for the ratio σ_2^2/σ_1^2. of the population variances are

$$\frac{2\cdot34}{6\cdot32 \times 3\cdot20},\quad \frac{2\cdot34 \times 2\cdot59}{6\cdot32} = 0\cdot12,\ 0\cdot96.$$

$$[F_{0\cdot95}(16,\ 8) = 3\cdot20,\quad F_{0\cdot95}(8,\ 16) = 2\cdot59] \qquad \square\,\square\,\square$$

A significance test of the hypothesis $H_0 : \sigma_1^2 = \sigma_2^2$ is obtained by calculating the variance ratio

$$V = (\text{larger variance})/(\text{smaller variance})$$

and referring to F-tables with ν_1 and ν_2 equal to the degrees of freedom of the numerator and denominator respectively.

Example 6.5. For Example 6.3 a significance test that $\sigma_1^2 = \sigma_2^2$ is as follows.

$$V = 6\cdot32/2\cdot34 = 2\cdot70\quad \text{on } (8,\ 16)\text{ d.f.}$$

Now $F_{0\cdot95}(8,\ 16) = 2\cdot59$, and $F_{0\cdot975}(8,\ 16) = 3\cdot12$, so that this result is just significant at the one-sided 5% level, or only the 10% level on a two-sided test. For Example 6.3 we therefore conclude that the F-ratio is just significant at the 10% level, and that there is some evidence that the two groups sampled in Example 6.3 may have different variances, but more evidence is required to settle the question. $\qquad \square\,\square\,\square$

The methods of this section assume independence and normality of the observations and because of the arguments of § 5.7, we expect the methods to be sensitive to these assumptions.

Exercises 6.3

1. Look up the tables of percentage points of the F-distribution given in Biometrika Tables or in Fisher and Yates, and construct a short table similar to Table 5.2, for values of ν_1 and ν_2 at about 5, 15, 50, including the combinations (5, 15), (5, 50), etc.

2. Write out in full the correct interpretation of the confidence intervals (6.10).

3. The t-test calculated in Exercise 6.2.1 assumed that the variances of observations on cut shoots and rooted plants are equal. Carry out an F-test of the hypothesis that the variances are equal, and also calculate 90% confidence intervals for the ratio of the two (population) variances.

4. Use a method similar to that indicated in Exercise 5.3.3 to find the approximate expectation and variance of (6.6).

5.* One difficulty about comparisons between variances is that the standard error of an estimate of σ^2 is proportional to σ^2; see Exercise 5.2.3. This difficulty can be avoided by taking a transformation.

If s^2 is an estimate of σ^2 based on ν degrees of freedom, write

$$\log_e s^2 = \log_e (\sigma^2 + s^2 - \sigma^2) = \log_e \sigma^2 + \log_e \left(1 + \frac{s^2 - \sigma^2}{\sigma^2}\right)$$

Then expand the last logarithm on the right-hand side in a series, and take expectations, assuming normality, to obtain

$$E(\log_e s^2) \simeq \log_e \sigma^2$$

$$V(\log_e s^2) \simeq 2/\nu,$$

see Exercise 5.3.3. The important feature here is that $V(\log_e s^2)$ is independent of σ^2. In addition, the distribution of $\log_e s^2$ is much more nearly normal than the distribution of s^2. Use the results of this exercise to suggest another method of comparing variances which are estimated on equal numbers of degrees of freedom.

6.4 Analysis of paired samples

A very common problem, especially in experimental work, is the comparison of two groups, for example treated and control groups, as in Example 6.1.

The best way of doing such an experiment is often to pair the experimental units, the two in a pair being as alike as possible, and then allocate the two treatments randomly to the individuals in each pair. The following example illustrates the method.

Example 6.6. Fertig and Heller (1950). In comparing two methods of chlorinating sewage, eight pairs of batches of sewage were treated. Each pair was taken on a different day, the two batches on any one day being taken close together in time, and the two treatments were randomly assigned to the two batches in each pair. Treatment A involved an initial period of rapid mixing, while treatment B did not. The results, in log coliform density per ml, were as follows.

	Day							
	1	2	3	4	5	6	7	8
	A 2·8	B 3·1	B 3·4	A 3·0	B 2·7	B 2·9	B 3·5	A 2·6
	B 3·2	A 3·1	A 2·9	B 3·5	A 2·4	A 3·0	A 3·2	B 2·8
Difference B − A	0·4	0·0	0·5	0·5	0·3	−0·1	0·3	0·2

There are three main sources of error in Example 6.6:

(*i*)   Differences in sewage on different days of the experiment, which could be systematic variations with time.

(*ii*)   Differences between individual batches of sewage selected on any day. The chemical content of the sewage will differ between the two batches due partly to variation arising from sampling the whole stream of sewage, and possibly partly to variations in the content of the sewage with time, if the batches were selected successively, etc.

(*iii*)   Measurement errors, and possible bias in measurement. Also, since the batches of sewage will be sampled for measurement, there will be a sampling error. For Example 6.6 such errors are liable to be small in comparison with others.

The advantage of the paired comparisons design is that the data can be analysed by taking the treatment differences, B − A, within each pair, and this will exclude the source of error listed under (*i*) from the treatment comparisons.

We may therefore assume that the differences (B − A) for each day are unbiased estimates of the difference in log coliform density caused by the treatment, which we call the *treatment effect*. Confidence intervals and significance tests for this treatment effect can be obtained by using the methods of § 5.3 and § 5.5 on the differences within pairs.

Example 6.7 The analysis of Example 6.6 data is as follows. We base calculations on the differences between the treatments within pairs.

Total	2·1
Mean	0·2625
Unc. SS	0·8900
Corr.	0·5512
CSS	0·3388 (on 7 d.f.)
Est. var.	0·0484
Est. var. mean	0·00605
Est. standard error	0·0778

The 95% confidence intervals for the treatment effect $B - A$ are

$$0·2625 \pm 2·36 \times 0·0778 = 0·079, 0·446,$$

where 2·36 is the two-sided 95% point of the t-distribution for 7 d.f.

A significance test that the treatments are identical in their effect is obtained by testing the hypothesis that the mean difference is zero. We therefore refer

$$t = 0·2625/0·0778 = 3·37$$

to the t-tables on 7 d.f. This is significant at the 2% level for a two-sided test, and at the 1% level for a one-sided test. If we assume that treatment A cannot yield a higher log coliform density than B, then a one-sided significance test is appropriate here; see the discussion on one- and two-sided tests in § 5.5.

Conclusion. The data give quite strong evidence (significant at the 1% level), that method A gives a lower coliform density than method B. The mean reduction in log coliform density was 0·26, with a standard error of 0·08, and 95% confidence intervals for the true reduction in log coliform density are (0·08, 0·45).   □ □ □

An alternative design for Example 6.6 would have been to draw two batches of sewage per day, not necessarily close together, for each of seven days, and randomly allocate seven of the fourteen batches to each treatment. This design would have to be treated by the methods of § 6.2, and the experimental error contains contributions from all three sources listed above.

Exercises 6.4

1. Analyse the data of Example 6.6 as if the treatments were not paired, but just seven of the fourteen batches were randomly allocated to each treatment. Compare the analysis with Example 6.7, and discuss the differences in the standard error of the treatment effect, and in the degrees of freedom for estimation of error.

When would a paired comparisons design be advisable, and when would it be inadvisable?

2. Why is it necessary to allocate the treatments within each pair of Example 6.6 randomly, rather than, for example, using B first on all days?

3. (*i*) The data given below are part of the results of an experiment (due to Drew, 1951) designed to investigate any possible relationship between the reflex blink rate and the difficulty of performing certain visual motor tasks.

Subjects were asked to steer a pencil along a moving track. Each trial lasted eight minutes, and involved alternating periods of straight and oscillating track. Give a 95% confidence interval for the true difference between

Blink rate (per minute)

Subject	Track Straight	Oscillating	Subject	Track Straight	Oscillating
1	14·0	5·0	7	8·2	0·6
2	19·5	6·6	8	10·1	0·5
3	8·2	1·9	9	5·5	0·5
4	8·5	1·5	10	10·1	3·1
5	12·1	1.1	11	7·2	2·1
6	8·0	2·5	12	5·6	1·6

blink rates for straight and oscillating tracks, stating any assumptions you make.

(*ii*) If when this experiment was being planned, it was guessed that the standard deviation of the difference in blink rates was 3 blinks per minute, how many subjects should have been used in the experiment in order to determine a 95% confidence interval of width of about 2 blinks per minute?

(B.Sc. Special, London, 1960. One Year Ancillary Statistics)

6.5 An example
As an illustration of the use of methods developed so far on the analysis of data, we consider the following example, which is taken from B.Sc. General, London, 1959 (and slightly modified).

Example 6.8. From among a large number of patients suffering from a rheumatic illness, two sets of ten volunteers were selected. The first patient in group 2 was chosen to match the first patient in group 1 as closely as possible for age and duration of illness. Similarly, the second patients in each group were paired, and so on.

Measurements of the angular movement of the knee were made on all twenty patients. Afterwards, drugs A and B were administered to patients in groups 1 and 2 respectively; the drugs were contained in pills similar in appearance. After a period of general treatment, measurements were again made on the knees of all patients, and the results are shown in Table 6.1 below. Improvement in a patient's condition should result in an increased measurement.

Table 6.1 *Angular measurements of the knee on 20 patients*

Patient no.	1st measurement		2nd measurement	
	Group 1	Group 2	Group 1	Group 2
1	9·53	9·47	12·03	12·36
2	12·86	10·26	12·47	12·49
3	10·51	6·66	8·59	11·04
4	10·31	11·70	10·61	12·11
5	14·00	13·39	14·59	17·65
6	7·87	7·24	8·48	9·51
7	4·78	4·79	3·82	4·79
8	10·63	9·71	12·24	12·11
9	11·19	10·52	12·17	12·88
10	9·86	11·92	12·39	13·10

The main aim of the analysis of this data should clearly be to examine for any difference in the effect of the two treatments, but other points can be studied. For example, it is possible to examine whether the pairing has been effective in increasing precision.

Firstly, any differences between the first two columns of Table 6.1 reflect differences between the conditions of patients selected for the experiment, and this variation is presumably of little interest. Most of the analysis can be based on the differences between the two measurements made on any patient. These differences are given in Table 6.2.

Table 6.2 *Differences between first and second observations on 20 patients*

Patient	Group 1 (drug A)	Group 2 (drug B)	Group 2 − Group 1
1	2·50	2·89	0·39
2	−0·39	2·23	2·62
3	−1·92	4·38	6·30
4	0·30	0·41	0·11
5	0·59	4·26	3·67
6	0·61	2·27	1·66
7	−0·96	0·00	0·96
8	1·61	2·40	0·79
9	0·98	2·36	1·38
10	2·53	1·18	−1·35

The data in columns 1 and 2 of Table 6.2 are measures of the change of the patient's condition during the period in which the drugs were administered. Since the patients in the groups are paired, we can subtract column 1 from column 2 to obtain measures of the difference between the amount of improvement caused by the two drugs; these differences are given in column 3 of Table 6.2. Calculations of the means and variances for the three columns of Table 6.2 are given in Table 6.3.

Table 6.3 *Means and variances of Table 6.2 data*

	Group 1	Group 2	2 − 1
Total	5·85	22·38	16·53
Mean	0·585	2·238	1·653
Unc. SS	21·7737	68·7000	68·2157
Corr.	3·4223	50·0864	27·3241
CSS	18·3514	18·6136	40·8916
Est. var.	2·0390	2·0682	4·5435

Several conclusions are immediately obvious from Tables 6.2 and 6.3. First the average improvement by patients treated by drug B is considerably more than those treated by drug A. Of the ten pairs of patients, only in one pair was the improvement due to drug A greater than that for drug B. Secondly, while the differences for drug A patients are well spread, the differences for drug B are heavily grouped at about 2·50, with stragglers. The ordered differences are as follows:

$$0·00, \quad 0·41, \quad 1·18, \quad 2·23, \quad 2·27, \quad 2·36, \quad 2·40, \quad 2·89, \quad 4·26, \quad 4·38.$$

The gap between 2·89 and 4·26 is 1·37, or about a standard deviation, and the gaps at the bottom are almost as great. Thus although most patients did moderately well on drug B, two were not improved much at all, while two others did extremely well. This fact should be explored, if possible, by studying further details of the particular patients, or by further trials.

Significance tests that drugs A and B have on average led to an improvement in patients' conditions are carried out by calculating paired sample t-tests separately for each drug. The details are as follows.

Drug A: Standard error of mean $= \sqrt{(2·039/10)} = 0·4516$

$$t = 0·585/0·4516 = 1·30 \text{ on } 9 \text{ d.f.}$$

This value of t is only just beyond the 25% level, and there is therefore no clear evidence that drug A does on average lead to an improvement. The 95% confidence intervals for the amount of improvement are

$$0·585 \pm 2·26 \times 0·452 = -0·437, 1·607.$$

Drug B: Standard error of mean $= \sqrt{(2·068/10)} = 0·4548$

$$t = 2·238/0·4548 = 4·92 \text{ on } 9 \text{ d.f.}$$

The 0·1% value of t is 4·78, so that this result is significant at beyond the 0·1% level, showing very clear evidence that drug B does on average lead to an improvement. The 95% confidence intervals for the amount of improvement are

$$2·238 \pm 2·26 \times 0·455 = 1·210, 3·266.$$

A comparison between the two drugs can be made by calculating a paired sample t-test on column 3 of Table 6.2. The calculations are as follows.

Drug B — Drug A (based on column 3 of Table 6.2)

$$\text{Standard error of mean} = \sqrt{(4·544/10)} = 0·6741$$

$$t = 1·653/0·6741 = 2·45 \text{ on 9 d.f.}$$

This is just significant at the 5% level, showing moderate evidence that drug B improves patients' condition on average more than drug A. The 95% confidence intervals for the difference in improvement are

$$1·653 \pm 2·26 \times 0·674 = 0·130, 3·176.$$

As an alternative method of comparing the two drugs we might consider applying the two-sample t-test to columns 1 and 2 of Table 6.2; but since the patients were paired, there is likely to be a correlation between the results in the two columns, and this procedure is not valid. Clearly, the difference in the means of columns 1 and 2 is equal to the mean of column 3, but the estimated standard error attributed to this mean difference is in general different when using the two sample and paired t-tests; see the calculations below.

Comparison of drugs by two sample t-test
The calculations are as follows

Pooled CSS	36·9650 (18 d.f.)
Est. Var.	2·0536
$V(\bar{x}_B - \bar{x}_A)$	0·4107
Standard error	0·6409

$$t = 1·653/0·6409 = 2·58 \text{ on 18 d.f.}$$

This is significant at the 5% level, showing moderate evidence that drug B improves patients' conditions on average more than drug A. The 95% confidence intervals for the difference in improvement are

$$1·653 \pm 2·10 \times 0·641 = 0·307, 2·999.$$

Effectiveness of the pairing

The mean difference between the results when using drug B and those when using drug A is 1·653, with a standard error of 0·674 as calculated by the pairing method, and 0·641 as calculated by the two sample method which ignores the pairing. That is, the standard error is *greater* when calculated using the pairing, although it is clearly not significantly greater.

The aim of pairing of patients is to cut out of the experimental comparisons a certain amount of extraneous variation between patients. The effectiveness of pairing would therefore be shown by a *reduced* standard error and in this instance nothing has been gained by the pairing.

It is important to notice that the degrees of freedom are 9 by the paired method, and 18 by the two sample method. Thus the width of 95% confidence intervals is

$2 \cdot 26 \times$ standard error   for the paired samples case, and
$2 \cdot 10 \times$ standard error   for the two sample case.

Thus unless the paired standard error is less than $2 \cdot 10 / 2 \cdot 26 = 0 \cdot 93$ of the two sample standard error, there is no gain in apparent precision as measured by the width of 95% confidence intervals.

The comparison between the paired and two sample methods of analysis given above has important implications when designing an experiment; the design aspects are discussed more fully in Chapter 12.   ☐ ☐ ☐

Exercises 6.5

1. Is there any evidence of a systematic difference in patients' conditions in the two groups of Example 6.8, as indicated by the first measurements (before treatment)?

Can you suggest an alternative design for the experiment, which may yield a more precise estimate of the treatment difference?

2. For an investigation of the effects of dust on tuberculosis, two laboratories each used sixteen rats. The set of rats at each laboratory was divided at random into two groups, A and B, the animals in group A being kept in an atmosphere containing a known percentage of dust while those in group B were kept in a dust-free atmosphere. After three months the animals were killed, and their lung weights measured, the results being given in the following table.

Lung weights (gm)

Laboratory 1		Laboratory 2	
A	B	A	B
5·44	5·12	5·79	4·20
5·36	3·80	5·57	4·06
5·60	4·96	6·52	5·81
6·46	6·43	4·78	3·63
6·75	5·03	5·91	2·80
6·03	5·08	7·02	5·10
4·15	3·22	6·06	3·64
4·44	4·42	6·38	4·53

Analyse these results, bearing in mind the possibility of differences both in mean and in dispersion. Also give 95% confidence intervals for σ^2 for each set of data.

(London, Anc. Stats., 1959)

6.6 Departures from assumptions

The discussion of assumptions given in § 5.7 applies with little change to the problems discussed in this chapter, and the general conclusions can be stated as in § 5.7(c). One further problem which arises in this chapter is the sensitivity of the two sample t-test to equality of variances in the two samples. Two fairly simple arguments throw some light on this point.

Firstly, the problem can be examined by asymptotic arguments, and this is outlined briefly in Exercise 6.2.4; see also Scheffé (1959) and references. The conclusion of this is that if the two variances are unequal, and if the sample sizes are also unequal, then the denominator of the two sample t-test (6.3), is not an estimate of the standard error of the numerator. In this situation therefore, use of (6.3) may lead to serious errors.

Secondly, let the population variance of a sample of n_1 observations be σ_1^2, and of an independent sample of n_2 observations be σ_2^2. Now if σ_1^2 is very small, most of the error will be due to σ_2^2, and the effective degrees of freedom of t would be nearer to $(n_2 - 1)$ rather than to $(n_1 + n_2 - 2)$. Hence we might expect that if $\sigma_1^2 \neq \sigma_2^2$, the distribution of quantities like (6.3) is still approximately as a t-distribution, but with degrees of freedom depending on the ratio σ_1^2/σ_2^2. Thus for example, if we have two samples of size seven, the degrees of freedom for (6.3) must be between 6 and 12. This information is often sufficient to enable us to draw valid conclusions; see Exercise 6.6.2.

The second argument can be extended to obtain a test more suitable for the situation where $\sigma_1^2 \neq \sigma_2^2$. Let s_1^2 and s_2^2, on $(n_1 - 1)$ and $(n_2 - 1)$ degrees of freedom respectively, be the estimates of variance based on the first and second samples. Then consider

$$t' = \frac{\bar{x}_1 - \bar{x}_2}{\sqrt{\left\{\dfrac{s_1^2}{n_1} + \dfrac{s_2^2}{n_2}\right\}}}. \tag{6.11}$$

If the population means of the two samples are equal, this statistic has expectation zero. Further, the term within brackets of the denominator is always an unbiased estimate of the variance of the numerator.

If the observations are all independently and normally distributed, the distribution of (6.11) is approximated by a t-distribution, with degrees of freedom f, where

$$\frac{1}{f} = \frac{\sigma_1^4}{k^2 n_1^2 (n_1 - 1)} + \frac{\sigma_2^4}{k^2 n_2^2 (n_2 - 1)}, \quad k = \frac{\sigma_1^2}{n_1} + \frac{\sigma_2^2}{n_2}. \tag{6.12}$$

When the test is used in practice, σ_1^2 and σ_2^2 in (6.12) must be replaced by the sample estimates s_1^2 and s_2^2.

Exercises 6.6

1. Derive formula (6.12) as follows. Write

$$\left\{\frac{s_1^2}{n_1} + \frac{s_2^2}{n_2}\right\}^{-1} = \left\{\frac{\sigma_1^2}{n_1} + \left(\frac{s_1^2 - \sigma_1^2}{n_1}\right) + \frac{\sigma_2^2}{n_2} + \left(\frac{s_2^2 - \sigma_2^2}{n_2}\right)\right\}^{-1},$$

and hence by expansion, find $E(t'^2)$ approximately. Hence find (approximately) the number of degrees of freedom on a t-distribution which yields the same variance as $V(t')$. [The variance of the t-distribution with f degrees of freedom is approximately $1 + 2/f$.]

2. In the discussion above we stated that when testing the difference between the means of two groups which have unknown and possibly unequal variances, then the test statistic has approximately a t-distribution but with unknown degrees of freedom. Look up the 5%, 2% and 1% values for t on 6 and 12 degrees of freedom, and then continue the discussion of the section to consider the importance you would attach to such inequality of variances.

CHAPTER SEVEN

Non-parametric Tests

7.1 Introduction

In the last two chapters we have seen how the t-test can be used to derive confidence intervals and significance tests for an unknown mean. In this chapter we present some alternative methods of analysis which do not make the assumption that the observations are normally distributed. For various reasons, discussed briefly at the end of this chapter, the methods are not of great importance, although with certain types of data these methods must be used. Further the logic underlying non-parametric methods is easily explained without elaborate mathematics, so that this chapter serves as an opportunity to study the basic principles underlying methods of statistical inference in a very simple setting. Those interested in following up the subject should consult Siegel (1956), Savage (1953) and references therein.

Nearly all the methods are based on test statistics which are functions of the ranks of the observations, and the distributions of the test statistics are all discrete. For the most part there is a normal approximation to the exact distribution of the test statistic under the null hypothesis, which can be used satisfactorily even in quite small sample sizes, and improves as the sample size increases. One particular case which is required is the normal approximation to the binomial distribution, and this is discussed in the next section as an example of the method of approximating discrete distributions by a continuous one. The methods of inference are set out briefly in succeeding sections, following the general outline of § 4.5.

7.2 Normal approximation to the binomial distribution

The binomial distribution is given in § 2.5(iii), and in § 2.6(iv) it is stated that the central limit theorem implies that the distribution tends to the normal as a limit, when n becomes large.

Suppose we have a binomial distribution with parameter θ and index n, then

$$\Pr(x = r) = {}^nC_r\, \theta^r (1 - \theta)^{n-r}, \tag{7.1}$$

for $r = 0, 1, \ldots, n$ and

$$E(x) = n\theta, \quad V(x) = n\theta(1 - \theta). \tag{7.2}$$

This distribution can be represented graphically by drawing a histogram with the rectangles centred on points $x = r$, and having heights determined

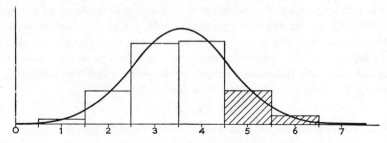

Figure 7.1 *Normal approximation to the binomial distribution*

by (7.1). Let us also plot on the same figure, an approximating normal distribution having mean and variance given by (7.2),

$$\frac{1}{\sqrt{(2\pi)}\sqrt{\{n\theta(1-\theta)\}}} \exp \left\{ -\frac{(x-n\theta)^2}{2n\theta(1-\theta)} \right\} \qquad (7.3)$$

Suppose we want to calculate the probability

$$\Pr(x \geqslant x_0) = \sum_{r=x_0}^{n} {}^nC_r\, \theta^r(1-\theta)^{n-r},$$

which corresponds to the shaded area on the histogram in Figure 7.1. We could clearly approximate to this probability by calculating the area under the approximating normal curve. We might think first of trying as an approximation the integral of (7.3) over the range (x_0, ∞).

$$\Pr(x \geqslant x_0) \simeq \int_{x_0}^{\infty} \frac{1}{\sqrt{(2\pi)}\sqrt{\{n\theta(1-\theta)\}}} \exp \left\{ -\frac{(x-n\theta)^2}{2n\theta(1-\theta)} \right\} dx$$

$$= 1 - \Phi\left[\frac{x_0 - n\theta}{\sqrt{\{n\theta(1-\theta)\}}} \right]. \qquad (7.4)$$

However, reference to Figure 7.1 indicates that a better approximation would be obtained by integrating over the range $(x_0 - \frac{1}{2}, \infty)$, since this will incorporate the whole range of x covering the shaded region. That is we try

$$\Pr(x \geqslant x_0) \simeq 1 - \Phi\left[\frac{x_0 - \frac{1}{2} - n\theta}{\sqrt{\{n\theta(1-\theta)\}}} \right] \qquad (7.5)$$

and calculations readily confirm that (7.5) is a better approximation than (7.4). The correction of $\frac{1}{2}$ is called the *continuity correction*.

In the same way, an approximation for the probability that $x \leqslant x_0$ is

$$\Pr(x \leqslant x_0) \simeq \Phi\left[\frac{x_0 + \frac{1}{2} - n\theta}{\sqrt{\{n\theta(1-\theta)\}}} \right] \qquad (7.6)$$

The accuracy of these approximations is a function of the parameters n and θ, and of the range of x over which the probability is calculated. The binomial distribution is only symmetrical if $\theta = \frac{1}{2}$, and the skewness of the distribution increases as θ approaches zero or unity. If the range of x includes the mean $n\theta$, the errors due to approximating a skew distribution by a symmetrical one tend to cancel each other out, but if single tail probabilities are required, as in (7.5), appreciable errors can arise due to skewness. However, the effect of skewness will not be large provided the minimum of $n\theta$ and $n(1 - \theta)$ is considerably larger than unity,

$$\text{Min}\{n\theta,\, n(1 - \theta)\} \gg 1. \tag{7.7}$$

Some restrictions for application of the normal approximation can be derived as follows. The normal distribution becomes negligible at about three standard deviations from the mean, and we would like these to be within the limits $(0, n)$ for the range of the binomial variable. That is, we want

$$n\theta + 3\sqrt{\{n\theta(1 - \theta)\}} < n$$

and

$$n\theta - 3\sqrt{\{n\theta(1 - \theta)\}} > 0.$$

These reduce to

$$n > 9\theta/(1 - \theta) \tag{7.8}$$

and

$$n > 9(1 - \theta)/\theta \tag{7.9}$$

respectively. Clearly if $\theta > \frac{1}{2}$, (7.8) is the important restriction, whereas if $\theta < \frac{1}{2}$, (7.9) is the important restriction. Some values of (7.8) and (7.9) are given in Table 7.1. The accuracy of the approximation depends on the per-

Table 7.1 *Minimum values of n for use of the normal approximation to the binomial distribution*

| | Values of θ in (7.8) or of $(1 - \theta)$ in (7.9) | | | | | |
	0·50	0·60	0·70	0·80	0·90	0·95
n	9	14	21	36	81	171

centage point being examined, and at 1% or 0·1% levels, values of n considerably in excess of the minimum given above are required for good approximations. [See Feller (1957) for a discussion of the normal approximation to the binomial distribution.]

Example 7.1. Suppose we have a binomial distribution with $n = 16$, $\theta = 0.50$, and we wish to find $\text{Pr}(x \geqslant 13)$, then we calculate

$$\Phi\left(\frac{x_0 - \frac{1}{2} - n\theta}{\sqrt{\{n\theta(1 - \theta)\}}}\right) = \Phi\left(\frac{13 - \frac{1}{2} - 8}{\sqrt{\{16 \times \frac{1}{2} \times \frac{1}{2}\}}}\right) = \Phi\left(\frac{4\frac{1}{2}}{2}\right)$$

$$= \Phi(2.25) = 0.9878.$$

The probability is therefore approximately 0·0122: the exact value is 0·0106.   □ □ □

Exercises 7.2

1. Examine the accuracy of the normal approximation to the binomial distribution for various probabilities for sets of parameters given in Table 7.1. For example, using $n = 9$, $\theta = 0.50$, calculate true and approximate values for

$$\Pr(x \leqslant 1), \Pr(x \leqslant 2), \Pr(x \geqslant 8), \Pr(x \geqslant 9), \Pr(3 \leqslant x \leqslant 7).$$

Also examine the value of the continuity correction in these cases.

2. Give a method for approximating to probabilities defined for Poisson random variables. Under what circumstances will skewness of the distribution give rise to large errors?

7.3 The sign test

We shall base our discussions of the sign test on the paired samples experiment described in Example 6.6. The results of this experiment can be considered as the differences $(B - A)$ between the individuals of each pair. The distribution of the differences is symmetrical about some value μ under the null hypothesis, since the treatments were randomly assigned within each pair. The analysis of § 6.4 applies if we assume that this symmetrical distribution is normal. In this section and § 7.4, methods of analysing this data are described which assume no more than that the differences have a symmetrical distribution about some mean μ.

Now if the null hypothesis is true, $\mu = 0$, and the probability p of a positive difference is $\frac{1}{2}$, whereas, say, if $\mu > 0$, then $p > \frac{1}{2}$. Therefore large or small numbers of positive signs indicate differences from the null hypothesis. This is the basis of the sign test, which proceeds as follows.

Null hypothesis. The probability of a positive sign is $\frac{1}{2}$, that is, the median of the distribution of differences is zero. For Example 6.6 the design ensures that the distribution of observations is symmetrical, so that this is equivalent to the hypothesis that the mean is zero in this case.

Alternative hypothesis. The probability of a positive sign is $p \neq \frac{1}{2}$. For Example 6.6 we are interested only in one-sided alternatives, so that if we take differences $(B - A)$, interest centres in alternatives $p > \frac{1}{2}$.

Test statistic. Observe the number n_+ and n_- of positive and negative signs. For this purpose zero results will have to be ignored, and n is defined as the number of non-zero results.

Distribution of test statistic. If the observations are independent the number of positive signs is binomially distributed with parameters (n, p) so that

$$\Pr(n_+ = r) = {}^nC_r\, p^r\, (1 - p)^{n-r}.$$

If the null hypothesis is true this becomes

$$\Pr(n_+ = r) = {}^nC_r\, 2^{-n}.$$

Example 7.2. For the data of Example 6.6 there is one zero result, which must therefore be ignored, and $n = 7$, $n_+ = 6$, $n_- = 1$. The probability distribution of n_+ under H_0 is given in Table 7.2. Thus $\mathrm{Prob}(n_+ \geqslant 6) = 8 \cdot 2^{-7}$ $= 1/16$, which is the significance level of the observed result for a one-sided

Table 7.2 *Distribution of* n_+ *under* H_0 *for* $n = 7$

	n^+							
	0	1	2	3	4	5	6	7
Prob. $\times\ 2^7$	1	7	21	35	35	21	7	1

test. This significance level is not high enough to draw any firm conclusions, though it is suggestive that treatment B may be better. On a two-sided test the significance level of the result would be $2 \times 8 \times 2^{-7} = \tfrac{1}{8}$. ☐ ☐ ☐

The sign test can only achieve certain significance levels, determined by the values of the corresponding binomial distribution. If n is large enough this cannot be considered a serious disadvantage, but it may lead to difficulties in small samples. For example, if $n = 9$, we have

$$\Pr(n_+ = 0 \mid H_0) = 0 \cdot 002$$
$$\Pr(n_+ = 1 \mid H_0) = 0 \cdot 020$$
$$\Pr(n_+ = 2 \mid H_0) = 0 \cdot 090$$

so that we have only the $0 \cdot 2\%$, $2 \cdot 2\%$ or $11 \cdot 2\%$ (one-sided) significance levels available in the range commonly used.

For large values of n it is satisfactory to use the normal approximation to the binomial distribution to calculate the significance level. If, for example, we have a paired comparisons experiment with $n = 16$, and 13 positive differences are observed, then the calculation of a one-sided significance level is set out in Example 7.1. (Double the result for a two-sided test.)

Extensions to the sign test

Denote by μ_{iA}, μ_{iB} the means of treatments A and B on day i, then the hypothesis tested above is that $\mu_{iA} = \mu_{iB}$. Two other null hypotheses are easily tested:

(i) $\mu_{iB} - \mu_{iA} = d$. That is, method B is better than method A by d units, and this hypothesis is tested by considering differences $(x_{iB} - x_{iA} - d)$, and proceeding as above.

(ii) $\mu_{iB} = \left(1 + \dfrac{d}{100}\right)\mu_{iA}$. That is, method B is better than method A by $d\%$,

and this is tested by considering differences $\left\{x_{iB} - x_{iA}\left(1 + \dfrac{d}{100}\right)\right\}$ of the observations. In both cases the resulting differences are assumed to have a symmetrical distribution.

The sign test can also be used for tests about the mean of a single sample (§ 5.3) by considering the differences $(x_i - \mu_0)$ and proceeding as above. However, the distribution of the observations x_i must be assumed symmetrical for this to be valid as a test of means.

One advantage of the sign test is that it can be used when the results of trials are strictly dichotomous, such as cure–no cure in a medical trial, preferences for one or other of a pair of foods, etc.

Confidence intervals with the sign test

The procedures (i) and (ii) of § 7.3 can be used to obtain confidence intervals. The ordered differences of Example 6.6 results are

$$-0\cdot1,\ 0\cdot0,\ 0\cdot2,\ 0\cdot3,\ 0\cdot3,\ 0\cdot4,\ 0\cdot5,\ 0\cdot6.$$

If we test hypothesis (i) of § 7.3, with $d = -0\cdot2$, all eight observations are positive, and the one-sided significance level is 2^{-8}. We thus have results of Table 7.3.

Table 7.3 *Test of hypothesis (i) of § 7.3 for various values of d*

d	n_+	n_-	Prob. element	One-sided sig. level
$< -0\cdot1$	8	0	2^{-8}	2^{-8}
$-0\cdot1 < d < 0$	7	1	8.2^{-8}	9.2^{-8}
$0 < d < 0\cdot2$	6	2	28.2^{-8}	37.2^{-8}
$0\cdot2 < d < 0\cdot3$	5	3	56.2^{-8}	93.2^{-8}
$0\cdot3 < d < 0\cdot4$	3	5	56.2^{-8}	219.2^{-8}
$0\cdot4 < d < 0\cdot5$	2	6	28.2^{-8}	247.2^{-8}
$0\cdot5 < d < 0\cdot6$	1	7	8.2^{-8}	255.2^{-8}
$0\cdot6 < d$	0	8	1.2^{-8}	$256.2^{-8} = 1$

Suppose that in (imaginary) repetitions of experiments of the form of Example 6.6 we denote the ordered observations (differences)

$$x_{(1)} < x_{(2)} < \ldots < x_{(8)}.$$

Then if, say $x_{(2)} < \mu$, where μ is the true mean of the differences, we have at least 2 negative signs. From Table 7.3 we have

$$\begin{aligned}
\Pr\{x_{(2)} < \mu\} &= \Pr\{\text{at least 2 negative signs}\} \\
&= 1 - \Pr\{\text{1 or 0 negative signs}\} \\
&= 1 - 9.2^{-8}.
\end{aligned}$$

Inverting this statement, $x_{(2)}$ is a lower confidence bound for μ with confidence $100(1 - 9.2^{-8})\%$. This means that in repeated application of this rule in experiments of 8 observations from similar experiments to Example 6.6, $x_{(2)} > \mu$ only with probability 9.2^{-8}.

Similarly we can show that $x_{(7)}$, for $n = 8$, is an upper $100(1 - 9.2^{-8})\%$ confidence bound, and $(x_{(2)}$ to $x_{(7)})$ is a $100(1 - 18.2^{-8})\%$ confidence interval.

Banerjee and Nair (1940) published tables of the confidence intervals obtained in this way, and their tables are reproduced in Siegel (1956).

Example 7.3. For the data of Example 6.6 we have

lower $100(1 - 2^{-8})\%$ confidence bound for μ is -0.1
lower $100(1 - 9.2^{-8})\%$ confidence bound for μ is 0
$100(1 - 9.2^{-7})\%$ confidence interval for μ is $0 < \mu < 0.5$. □□□

Two difficulties arise with this procedure. Firstly, we observe that for the numerical example given above, a test of the null hypothesis $\mu = 0$ gave a non-significant result at 5% level, yet zero is a lower $9.2^{-8} = 96.6\%$ confidence bound! This anomaly arises from the fact that the usual practice of ignoring zeros for a significance test has been followed. This difficulty is only serious with small numbers of observations; one possible way round it is to interpret data using the confidence bound or confidence interval argument when zero observations arise. [See Pratt (1959) and Lehman (1961).]

The second apparent anomaly arises as follows. Pressing the confidence interval argument to the limit, for $n = 8$, $(x_{(4)}, x_{(5)})$ is a $100(70.2^{-8})\%$ confidence interval. For the above data this interval is $(0.3, 0.3)$! However, we must remember that a confidence statement is true over repetitions of the experiment, using the same rule, and in repetitions $x_{(4)}$ and $x_{(5)}$ would not always be equal. Furthermore, the difficulty in this sample arises in that the observations have been approximated to the nearest 0.1 unit.

We emphasize that in the above connection, a confidence interval is a statement of the probability that an interval $(x_{(r)}, x_{(s)})$ contains μ, where $x_{(r)}$ and $x_{(s)}$ are random variables. It is not a statement of the probability that μ lies between two actual numerical values, such as $(0, 0.5)$. However, if the confidence coefficient is high enough, we have a measurable 'confidence' that the interval does contain μ.

Exercises 7.3

1. Analyse the differences given in column 3 of Table 6.2, § 6.5, by the sign test. Also give an approximately 95% confidence interval for the mean difference.

2. In a paired comparisons experiment similar to Example 6.6, 40 pairs were measured, yielding 29 positive and 11 negative differences. Find the significance level of this result on a two-sided test by the normal approximation to the binomial distribution, set out in § 7.2.

7.4 The signed rank test (Wilcoxon one sample) test

Let us consider again the results of Example 6.6. The one negative observation is smaller in absolute magnitude than any positive observation, but this information is wasted when analysis proceeds by the sign test. It would appear that a more efficient test could be obtained by using not only the sign of an observation but also its rank in order of absolute magnitude.

Suppose we have data from an experiment such as Example 6.6 as follows (written out in increasing order of absolute magnitude).

$$0.22 \quad 0.60 \quad -0.70 \quad 0.87 \quad 1.23 \quad 1.33 \quad 1.67 \quad 2.17$$

Assign to each observation its rank order, with an appropriate sign, and add, thus

$$1 + 2 - 3 + 4 + 5 + 6 + 7 + 8 = 30.$$

Denote this sum of signed ranks by T, then if the null hypothesis states that the distribution of the observations is symmetrical about zero, T would on average be zero under the null hypothesis. Large values of $|T|$ indicate probable deviations from the null hypothesis. The maximum possible value of T for n observations is

$$1 + 2 + 3 + \ldots + n = n(n + 1)/2.$$

Denote the (absolute) sum of positive and negative ranks by S_+ and S_- respectively. Then

$$S_+ - S_- = T$$
$$S_+ + S_- = n(n + 1)/2,$$

and for a given n, either S_+ or S_- are equivalent test statistics to T. [For the above example, $S_+ = 1 + 2 + 4 + 5 + 6 + 7 + 8 = 33$, $S_- = 3$, and $S_+ - S_- = T = 30$.] In the remainder of this section we shall use S_+.

We proceed formally as follows.

Null hypothesis. The distribution of observations is symmetrical about zero.

Test statistic. Observe the sums S_+ and S_- of ranks of positive and negative observations respectively. For this purpose, zero observations will have to be ignored. [Note that similar difficulties arise over this point to those already discussed in the sign test.]

Distribution of test statistic under H_0. Under H_0 each rank is equally likely to be positive or negative. If the experiment is a paired comparison design, the sign of a rank can be regarded as the result of the random allocation of treatments to units in a given pair. Thus with n observations there are 2^n different and equally likely arrangements of ranks.

Table 7.4 *Some extreme samples for* $n = 8$

Rank								
1	2	3	4	5	6	7	8	S_+
+	+	+	+	+	+	+	+	36
−	+	+	+	+	+	+	+	35
+	−	+	+	+	+	+	+	34
+	+	−	+	+	+	+	+	33
−	−	+	+	+	+	+	+	32
−	−	−	−	−	−	−	−	⋮

The probability of each arrangement is 2^{-8} under H_0

All 256 possibilities could easily be written down, as in Table 7.4, and Table 7.5 obtained.

Table 7.5 *Probabilities of values of* S_+ *under* H_0

	S_+											
	36	35	34	33	32	31	30	29	28	27	26	...
$256 \times Prob.$	1	1	1	2	2	3	4	5	5	6	7	...
$256 \times Cum.\ prob.$	1	2	3	5	7	10	14	19	24	30	37	...

Significance level. The significance level of the result above, $S_+ = 33$, $n = 8$, is 5/256 one-sided, or 10/256 two-sided.

Mathematicians have devised more convenient ways of generating distributions such as Table 7.5 and tables of percentage points of S_+ are now available; see Siegel (1956). Many of these tables are indexed by the minimum sum of like signed ranks, $\text{Min}(S_+, S_-)$.

If we carry out the above procedure on the data of Example 6.6, we shall notice that sometimes a difficulty arises over tied observations. If the tied observations are of the same sign, the tie makes no difference to the significance level. If the tied observations have different signs a more difficult problem arises, which we shall not deal with here; if the ties are not too numerous it will be satisfactory to assign to each observation the average of the tied ranks.

The extensions described in § 7.3 for the sign test can also be applied to the signed rank test, but it is more usual to consider extensions of type (i), (slippage). Throughout this discussion we have said nothing of the alternative hypothesis. It is usual to consider that the distribution of the observations under the alternative hypothesis differs only from the distribution under the null hypothesis in its location. However, any departure from the null hypothesis, such as lack of symmetrical form, may show up with the signed-rank test. The concept of alternative hypothesis becomes rather difficult with this type of test, and the distribution of S_+ under particular alternatives is almost impossible to evaluate exactly.

Approximate distribution of S_+ under H_0
Write the ranks $r = 1, 2, 3, \ldots, n$, and the signs d_r,

$$d_r = 1 \text{ if } r \text{ +ve}$$
$$= 0 \text{ otherwise}$$

and consider the sum of positive ranks

$$S_+ = \sum r \, d_r.$$

In S_+ the d_r are the random variables, with properties

$$E(d_r) = \tfrac{1}{2}, \quad E(d_r^2) = \tfrac{1}{2}, \quad V(d_r) = \tfrac{1}{4},$$

and the d_r's are independent.

Thus we have

$$E(S_+) = \sum r \, d_r = \tfrac{1}{2} \sum r = n(n + 1)/4 \tag{7.10}$$

$$V(S_+) = V\left(\sum r \, d_r\right) = \sum r^2 \, V(d_r) = n(n + 1)(2n + 1)/24 \tag{7.11}$$

Now since S_+ is a sum of independent quantities, none predominant, we expect S_+ to be closely normally distributed, and in fact the normal distribution very closely approximates to the distribution of S_+ for even quite small n.

In constructing the normal approximation to the sampling distribution of S_+, we can include a continuity correction, as for the sign test. However, S_+ changes in steps of two, and the continuity correction will therefore be unity, not $\tfrac{1}{2}$ as for the binomial distribution. We therefore use the approximations

$$\Pr(S_+ \geqslant A) \simeq 1 - \Phi\left(\frac{A - 1 - n(n + 1)/4}{\sqrt{\{n(n + 1)(2n + 1)/24\}}}\right)$$

and $$\Pr(S_+ \leqslant A) \simeq \Phi\left(\frac{A + 1 - n(n + 1)/4}{\sqrt{\{n(n + 1)(2n + 1)/24\}}}\right),$$

where A is the observed score S_+.

Example 7.4. Suppose we observe $S_+ = 70$ for $n = 12$, then the significance level of this result is calculated as follows.

$$E(S_+) = n(n + 1)/4 = 39$$
$$V(S_+) = 162\cdot5$$

$$\Phi\left(\frac{70 - 1 - 39}{\sqrt{(162\cdot5)}}\right) = \Phi(2\cdot35) = 0\cdot9906.$$

The significance level is therefore $0\cdot9\%$ (one-sided) or $1\cdot9\%$ (two-sided).

□ □ □

Example of extension of the test

Suppose the results discussed above in § 7.4 are from a paired comparisons experiment with measurements x_A, x_B on two treatments. We wish to test the null hypothesis that $(x_B - d)$ has the same distribution as x_A, so that $\mu_B - \mu_A = d$, where the μ's are population means.

Suppose that we wish to test the null hypothesis that $d = 1$, then we consider differences $(x_B - x_A - 1)$, which are as follows,

Diff.	$-0{\cdot}78$	$-0{\cdot}40$	$-1{\cdot}70$	$-0{\cdot}13$	$0{\cdot}23$	$0{\cdot}33$	$0{\cdot}67$	$1{\cdot}17$
Signed rank	-6	-4	-8	-1	2	3	5	7

We have $S_+ = 17$, so that the data are in good agreement with this hypothesis.

Confidence intervals for d

If we assume that x_A and $(x_B - d)$ have identical distributions it is possible to calculate confidence intervals for d. The method is to find the values of d for which the observed data are just significant at a given level, and basically this involves trial and error calculations of the type illustrated above for the case $d = 1$.

Figure 7.2 below shows the above data represented by points on a line, the solid vertical line representing the null hypothesis value of d (zero as shown). The ranks assigned to observations are determined by the rank order of distance from the vertical line.

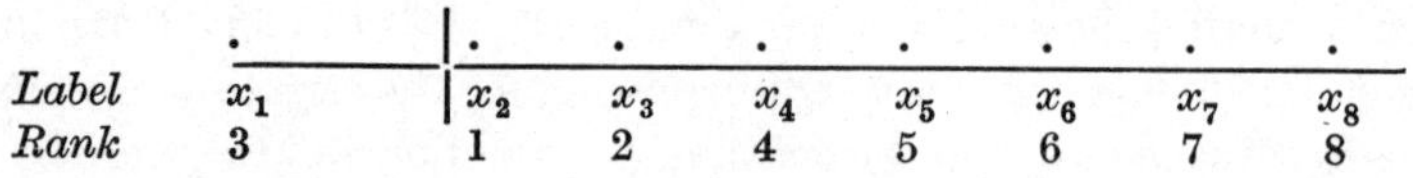

Label	x_1	x_2	x_3	x_4	x_5	x_6	x_7	x_8
Rank	3	1	2	4	5	6	7	8

Figure 7.2 *Confidence intervals by the Wilcoxon test*

Altering d is equivalent to moving the vertical line, and a confidence interval is determined by the two positions of the line which are the furthest possible distance apart, and for which the corresponding values of d are just not significant at a given level.

As the d-line is moved to the left, x_1 and x_3 change their ranks. Since $x_1 = 0{\cdot}70$, $x_3 = 0{\cdot}60$, the ranks of x_1 and x_3 are tied when $d = -0{\cdot}05$. As the d-line is moved to the right the ranks of x_1 and x_4 change, and these are tied when $d = \frac{1}{2}(-0{\cdot}70 + 0{\cdot}87) = 0{\cdot}085$.

Now suppose we want a lower $\frac{251}{256} \times 100\%$ confidence bound, corresponding to $S_- = 3$, $n = 8$. Any value of d in the range $-0{\cdot}05 < d < 0{\cdot}085$ gives $S_- = 3$, while $d < -0{\cdot}05$ gives a significance level higher than $100 \times \frac{5}{256}\%$. Therefore $d = -0{\cdot}05$ is a lower $\frac{251}{256} \times 100$ confidence bound.

Similarly, we can show that $d = 1{\cdot}75$ is an upper $100 \times \frac{251}{256}\%$ confidence bound, and hence

$$-0{\cdot}05 < d < 1{\cdot}75$$

is a $100 \times \frac{246}{256}\%$ confidence interval for d. The reader will notice that the calculation of confidence intervals with the signed-rank test is very tedious, and this is a very severe disadvantage to practical use of the test.

Exercises 7.4

1. Generate the sampling distribution of S_+ under the null hypothesis for a sample of six unequal observations.

2. Analyse the data indicated in Exercise 7.3.1 by the signed-rank test.

3. By a method similar to the one used in the binomial case (which led to Table 7.1), find the lowest value of n for which the normal approximation to the distribution of S_+ is liable to be sufficiently accurate. For this value of n, compare the tails of the exact and approximate distributions, given H_0 true. (Examine for beyond the 5% points.)

7.5 Two sample rank (Wilcoxon) test

So far in this section we have given two tests suitable for essentially single sample problems. These methods can be generalized to deal with some more complicated situations, and here we consider a two sample test of location, suitable for problems such as Example 6.1 of § 6.1.

Suppose we have two samples,

$$x\text{'s} \quad 1\cdot2 \quad 2\cdot3 \quad 2\cdot4 \quad 3\cdot2$$
$$y\text{'s} \quad 2\cdot8 \quad 3\cdot1 \quad 3\cdot4 \quad 3\cdot6 \quad 4\cdot1$$

and suppose we wish to test the null hypothesis that both of these samples are random samples from the same population.

Let us rank these observations in one group.

Obs.	1·2	2·3	2·4	2·8	3·1	3·2	3·4	3·6	4·1
Source	x	x	x	y	y	x	y	y	y
Rank	1	2	3	4	5	6	7	8	9

Now if these two samples are in fact random samples from one population, the rank sum of, say, the x's, is distributed in such a way that any combination of four out of the nine observations in all are equally likely. Thus the rank sums

$$
\begin{array}{ll}
1 + 2 + 3 + 4 & 10 \\
1 + 2 + 3 + 5 & 11 \\
1 + 2 + 3 + 6 & 12 \\
1 + 2 + 4 + 5 & 12 \\
\cdots\cdots\cdots\cdots\cdots & \\
6 + 7 + 8 + 9 & 30
\end{array}
$$

are all equally likely, and a distribution could be computed as in § 7.4 above for the one sample rank test.

It is intuitively obvious that if the rank sum of one group is used as a test, statistic, this sum will indicate differences from the null hypothesis.

The formal statement of the test is very similar to the cases already discussed.

Null hypothesis. The x's and y's are identically distributed.

Test statistic. The rank sum of one group, when the x's and y's are ranked together.

Distribution of test statistic under H_0. If H_0 is true, and there are m x's and n y's, the rank sum of the x's is distributed as a sum of m integers chosen randomly from $1, 2, \ldots, (m + n)$.

Again, mathematicians have obtained more suitable ways of tabulating the distribution under H_0 than that indicated above; see Mann and Whitney (1947).

Example 7.5. The rank sum of the x's in the above example is $1 + 2 + 3 + 6 = 12$, and there are four ways in which a rank sum of 12 or less could be obtained. The total number of different rankings is ${}^9C_4 = 126$. The observed results therefore have a significance level of 4/126. Ordinarily this would be only moderate evidence that a real difference exists, and depending on the situations, further experimentation may or may not be required. □ □ □

It is usual to consider only slippage alternatives here, as in § 7.4, but it is important to notice that any difference in the shape of the population distribution of the x's and y's, such as skewness, could throw up a significant result. Confidence intervals could be constructed using this test in a manner analogous to that described in § 7.4, but this procedure would be computationally involved. In cases of tied ranks, average ranks can be assigned, and this will not seriously affect results provided ties are not too numerous.

For large samples, the distribution of the rank sum S_x of a group with m observations approaches a normal distribution, with mean and variance

$$E(S_x) = m(m + n + 1)/2 \qquad (7.12)$$

$$V(S_x) = mn(m + n + 1)/12, \qquad (7.13)$$

under the null hypothesis. It is left to the reader to check these formulae, and obtain the normal approximation to the sampling distribution of S_x.

Exercises 7.5

1. Derive formulae (7.12) and (7.13) as follows. Write

$$S_x = \sum_{r=1}^{m+n} r\, d_r,$$

where
$$d_r = \begin{cases} 1 & \text{if } r \text{ is chosen} \\ 0 & \text{otherwise} \end{cases}$$

and exactly m of the d_r's are positive in any random selection of the $(m + n)$ ranks. The formulae now easily follow, once the quantities $E(d_r)$, $V(d_r)$, $C(d_r, d_s)$ are obtained.

2. Analyse the data of Exercise 6.2.1 by the two sample rank test.

7.6 Discussion

Suppose that either the sign test or the signed-rank test is used in a situation such as Example 7.2 when normal distribution theory might apply, then one question we would ask is how much waste of information is involved. Discussion of this sort is based upon the *power* of a test, defined,

Power = probability of rejecting the null hypothesis at a given significance level, when some specific alternative hypothesis is true.

Calculations of power are outlined in Exercise 4.5.2 and one calculation for this example is given below.

Example 7.6. From Exercise 4.5.2, we have that x_i have a distribution $N(\mu, 9)$, so that the distribution of $\bar{x}$ (with 9 observations) is $N(\mu, 1)$. A one-sided test of $H_0 : \mu = 0$ would reject H_0 when $\bar{x} > 1 \cdot 65$ for the 5% level. The probability that H_0 is rejected is therefore

$$\text{Power} = \Pr\{\bar{x} > 1 \cdot 65 \mid \bar{x} \text{ is } N(\mu, 1)\}$$

$$= \int_{1 \cdot 65}^{\infty} \frac{1}{\sqrt{(2\pi)}} \exp\left\{-(t - \mu)^2/2\right\} dt$$

$$= 1 - \Phi(1 \cdot 65 - \mu) = \Phi(\mu - 1 \cdot 65).$$

If $\mu = 0 \cdot 5$, this is $\Phi(-1 \cdot 15) = 0 \cdot 1251$. That is, if $\mu = 0 \cdot 5$, the probability that the test rejects H_0 is 12·51%, and this is the power of the test for this case. □ □ □

The power of the sign test for the same set-up can be calculated, and a graph of the power curves for the two tests for varying μ, but μ_0, n, σ as given, would be one way of comparing the tests. However, the graphs differ for different n.

Often a single measure is obtained for comparing two tests, in the following way. Consider a value μ very slightly different from μ_0, then choose numbers of observations n_s, n_t for, say, the sign test and the t-test respectively, so that the same power is obtained for the two tests. As we let μ become nearer to μ_0, and the number of observations increase, the ratio n_t/n_s tends to a limit which is called the asymptotic relative efficiency (A.R.E.). This limit is often easy to calculate mathematically.

Test	A.R.E. relative to the t-test
Sign test	66·7%
One or two sample rank test	96·4%

These A.R.E. mean roughly, that in very large samples, the same power is obtained by using $67k$ observations with the t-test as $100k$ observations on the sign test. Calculations made in particular cases tend to show that the A.R.E. is often a good measure of comparison of two tests, even in small samples.

A further general point to examine is how sensitive various tests are to deviations in assumptions. The tests given in this section do not assume normality, but they make other assumptions, such as assuming a symmetrical distribution of differences (see § 7.3). Theoretical investigations have shown that some rank tests are more sensitive to certain types of deviation from assumption than is the t-test; see Wetherill, 1960.

Exercises 7.6

1. Suppose x_i are independent $N(\mu, 1)$, and it is required to test $H_0: \mu = 0$ against one-sided alternatives, $H_1: \mu > 0$. Instead of using a normal theory test, the sign test is used, with $n = 10$, H_0 is rejected for two or less negative signs. Calculate the power for $\mu = 0$, 0·4, 0·8, 1·2, and plot the power curve.

Consider how the power function might help you choose an appropriate sample size for an experiment. (See Exercise 4.5.3.)

CHAPTER EIGHT

The Analysis of Discrete Data

8.1 Introduction

One consideration which has a very large influence on the methods of analysis which we adopt is the question of the type of measurement of which our observations are composed. Some possibilities are as follows:

(i) *Continuous data.* This includes data such as the weight per unit length of plastic pipe, Example 1.3; measurements of height such as in Example 1.6, etc. In these cases it is convenient to think of the variables as being continuous, even though they must be recorded to a finite number of decimal places.

(ii) *Discrete data.* This is when the observed values of our variables can only be one of a discrete set of values. Examples include the number of seeds germinating out of a box of 100, Example 1.1; the number of particles emitted from a radioactive source in a 5-second period, Example 1.2, etc.

(iii) *Ranked data.* This is when, for example, a subject is asked to rank some drawings in order of preference; if the subject is asked to rank n items the observation on any item is one of the numbers, $1, 2, \ldots, n$, representing the rank, and each of the numbers $1, 2, \ldots, n$ appears just once in any ranking of n items. Methods of analysis for ranked data include those given in Chapter 7. (The methods of Chapter 7 can also be used on continuous data.)

There are other types of measurement, but we shall not consider them; see Stevens (1958) for a more detailed discussion of measurement. Most of this book, Chapters 4–6 and 9–12, deals with the analysis of continuous data. This chapter is devoted to the methods of analysis appropriate for discrete data.

We shall begin by considering very simple problems, which may appear a little unrealistic. However, it is necessary to treat the simple cases first. It is important to note that raw observations are not often presented in a form suitable for simple techniques to be used directly, and it is part of the art of the statistician to calculate the statistics which are appropriate to a given problem, and which are also simple to analyse. In § 8.12, the results of a taste testing experiment are discussed, as an illustration of this point.

8.2 Distributions and approximations

It is convenient to collect together here some useful distributional results. Although we are mainly concerned with the binomial and Poisson distributions, it is important to realize that these are not the only discrete distributions. In some applications, the *negative binomial* distribution is very common and occasionally very much more complicated distributions arise.

Binomial distribution

For the binomial distribution,

$$\Pr(x = r) = {}^nC_r\theta^r(1 - \theta)^{n-r} \quad (r = 0, 1, \ldots, n), \tag{8.1}$$

one normal approximation has been discussed in § 7.2. Another normal approximation can be based on the transformation

$$z = \sin^{-1}\sqrt{(r/n)}. \tag{8.2}$$

We find from Exercise 8.2.2 that

$$E(z) \simeq \sin^{-1}\sqrt{\theta}$$

and

$$V(z) \simeq 1/(4n),$$

so that the variance of z does not depend on θ to this order of approximation. This is the main value of the transformation and it is called a *variance stabilizing transformation*. It is used, for example, when we wish to compare several binomial variables, which have the same or very similar values of n, but possibly differing values of θ. More complicated transformations of the same kind are given by Freeman and Tukey (1950). Freeman and Tukey show that in order to stabilize the variance, the transformation

$$\frac{1}{2}\left\{\sin^{-1}\sqrt{\left(\frac{r}{n+1}\right)} + \sin^{-1}\sqrt{\left(\frac{r+1}{n+1}\right)}\right\} \tag{8.3}$$

is to be preferred. Some tables of (8.3) for use in desk machine calculations are given by Mosteller and Youtz (1961). Tables of $\sin^{-1}\sqrt{\theta}$ are given, for example, in Table 9 of Lindley and Miller (1966) or Fisher and Yates (1963).

Poisson distribution

The Poisson distribution is

$$\Pr(x = r) = e^{-m}m^r/r! \quad (r = 0, 1, 2, \ldots), \tag{8.4}$$

where

$$E(x) = V(x) = m.$$

A normal approximation is therefore

$$\left.\begin{array}{l} \Pr(x \leqslant r) = \Phi\left(\dfrac{r + \frac{1}{2} - m}{\sqrt{m}}\right) \\[3mm] \Pr(x \geqslant r) = 1 - \Phi\left(\dfrac{r - \frac{1}{2} - m}{\sqrt{m}}\right) \end{array}\right\} \tag{8.5}$$

In order to stabilize the variance we use the square root transformation,

$$z = \sqrt{x},$$

and we have $\qquad\qquad E(z) \simeq \sqrt{m}, \quad V(z) \simeq 1/4,$

see Exercise 8.2.1. Freeman and Tukey (1950) show that a better transformation for stabilizing the variance is

$$\{\sqrt{x} + \sqrt{(x+1)}\} \tag{8.6}$$

which has a variance of (very nearly) unity provided $m > 1$. Tables of (8.6) are given by Mosteller and Youtz (1961). There is a serious difficulty about this transformation, which also applies to (8.3) for the binomial distribution. Most methods of statistical analysis are based on an underlying model, as described briefly in § 3.5. The analysis of data using (8.3) and (8.6), is equivalent to assuming a model in which the expectations of observations are additive in a very peculiar scale.

Negative binomial distribution

Suppose a random variable x has a Poisson distribution (8.4) where m has a probability density

$$\alpha^k m^{k-1} e^{-\alpha m}/(k-1)! \quad (0 < m < \infty), \tag{8.7}$$

where α and k are given parameters, both positive, then it is proved in Exercise 8.2.3 that the overall distribution of x is

$$\Pr(x = r) = \frac{(k+r-1)!}{(k-1)!\,r!}\theta^k(1-\theta)^r \quad (r = 0, 1, 2, \ldots), \tag{8.8}$$

where $\theta = \alpha/(1 + \alpha)$. This distribution is called the negative binomial distribution, and we have

$$E(x) = k(1-\theta)/\theta, \quad V(x) = k(1-\theta)/\theta^2.$$

One way in which the negative binomial distribution could arise is as follows. Suppose that in a large factory we record the number of accidents by each employee in a six-month period. Suppose that the distribution of the number of accidents for each individual during the period has a Poisson distribution, with a value m_i typical of the ith individual. If the distribution of m_i among individuals in the factory is approximately (8.7), then the distribution of the total number of accidents in the factory during the six-month period will be approximately negative binomial (8.8). [It is not possible to determine the mechanism which generates the distribution merely by fitting the marginal distribution of accidents in a given six-month period; the data relevant for studying the mechanism of the process are the numbers of accidents sustained by given individuals in successive time periods.] For examples of situations where the negative binomial distribution has been found a reasonable fit see, for example, Oakland (1950) and Waters (1955). The negative binomial distribution is an example of a general class of distributions called contagious distributions and for a brief treatment and references on these distributions see Douglas (1955) and Evans (1953).

Addition of random variables

Consider again the radioactive emissions experiment, Example 1.2, and suppose that we form a series of counts based on 10 seconds by adding together appropriate figures. If, as we expect, the original counts based on 5-second periods have a Poisson distribution, what is the distribution of counts based on a 10-second period? In fact it is obvious physically here that the distribution is also Poisson, but this is easily proved.

Write X and Y for independent random variables having a distribution (8.4) and put $Z = X + Y$. Then we have

$$\Pr(Z = z) = \sum \Pr(X = x) \Pr(Y = z - x)$$

$$= \sum_{x=0}^{z} e^{-m} \frac{m^x}{x!} \cdot e^{-m} \frac{m^{(z-x)}}{(z - x)!}$$

$$= e^{-2m} \frac{(2m)^z}{z!}, \tag{8.9}$$

this last result following from the binomial expansion

$$(a + b)^z = a^z + {}^zC_1 a^{z-1}b + \ldots + b^z,$$

with $a = b = \frac{1}{2}$. Therefore Z has a Poisson distribution with expectation $2m$. A much more general result holds, that if $x_1, \ldots, x_n$ have Poisson distributions with expectations $m_1, m_2, \ldots, m_n$, then the distribution of

$$z = x_1 + \ldots + x_n$$

is Poisson with expectation
$$E(z) = m_1 + \ldots + m_n,$$

see Exercise 8.2.6. A similar result holds for the binomial distribution, see Exercise 8.2.7, and also for the normal distribution, see Exercise 4.3.4. However, this is *not* a general property of probability distributions, and does not hold, for example for the negative binomial, or exponential distributions, nor for the binomial distribution if the probabilities are different.

Exercises 8.2

1. Let x be a Poisson variable with mean m, and write

$$\sqrt{x} = \sqrt{(x - m + m)} = \sqrt{m}\sqrt{\left(1 + \frac{x - m}{m}\right)}$$

$$= \sqrt{m}\left\{1 + \frac{1}{2}\left(\frac{x - m}{m}\right) - \frac{1}{8}\left(\frac{x - m}{m}\right)^2 + \frac{1}{16}\left(\frac{x - m}{m}\right)^3 - \ldots\right\}$$

then by taking expectations, show that

$$E(\sqrt{x}) = \sqrt{m}\left(1 - \frac{1}{8m} - \frac{7}{128m^2} + \cdots\right).$$

Also show that

$$V(\sqrt{x}) = E(x) - E^2(\sqrt{x}) = \frac{1}{4} + \frac{3}{32m} + \cdots$$

[Compare with the method used in Exercise 5.3.3.]

2.* Use an expansion method similar to that used in Exercise 8.2.1 to find the approximate expectation and variance of $z = \sin^{-1}\sqrt{(r/n)}$ given in (8.2).

3.* Given that the conditional distribution of x given m is Poisson with parameter m, where m has a distribution (8.7), then show that the marginal probability distribution of x is

$$\Pr(x = r) = \int \{e^{-m}\, m^r \alpha^k m^{k-1}\, e^{-\alpha m}/(k-1)!r!\}dm.$$

Prove that this is (8.8).

4.* A normal approximation to the Poisson distribution can be based on the square root transformation. That is $2(\sqrt{x} - \sqrt{m})$ is approximately $N(0, 1)$. Suggest a continuity correction, and examine numerically the accuracy of this normal approximation in comparison with (8.5). [Similarly a normal approximation to the binomial distribution could be made on the basis of (8.2).]

5.* Generalize the result given in Exercise 8.2.1. as follows: Let x be a random variable with $E(x) = \mu$, $V(x) = \sigma^2$, and let $f(x)$ be any function which can be expanded in a Taylor's series about μ,

$$f(x) = f(\mu) + (x - \mu)f'(\mu) + \tfrac{1}{2}(x - \mu)^2 f''(\mu) + \cdots$$

By taking expectations show that approximately,

$$E\{f(x)\} = f(\mu)$$

and

$$V\{f(x)\} = \sigma^2 f'(\mu)^2.$$

Check the result given above for $f(x) = \sqrt{x}$.

6. Let x_1 and x_2 have a Poisson distribution with expectations m_1, m_2 respectively. Show that the distribution of $z = x_1 + x_2$ is Poisson with expectation $(m_1 + m_2)$. [Follow the method leading to the result (8.9).]

7. Let x_1 and x_2 be binomial distributions with different indexes n_1, n_2, but the same probability θ of a success at any trial. Follow the method leading to

(8.9) to show that the distribution of $x_1 + x_2$ is binomial with index $(n_1 + n_2)$ and probability θ. Obtain by induction the general result for the distribution of

$$z = x_1 + \ldots + x_n,$$

where the distribution of x_i is independently binomial with index n_i and probability θ.

8.3 Inference about a single Poisson mean
Consider the following problem.

Example 8.1. The distribution of the number of stoppages on a loom due to warp breakages is often, but not inevitably, Poisson. Assume that the number of stoppages in, say, one hour's running has a Poisson distribution with a mean μ, where μ is the stoppage rate. In 75 running hours on a particular loom, 30 warp breakages are observed. The following questions now arise:
(*i*)   What is the best point estimate of μ?
(*ii*)   Between what limits is μ likely to lie?
(*iii*) Is the observed stoppage consistent with a rate of $\frac{1}{2}$ per hour observed on other looms?   □ □ □

We can deduce from the assumptions given in Example 8.1 and statements on addition of random variables in § 8.2, that the distribution of the number of stoppages in 75 hours is Poisson with a mean μ; we shall clearly take 30 as an estimate of μ from Example 8.1. In general, when we have a whole set of observations which can be assumed to be from the same Poisson distribution, as in the radioactive emissions experiment, Example 1.2, we take the sample mean as an estimate of the parameter μ. Clearly, when we have a set of observations, rather than a single observation, then we can examine whether the Poisson distribution fits the data, and look for outliers, etc.

Confidence intervals and significance tests can be constructed either on the normal approximation or on the exact distribution. For most applications the normal approximation theory is satisfactory, and the exact theory is left as an exercise.

Provided μ is not too small, the quantity $(x - \mu)/\sqrt{\mu}$ is approximately distributed as $N(0, 1)$, so that

$$\Pr\{-\lambda_{\alpha/2} \leqslant (x - \mu)/\sqrt{\mu} \leqslant \lambda_{\alpha/2}\} \simeq 1 - \alpha, \qquad (8.10)$$

where
$$\alpha/2 = \int_{\lambda_{\alpha/2}}^{\infty} \frac{1}{\sqrt{(2\pi)}} e^{-\frac{1}{2}t^2} \, dt.$$

This normal approximation is adequate at the 5% level for $\mu \geqslant 10$, and at the

1% level for $\mu \geqslant 20$; see Cox and Lewis (1966, p. 21) for a table comparing exact and approximate probabilities. To obtain a confidence interval we observe an x, and then invert (8.10) to form an inequality on μ. There are two ways of doing this; firstly we could replace $\sqrt{\mu}$ in the denominator of the central term of the inequality by $\sqrt{x}$, from which we obtain rough confidence intervals

$$x \pm \lambda_{\alpha/2}\sqrt{x}. \tag{8.11}$$

This method is based on a very approximate argument, and a second, more exact, method is to solve the equations

$$-\lambda_{\alpha/2} = (x - \mu)/\sqrt{\mu}, \quad +\lambda_{\alpha/2} = (x - \mu)/\sqrt{\mu}$$

for μ. By squaring to eliminate the square root, and solving the resulting quadratic equation in μ, we have confidence intervals

$$x + \tfrac{1}{2}\lambda_{\alpha/2}^2 \pm \lambda_{\alpha/2}\sqrt{(x + \tfrac{1}{4}\lambda_{\alpha/2}^2)}. \tag{8.12}$$

Unless x is very large, this interval would ordinarily be used in preference to (8.11).

Example 8.2. Approximate 95% confidence intervals for Example 8.1 are as follows. Here $x = 30$, $\lambda_{0.025} = 1.96$, so that rough confidence intervals (8.11) are

$$30 \pm 1.96\sqrt{30} = 19.3, 40.7.$$

Improved confidence intervals (8.12) are

$$30 + \tfrac{1}{2}(1.96)^2 \pm 1.96\sqrt{\{30 + \tfrac{1}{4}(1.96)^2\}} = 21.0, 42.8. \qquad \square\ \square\ \square$$

Significance tests can also be based on the normal approximation. Suppose that our null hypothesis is that the mean $\mu = \mu_0$, then if this is true $(x - \mu_0)/\sqrt{\mu_0}$ is approximately distributed as $N(0, 1)$, and we use (8.5), with $m = \mu_0$ to test the significance of an observed result. We therefore compare

$$z = (|\, x - \mu_0\,| - \tfrac{1}{2})/\sqrt{\mu_0} \tag{8.13}$$

with $N(0, 1)$ tables.

Example 8.3. In Example 8.1 the null hypothesis is the previously established stoppage rate of $\tfrac{1}{2}$ per hour, or 37.5 in 75 hours. The observed result being 30, we calculate

$$z = \frac{|\,30 - 37.5\,| - 0.5}{\sqrt{(37.5)}} = 1.14,$$

which corresponds to a probability 0.873 on $N(0, 1)$ tables. The significance level is therefore 13% for a one-sided test, or 25% for a two-sided test. We therefore conclude that the observed stoppage rate of 30 in 75 hours is consistent with the previously established rate of $\tfrac{1}{2}$ per hour. $\qquad \square\ \square\ \square$

The methods given above are for a single observation. However, if we observe $x_1 \ldots, x_n$, all Poisson variables with mean μ, then $\sum x_i$ can be considered as a single observation from a Poisson distribution with mean $n\mu$. Thus from the point of view of inference about μ, it is immaterial whether x is a single observation, or a sum of observations which have independent Poisson distributions with the same mean.

Exercises 8.3

1. Examine the data of the radioactive emissions experiment, Example 1.2, and Table 1.3. Assume that the observations have a Poisson distribution with some unknown mean μ, and obtain point and interval estimates for μ.

2. Consider how you would incorporate the continuity correction into the derivation of confidence intervals (8.12).

3. Another normal approximation to the Poisson distribution can be based on the square root transformation; see Exercise 8.2.4. The quantity $(\sqrt{x} - \sqrt{\mu})2$ is approximately distributed as $N(0, 1)$. Construct confidence intervals for μ on this basis, and compare your answers for Example 8.1, with those already given in Example 8.2.

4. The exact cumulative Poisson distribution is related to the χ^2-distribution as follows,

$$\Pr(x \leqslant c) = \sum_{r=0}^{c} e^{-\mu} \mu^r / r! = 1 - \Pr(\chi^2 < 2\mu),$$

where the χ^2 is calculated on $2(c + 1)$ degrees of freedom. Thus the χ^2-tables provide a means of obtaining exact confidence intervals for the Poisson distribution, (μ_l, μ_u), given by

$$\Pr(r \geqslant x \mid \mu_l) = \alpha/2$$
$$\Pr(r \leqslant x \mid \mu_u) = \alpha/2$$

where x is the observation. Find the exact confidence intervals for Example 8.1.
What difficulties arise here out of the discreteness of the distribution?

5. Estimate approximately how long it would be necessary to observe the experiment of Example 1.2 so that the 95% confidence intervals for the mean number of particles emitted per 5-second period, have a width equal to 10% of the mean. [For hint see Example 8.13.]

8.4 Inference about a single binomial probability

All of the discussions in § 8.3 for the Poisson distribution have an exact parallel for the binomial distribution, so that we only give a brief outline here.

If x is a binomial variable with parameters (n, θ), then

$$\frac{x - n\theta}{\sqrt{\{n\theta(1 - \theta)\}}}$$

is approximately distributed as $N(0, 1)$. Therefore we have approximately

$$\Pr\left\{-\lambda_{\alpha/2} < \frac{x - n\theta}{\sqrt{\{n\theta(1 - \theta)\}}} < \lambda_{\alpha/2}\right\} = 1 - \alpha.$$

To obtain confidence intervals, we observe an x, and invert the inequalities by solving

$$\lambda_{\alpha/2}^2 = \frac{(x - n\theta)^2}{n\theta(1 - \theta)}$$

for θ. This yields

$$\left[\frac{x}{n} + \frac{\lambda_{\alpha/2}^2}{2n} \pm \frac{\lambda_{\alpha/2}}{\sqrt{n}}\left\{\frac{x}{n}\left(1 - \frac{x}{n}\right) + \frac{\lambda_{\alpha/2}^2}{4n}\right\}^{\frac{1}{2}}\right] \bigg/ \left(1 + \frac{\lambda_{\alpha/2}^2}{n}\right) \qquad (8.14)$$

which are $100(1 - \alpha)\%$ confidence intervals for θ.

Example 8.4. Suppose that in Example 1.1, exactly 80 seeds germinated out of 100 seeds on trial. We calculate 99% confidence intervals for the true proportion germinating in the whole consignment as

$$\left[\frac{80}{100} + \frac{(2 \cdot 5758)^2}{200} \pm \frac{2 \cdot 5758}{10}\left\{\frac{80}{100}\frac{20}{100} + \frac{(2 \cdot 5758)^2}{400}\right\}^{\frac{1}{2}}\right] \bigg/ \left\{1 + \frac{(2 \cdot 5758)^2}{100}\right\}$$
$$= (0 \cdot 6777, \, 0 \cdot 8807)$$

Significance tests of a null hypothesis $\theta = \theta_0$ proceed by comparing

$$z = \frac{|x - n\theta_0| - \frac{1}{2}}{\sqrt{\{n\theta_0(1 - \theta_0)\}}} \qquad (8.15)$$

with $N(0, 1)$ tables. This significance test was used in § 7.3 for $\theta_0 = 0 \cdot 5$, as the sign test.   □ □ □

Example 8.5. Out of 112 children who have one parent with an opalescent dentine, exactly 52 also have the condition. Is this consistent with a single dominant gene?

If there is a single dominant gene, the number of children with the condition should be binomially distributed with parameters $n = 112$, $\theta = \frac{1}{2}$. Therefore we calculate (8.15) with $\theta_0 = \frac{1}{2}$, and we have,

$$z = \frac{|52 - \frac{1}{2} \cdot 112| - 0 \cdot 5}{\sqrt{\{112 \cdot \frac{1}{2} \cdot \frac{1}{2}\}}} = 0 \cdot 66.$$

This value is well to the centre of the normal distribution, and is clearly not significant. The data are therefore in good agreement with there being a single dominant gene. □ □ □

In order to be able to apply the methods of this section it is necessary to assume that the data have a binomial distribution. Sometimes the binomial distribution is assumed because of the physical set-up of the experiment, but it is important to note that the binomial distribution does not inevitably apply when at first sight it ought to. Sometimes, for some reason, data is more nearly negative binomial. There is no substitute for an empirical check on the distribution. Similar remarks apply elsewhere in this chapter, and also in connection with the Poisson distribution.

Exercises 8.4
1. In Example 8.4 confidence intervals for the percentage of seeds germinating were calculated based only on the results of one set of 100 seeds. Calculate confidence intervals based on the full set of data given in Example 1.1.

8.5 The comparison of two Poisson variates
Suppose we observe two independent Poisson variables, x_1 and x_2, with unknown means μ_1 and μ_2 respectively, then we may wish to compare the two means.

Example 8.6. In 75 running hours on loom 1 there are 30 stops, whereas in 26 running hours on loom 2 there are 21 stops. Is there evidence that the stopping rates of the two looms are not equal? □ □ □

Let the mean for 75 running hours on loom 1 be μ_1, and for 26 hours on loom 2 be μ_2, and let us write

$$\mu_1 = k\mu_2,$$

then the null hypothesis for Example 8.6 can be put

$$H_0 : k = 75/26. \tag{8.16}$$

This null hypothesis can be tested by obtaining the conditional distribution of x_1 for given values of $(x_1 + x_2)$. We proceed as follows.

Since x_1 and x_2 are independent Poisson variables, then

$$\Pr(x_1 = r, x_2 = s) = (e^{-\mu_1} \mu_1^r/r!)(e^{-\mu_2} \mu_2^s/s!). \tag{8.17}$$

Further, the distribution of $(x_1 + x_2)$ is Poisson with mean $(\mu_1 + \mu_2)$, so that

$$\Pr(x_1 + x_2 = r + s) = e^{-(\mu_1+\mu_2)} (\mu_1 + \mu_2)^{r+s}/(r + s)! \tag{8.18}$$

We now obtain the conditional probability of x_1 for given $(x_1 + x_2)$ by using (2.16) with the events A and B

$$A: (x_1 = r) \quad B: (x_1 + x_2 = r + s).$$

The event A & B is the event $(x_1 = r, x_2 = s)$, which has a probability (8.17). The conditional probability is therefore (8.17) divided by (8.18), which is

$$\Pr(x_1 = r \mid x_1 + x_2 = r + s) = \binom{r+s}{r} \frac{\mu_1^r \mu_2^s}{(\mu_1 + \mu_2)^{r+s}}$$

$$= \binom{r+s}{r} \left(\frac{k}{1+k}\right)^r \left(\frac{1}{1+k}\right)^s. \quad (8.19)$$

This probability depends on k, but not otherwise on the μ_i's. Under the null hypothesis (8.16) for Example 8.6, the conditional distribution of r given $(r + s)$ is binomial with index $x_1 + x_2 = 30 + 21 = 51$, and parameter

$$\theta = \frac{k}{1+k} = \frac{75}{26+75} = \frac{75}{101}.$$

We can therefore test the null hypothesis (8.16) by asking whether the data $r = 30$, $s = 21$, are consistent with the binomial distribution just given. This latter problem can be treated by the method given in § 8.4.

Example 8.7. Continuing the discussion of Example 8.6, we wish to know whether $r = 30$ is consistent with the binomial distribution $n = 51$, $\theta = 75/101$. The statistic (8.15) is

$$\frac{|30 - 51 \times \frac{75}{101}| - \frac{1}{2}}{\sqrt{(51 \times \frac{75}{101} \times \frac{26}{101})}} = 2 \cdot 4.$$

This result is statistically significant at about the 2% level on a two-sided test, showing quite strong evidence of a difference in stopping rate between the two looms. We therefore conclude that the data give quite strong evidence that loom 1 has a lower stopping rate than loom 2. □□□

Exercises 8.5

1. Obtain a significance test for Example 8.6 based on the normal approximation to the Poisson distribution.

2. Obtain approximate 95% confidence intervals for the ratio μ_1/μ_2 in Example 8.7.

3. Recalculate the significance test for Example 8.6 if there had only been 21 stops in the 75 running hours of loom 1.

8.6 The comparison of two binomial variates

Consider the following example, which is based upon data given by Barr (1957).

Example 8.8. A survey was carried out among young mothers of a certain town, relating to the need to provide additional maternity facilities. The town was fairly homogeneous, and was divided into two areas, two experienced interviewers A and B taking one area each, and doing all the interviews in that area. Table 8.1 below presents some of the results of interviews of mothers who had their babies at home. The responses of these mothers have been put into two groups, those who would have preferred a confinement in hospital, and those who preferred a confinement at home. Do the data give any

Table 8.1 *Results of interviews among mothers who had their babies at home*

	Interviewer		Total
	A	B	
Wanted to be in hospital	40	48	88
Wanted to be at home	44	31	75
	84	79	163

evidence of a systematic difference between the results of the two interviewers?

□ □ □

Table 8.1 is called a 2×2 *contingency table*. In the table, the totals for the columns, 84 and 79, are the numbers of mothers selected and interviewed by A and B respectively, and no particular importance can be attached to them. Each column can be considered as derived from binomial distribution with true proportions θ_A and θ_B of mothers who would have preferred to be in hospital. For our present purpose, therefore, interest centres in the difference between the observed proportions, $40/84 = 47.6\%$ for A, and $48/79 = 60.8\%$ for B. In any analysis of data of this kind, the calculation of these percentages is always the first step.

Let us represent a general 2×2 contingency table as in Table 8.2. Then the question we ask is either:

(i) That the true proportion of Row I's is the same for both columns. That is, we test the observed proportions $a/(a + c)$ and $b/(b + d)$; or

(ii) that the true proportion of column I's is the same for both rows. That is, we test the observed proportions $a/(a + b)$ and $c/(c + d)$.

Table 8.2 *General form of a 2 × 2 contingency table*

Row	Column		Totals
	I	II	
I	a	b	$a + b$
II	c	d	$c + d$
Totals	$a + c$	$b + d$	$N = a + b + c + d$

Both questions are arithmetically equivalent, since both sets of observed proportions are equal if $ad = bc$. However, often only one of the questions has a meaning, as in Example 8.8 above.

In Example 2.2 we presented some data on the heights of fathers and daughters, and Table 2.2 represents a different kind of 2 × 2 contingency table from that given above. A pair (father, daughter) is selected, and each individual classified as short or tall. Any selected pair may therefore fall in any of the four cells of the table, with probabilities $\theta_{11}, \theta_{12}, \theta_{21}, \theta_{22}$ as given in Table 8.3. Thus, for example, θ_{12} is the probability that a pair selected at

Table 8.3 *Population proportions for Table 2.2 data on heights of fathers and daughters*

Father	Daughter		Totals
	Short	Tall	
Short	θ_{11}	θ_{12}	$\theta_{11} + \theta_{12}$
Tall	θ_{21}	θ_{22}	$\theta_{21} + \theta_{22}$
Totals	$\theta_{11} + \theta_{21}$	$\theta_{12} + \theta_{22}$	1

random fall in the cell for father short, daughter tall. This type of sampling is called multinomial sampling. As indicated in Example 2.2, one problem with this type of table is to investigate whether the row and column classifications are associated or not. This reduces to questions (*i*) and (*ii*) given above for Table 8.2.

Both types of contingency table are analysed by the same methods, but there is an important difference in interpreting the results; see below. We can proceed to derive a test either on exact theory or, alternatively, on the basis of the normal approximation. Here we shall give only the normal approximation method, since this is satisfactory for most situations. We consider Table 8.2 and test the null hypothesis that the true proportion θ of row I's is the same in both columns. In Table 8.2 the proportion $a/(a + c)$ is approximately normal with a mean θ, and a variance $\theta(1 - \theta)/(a + c)$. Similarly the proportion $b/(b + d)$ is approximately normal with the same mean θ and variance $\theta(1 - \theta)/(b + d)$. Therefore the difference

$$\frac{a}{a + c} - \frac{b}{b + d} = \frac{ad - bc}{(a + c)(b + d)} \tag{8.20}$$

is approximately normal with zero mean under the null hypothesis, and variance

$$\theta(1 - \theta)\left\{\frac{1}{(a + c)} + \frac{1}{(b + d)}\right\} = \frac{\theta(1 - \theta)N}{(a + c)(b + d)}. \tag{8.21}$$

Therefore the expression on the right-hand side of (8.20) divided by the square root of the right-hand side of (8.21) is approximately $N(0, 1)$ under the null hypothesis, and by squaring the result we obtain a quantity which is approximately χ_1^2 under the null hypothesis. This is

$$\frac{(ad - bc)^2}{(a + c)^2(b + d)^2} \cdot \frac{(a + c)(b + d)}{\theta(1 - \theta)N}. \tag{8.22}$$

To eliminate θ we insert our best estimate of it, which is $(a + b)/N$, and (8.22) becomes

$$\frac{(ad - bc)^2 N}{(a + c)(b + d)(a + b)(c + d)}. \tag{8.23}$$

Again here, we are approximating a discrete distribution by a continuous one, and a continuity correction is used. Now if the marginal totals are fixed, (8.23) depends on $(ad - bc)$. If a is changed by one, say to $(a - 1)$, then to keep the marginal totals fixed, b, c, d become $(b + 1)$, $(c + 1)$, $(d - 1)$ respectively, and

$$(a - 1)(d - 1) - (b + 1)(c + 1) = ad - bc - N.$$

Therefore $(ad - bc)$ changes in steps of N, and a continuity correction must be $N/2$. The final test statistic is therefore to calculate

$$\frac{\{|\, ad - bc\, | - \tfrac{1}{2}N\}^2 N}{(a + b)(c + d)(a + c)(b + d)} \tag{8.24}$$

and refer to tables of χ_1^2. From the way in which the test has been derived we would expect χ_1^2 approximation to be poor when some of the cells have small frequencies.

Example 8.9. For Example 8.8 the test statistic is

$$\frac{\{|\, 1240 - 2112\, | - 81{\cdot}5\}^2 163}{88.75.84.79} = 2{\cdot}33,$$

which is not significant on χ_1^2. Thus although the observed proportion of mothers who would have preferred a hospital confinement is different for A and B, being 47·6% and 60·8% respectively, the difference is not statistically significant.   □ □ □

If Example 8.9 had been significant, we would have sought an explanation in that either

(a) a bias was introduced in the response of the mothers by the way the interviewers put the questions, and that the bias was different for A and B; or

(b) the proportions differed in the two areas; or perhaps partly in both explanations. Because of the design of the survey it is impossible to separate an area bias from an interviewer bias. A further question which arises here is whether the samples in each area were representative or not. Methods of selecting samples are not discussed in this book; see for example Moser (1958) or Cochran (1963). However, to ensure against possible bias some physical act of randomization is needed in selecting individuals.

A significance test can be carried out on Table 2.2 data exactly as in the manner described above. However, the explanation of a significant effect would be sought for in an association between the classification factors, and genetic theory predicts an association between the heights of fathers and daughters.

Combining evidence from contingency tables

Sometimes the problem arises that the conclusions from one particular contingency table are doubtful, and we wish to combine the evidence from two or more tables. The following example shows that adding appropriate elements of different contingency tables can produce spurious results.

Example 8.10. [Fictitious data.] A survey is carried out to find the percentage of individuals having lung cancer, among smokers and non-smokers. The results are as follows.

	Men		Women	
	LC	Not LC	LC	Not LC
Smokers	40	200	10	10
Non-smokers	2	10	200	200

	Men + Women	
	LC	Not LC
Smokers	50	210
Non-smokers	202	210

The ratio of the number of individuals with lung cancer to the number without is 1/5 for both smokers and non-smokers among men, but 1/1 for both smokers and non-smokers among women. By combining these tables over sex, the ratio is now 50/210 for smokers, but 202/210 for non-smokers! That is, the proportion with lung cancer is higher for non-smokers, even though the proportion with lung cancer is the same for smokers and non-smokers in both original tables. This curious result is produced by exploiting the difference in the proportion with lung cancer for men and women, by the numbers of individuals in various categories selected for test. This example is extreme, but it illustrates an important point – *very great care must be taken if contingency tables are added*. It also means that we should take care in interpreting data such as that in Example 8.8, and watch out for the possibility that the data has been added over another factor. For example, it is possible in Example 8.8 that the age distributions of the mothers in the two areas are different, and that there is a strong association between age and mothers' preferences.

A convenient way to combine evidence from 2×2 contingency tables is as follows. Take the square root of chi-square, and give this the sign of the quantity $(ad - bc)$; denote this quantity for the ith table by z_i. Then z_i is approximately $N(0, 1)$ under the null hypothesis, and by adding over k tables, the sum

$$S = \sum_{i=1}^{k} z_i \qquad (8.25)$$

will also be normally distributed, with expectation zero and variance k. Therefore S^2/k will be approximately χ_1^2, and can be referred to chi-square tables in the usual way.

Another way of combining evidence from contingency tables is to add the k χ_1^2's to obtain a chi-square on k degrees of freedom. By this method no account is taken of the direction of the difference in the proportions and, for example, equal and opposite differences in two tables will *not* cancel, whereas they would cancel by use of (8.25).

Exercises 8.6

1. Carry out a chi-square test on Examples 2.1 and 2.2. Are the details of the calculation really required in either case, in order to make conclusions from the data?

2. In a recent outbreak of poliomyelitis 12 of the 25 children who contracted the disease had been previously vaccinated. Two of the vaccinated and nine of the non-vaccinated developed symptons of paralysis. Do you consider that the data given support the claim that the vaccine is effective in helping to provide immunity?

8.7 Comparison of proportions in matched pairs

Sometimes we have data which can be presented in the same form as Table 8.1, and which must nevertheless be treated rather differently. Consider the following example.

Example 8.11. A certain psychological test can be presented to subjects either orally (O) or visually (V). Subjects are classified as either succeeding (S) or failing (F) the test, and it is required to know which form of presentation gives a higher success rate. Subjects used in the experiment were paired with respect to I.Q., age and background. The results can be written out as follows;

Subject pair	Method of presentation	
	O	V
1	S	S
2	S	F
3	F	F
4	S	S
5	F	S etc.

One method of presenting the results is shown in Table 8.4. However, this table ignores the fact that subjects have been paired, so that the two rows of this table are not independent and the method of testing used in the previous section does not apply.

Table 8.4 *Results of Example* 8.11

Method of presentation	S	F	
V	18	32	50
O	10	40	50

Another method of summarizing the results is shown in Table 8.5. This table reveals more clearly that there were just 50 pairs of subjects in all, and that for 38 pairs, both of the paired subjects gave the same response.

Table 8.5 *Results of Example* 8.11

Presentation O	Presentation V		
	S	F	
S	8	2	10
F	10	30	40
	18	32	50

Although Table 8.5 *can* be analysed by the method of the previous section, this analysis would test the hypothesis that the success rate for oral presentation is the same among pairs giving response S to visual presentation as among pairs giving response F to visual presentation. This hypothesis is very interesting (and manifestly untrue), but it is not the question asked!

To decide whether oral or visual presentation gives a significantly higher success rate we proceed as follows. The two cells $(S_V S_O, F_V F_O)$ give the same response to both visual and oral presentation. It can be shown that an appropriate way to analyse for a difference in success rates is to concentrate on the two diagonal cells $(S_V F_O, F_V S_O)$ in which there is a difference in responses. These diagonal cells contain 12 pairs of subjects in all, and if the success rates for visual and oral presentation are the same, results would be expected to fall in either of these two cells with equal frequencies. The null hypothesis of equal success rates can therefore be put in the form that among responses falling in cells $(S_V F_O, F_V S_O)$,

$$\text{Pr(obs. in } S_V F_O) = \text{Pr(obs. in } F_V S_O) = \tfrac{1}{2}.$$

This hypothesis can be tested by the binomial probability test of § 8.4.

Example 8.12. The null hypothesis of equal success rates can be tested on Example 8.11 data by testing whether $n = 12$, $r = 2$ is consistent with the binomial distribution for $\theta = \tfrac{1}{2}$. Formula (8.15) is

$$z = \frac{|2 - 6| - \tfrac{1}{2}}{\sqrt{(12 \times \tfrac{1}{2} \times \tfrac{1}{2})}} = \frac{7}{2\sqrt{3}} = 2 \cdot 02.$$

This result is just significant at the 5% level on a two-sided test. We therefore make the following conclusion. Among 38 of the 50 pairs of subjects, the response was identical to visual and oral presentation. Among the remaining 12 pairs of subjects there is some evidence, just significant at the 5% level, that visual presentation gives a higher success rate.

Exercises 8.7

1. In a medical trial two drugs A and B are being compared. As patients become available they are paired with respect to sex, age and state of the illness, and then the patients are randomly assigned to drug A or B. For each patient, it is noted whether his condition was improved (I), or was not improved (N) by the treatment. The results are given below. Is there any evidence that one drug is to be preferred to the other?

Pair no.	A	B		Pair no.	A	B
1	I	I		21	I	I
2	I	I		22	I	I
3	I	I		23	N	N
4	I	I		24	I	I
5	I	I		25	N	I
6	N	N		26	I	I
7	I	I		27	N	I
8	N	I		28	I	I
9	N	I		29	I	I
10	N	N		30	N	I
11	N	I		31	N	I
12	I	I		32	I	I
13	I	I		33	I	I
14	I	I		34	I	I
15	I	N		35	I	I
16	N	N		36	I	I
17	N	I		37	I	I
18	I	I		38	N	I
19	I	I		39	I	I
20	N	N		40	I	I

8.8 Examination of Poisson frequency table

Table 8.6 below reproduces the frequency table for the radioactive emissions experiment, Example 1.2 and Table 1.3. We have already indicated that we expect this data to follow the Poisson distribution, with some unknown and

Table 8.6 *Number of emissions per 5-second period from a radioactive source*

No.	0	1	2	3	4	5	6	7	8
Frequency	0	2	4	13	17	29	24	25	31

No.	9	10	11	12	13	14	Total
Frequency	27	13	5	5	4	1	200

constant value of the mean, m. In order to answer the question, 'Does the data agree with the Poisson distribution, with an unknown value of m?', we develop a test as follows.

One property of the Poisson distribution is that $E(X) = V(X) = m$. If therefore we calculate the ratio,

$$\frac{\text{sample variance}}{\text{sample mean}}$$

called the *index of dispersion*, then this should be close to unity. If the ratio is too high or too small, this will indicate that the Poisson distribution is unlikely to hold, or if it does, that m is not constant from trial to trial.

Most of the sampling variation in the index of dispersion will be in the numerator, since the denominator is the mean of the whole set of data, and will be estimated fairly precisely in large enough samples. The index of dispersion is therefore in the nature of a test of agreement of a sample variance with a theoretical variance; see § 5.4 and § 5.6.

Therefore if we calculate

$$\frac{\text{corrected sum of squares}}{\text{sample mean}} \tag{8.26}$$

then by analogy with § 5.4 we expect this statistic to be approximately distributed as χ^2 on $(n-1)$ degrees of freedom when the Poisson distribution is the true distribution of the data. If the distribution of (8.26) is $\chi^2_{(n-1)}$, then its expectation and variance would be $(n-1)$ and $2(n-1)$ respectively. Exercise 8.8.3 shows that the expectation is correct, and similarly it can be shown that the variance is approximately $2(n-1)$. In fact the distribution of (8.26) is very closely $\chi^2_{(n-1)}$ under the null hypothesis, even in quite small samples

The final method is therefore to calculate (8.26) and refer to tables of $\chi^2_{(n-1)}$. This test is called the χ^2 *dispersion test for Poisson variates*, and calculations for the radioactive emissions data are set out in Table 8.7.

Table 8.7 *Calculation of χ^2 dispersion test for the data of Table* 8.6

Total	1383·000	
Mean	6·915	$\chi^2 = \dfrac{1325 \cdot 555}{6 \cdot 915}$
Unc. SS	10,889·000	
Corr.	9563·445	
CSS	1325·555	$= 191 \cdot 7$ on 199 d.f.
Est. var.	6·661	

Index of dispersion $= 6 \cdot 661/6 \cdot 915 = 0 \cdot 963$.

The final value, 191·7 must be referred to the χ^2_{199} distribution, but very few tables extend this far. Now from (5·7) and (5.8) we have

$$E(\chi^2_{199}) = 199 \quad V(\chi^2_{199}) = 398$$

so that a normal approximation to the 5% points would be

$$199 \pm 1 \cdot 96\sqrt{(398)}$$

or roughly 199 ± 40. This normal approximation is rather poor, but quite adequate for our example here; the observed value of 191·7 is clearly nowhere near being significant at the 5% level. A more satisfactory approximation when the number of degrees of freedom exceeds 100 is to treat

$$\sqrt{(2\chi^2)} - \sqrt{\{(2 \times \text{degrees of freedom}) - 1\}} \tag{8.27}$$

as $N(0, 1)$. For our example therefore we have

$$\sqrt{(383\cdot4)} - \sqrt{(397)} = 19\cdot58 - 19\cdot92 = -0\cdot34,$$

which is not significant even at the 10% level, when referred to $N(0, 1)$ tables. We therefore conclude that the data are in excellent agreement with the Poisson distribution, and that there is no evidence of a variation in the mean; see Exercise 8.8.2.

One advantage of the χ^2 dispersion test is that it can be used with quite small numbers of observations, such as Exercise 8.8.1.

A brief discussion, with references, of the accuracy of the χ^2 dispersion test for Poisson variates is given in Cox and Lewis (1966, p. 234). The chief error arises from the variance of the statistic, which is $(1 - 1/S)$ times that of the χ^2-distribution on the appropriate number of degrees of freedom, where S is the *total* of the Poisson variates. We might expect the approximation to be reasonable then, when this total is, say, larger than 10.

If we are satisfied that our data are in good agreement with the Poisson distribution with some constant but unknown value of m, we shall require point and interval estimates for m. The methods of § 8.3 can be used directly here, since the sum of n Poisson variables with mean m is also Poisson, with mean nm. Thus if we put $X = \sum_1^n x_i$ into (8.12), we obtain confidence intervals for nm, and by dividing by n we obtain confidence intervals for m.

Example 8.13. For Example 1.2 data, approximate 99% confidence intervals for the mean number of emissions in 5-second periods are

$$\tfrac{1}{200}[1383 + \tfrac{1}{2}(2\cdot5758)^2 \pm 2\cdot5758 \sqrt{\{1383 + \tfrac{1}{4}(2\cdot5758)^2\}}]$$
$$= \tfrac{1}{200}(1386\cdot32 \pm 95\cdot85)$$
$$= \tfrac{1}{200}(1290\cdot47, 1482\cdot17) = (6\cdot45, 7\cdot41). \quad \square\,\square\,\square$$

Exercises 8.8

1. Each of five different looms are observed for 100 running hours, and the number of stops observed are respectively

$$25, 18, 34, 22, 37.$$

Is there any evidence of real differences in the stoppage rate?

2. In the radioactive emissions experiment, Example 1.2, we expect a gradual decline in the average over the period of observation, although in this particular case the decline will be very small. Will the dispersion test be a good test of whether such a trend exists or not? In what way would you examine the data in order to discover evidence, (a) of a linear trend, (b) of a more general trend?

3.* Derive the expectation of (8.2.6) as follows. If $x_1, \ldots, x_n$ are Poisson with mean m, then

$$\Pr(x_1 = r_1, \ldots, x_n = r_n) = e^{-nm} m^{\Sigma r_i} / \Pi r_i!,$$

if the observations are independent. Also $\sum x_i$ is Poisson with mean nm, so that

$$\Pr\left(\sum x_i = \sum r_i\right) = e^{-nm} (nm)^{\Sigma r_i} \bigg/ \left(\sum r_i\right)!.$$

Hence show that the conditional probability that $x_1 = r_1, \ldots, x_n = r_n$, given that $\sum x_i = \sum r_i$ is

$$\frac{\left(\sum r_i\right)!}{r_1! \ldots r_n!} \cdot \left(\frac{1}{n}\right)^{\sum r_i}$$

which is a multinomial distribution with $\sum r_i$ trials, and n equally likely events at each trial. Hence show that

$$E\left\{\sum (x_i - \bar{x})^2 \,\Big|\, \sum x_i = \sum r_i\right\} = (n-1) \sum r_i / n$$

and that

$$E\left\{\frac{\sum (x_i - \bar{x})^2}{\bar{x}} \,\Big|\, \sum x_i = \sum r_i\right\} = (n-1)$$

provided $\sum r_i \neq 0$.

8.9 Examination of binomial frequency tables

We now discuss a similar set of problems to those already given in § 8.8 for the Poisson distribution. Consider the following example.

Example 8.14. Laner *et al.* (1957) were concerned with visual space perception problems, and required a screen having random patterns. A screen of 27 by 40 in of $\frac{1}{12}$-in squares (about 155,000 elements) was desired, with black and white squares in the ratio 29/71. A series of pseudo random digits was generated, writing 1 for black and 0 for white, with the probability of a 1 of 0·29, and this series was used to fill in the squares. As a check, the screen was divided into blocks of 16 elements, and the numbers of black elements in a random set of 1000 such blocks were counted. The results are given in Table 8.8 below. Clearly we would expect the frequency table to be consistent with

Table 8.8 *Results of counts on a random screen pattern*

No. of black elements in block	No. of blocks
0	2
1	28
2	93
3	159
4	184
5	195
6	171
7	92
8	45
9	24
10	6
11	1
Total	**1000**

the binomial distribution, with $n = 16$, $\theta = 0 \cdot 29$. We consider four problems concerning this data, as follows.

(a) *Point estimates and confidence intervals for θ*
This problem does not really arise for Example 8.14, since a value of θ is given, and what is required is a test of significance; however, we illustrate the methods used on this example.

The total number of black squares in the sample is

$$2 \times 0 + 28 \times 1 + 93 \times 2 + \ldots 1 \times 11 = 4719,$$

and since there are 16,000 squares in the sample the observed proportion of black squares is $4719/16{,}000 = 0 \cdot 2949$.

Confidence intervals for θ are obtained by following the argument of § 8.4 directly, with $n = 16{,}000$, $X = 4719$. The reader should check for himself that 95% confidence intervals for θ are $(0 \cdot 2879, 0 \cdot 3020)$.

(b) *Tests of hypotheses about θ*
To answer the question 'Are the data consistent with the overall proportion of black squares being $0 \cdot 29$?', we use the method given in § 8.4. We calculate (8.15) with $\theta_0 = 0 \cdot 29$, $n = 16{,}000$ and $X = 4719$; we obtain $z = 1 \cdot 37$, which represents a significance level of about 17%.

(c) χ^2 *dispersion test for binomial variates*
To answer the question, 'Are the data consistent with there being a constant probability of a black square?' we use the dispersion test for binomial variates. The argument is parallel to that for the Poisson dispersion test, described in § 8.8.

If $x_1, \ldots, x_m$ each have a binomial distribution with index n and parameter θ, then $E(X) = n\theta$, and $V(X) = n\theta(1 - \theta)$. Thus we can obtain a test by comparing the observed sample variance with $n\theta(1 - \theta)$. Therefore we calculate the ratio

$$\frac{\text{corrected sum of squares}}{n\theta(1 - \theta)} \qquad (8.28)$$

and refer to χ^2 tables on $(m - 1)$ degrees of freedom, where m is the number of binomial variables in the sample. If a hypothesis value of θ is given, as in Example 8.14 above, then this is used in the denominator of (8.28). If no value of θ is given, a value is estimated as in paragraph (a) above, and this is used in the denominator of (8.28); the degrees of freedom of the test are not altered if an estimated value of θ is used. The test is approximate, but it can be used reliably even if only a few observations are available. See Cochran (1952, 1954) for some discussion of this and other χ^2-tests mentioned in this chapter. □ □ □

Example 8.15. In Example 8.14 we have $m = 1000$ observations from a binomial distribution for $n = 16$, $\theta = 0.29$. Therefore the sample variance should be close to

$$n\theta(1 - \theta) = 16(0.29)(0.71) = 3.2944.$$

The corrected sum of squares and sample variances are obtained as in Table 8.7 and we have

$$\text{corrected sum of squares} = 3590.04 \text{ on } 999 \text{ d.f.},$$

$$\text{sample variance} = 3.5936$$

The χ^2 dispersion statistic is therefore, from (8.28)

$$\chi^2 = 3590.04/3.2944 = 1089.7,$$

which is on 999 d.f. The test of significance must be done using (8.27), and this yields 1.99, which is a borderline result, just above 5%. If the estimated value of θ is used in the denominator instead of the theoretical value, the denominator is

$$16(0.2949)(0.7051) = 3.3269.$$

This yields a χ^2 value of 1079.1, which is equivalent to a standard normal variable of 1.77, or about 7.6% on a two-sided test.

There are therefore grounds for suspicion as to whether the screen really is a random pattern with the probability of a black square being $\theta = 0.29$. Even if it is allowed that θ may not be exactly 0.29, the χ^2 dispersion statistic, though non-significant, is still high enough to indicate some possible non-randomness. □ □ □

With problems such as Example 8.15 it is convenient to separate two questions. Firstly we can ask whether the distribution is binomial, and secondly we can ask whether θ has the hypothesis value. The second question is already considered in part (b) above. The first question is considered by using (8.28) with an *estimated* value of θ in the denominator. For practical purposes it is therefore sufficient to restrict the use of (8.28) with *estimated* values of θ in the denominator.

Exercises 8.9

1. Carry out the estimation of confidence intervals and test of significance for θ in Example 8.14, as indicated in paragraphs (a) and (b) above.

2. On four successive occasions twenty items are selected from a large batch, and the number of defectives are counted. The results are given below. Are they consistent with a constant probability of a defective?

| | Trial | | | |
	1	2	3	4
No. of defectives	3	4	7	8
No. of items inspected	20	20	20	20

8.10 Comparison of observed and expected frequencies

Many of the tests given in this chapter can be viewed as particular cases of a general method of comparing observed and expected frequencies. Instead of stating the method in a general way, we illustrate how it would be applied in some particular cases. We begin by discussing Example 8.14 of the previous section.

Table 8.9 *Calculation of χ^2 goodness of fit test*

No. black in block	Frequency observed	Frequency expected	$(Obs - Exp)^2$	$(O - E)^2/E$
0	2	4·2	4·84	1·15
1	28	27·2	0·64	0·02
2	93	83·4	92·16	1·11
3	159	159·0	0·0	0·0
4	184	211·1	734·41	3·48
5	195	206·9	141·61	0·68
6	171	154·9	259·21	1·67
7	92	90·4	2·56	0·03
8	45	41·5	12·25	0·30
9	24	15·1	79·21	5·25
10	6	4·3	2·89	0·67
11	1	1·0	0·0	0·0

The χ^2 goodness of fit test

For the frequency table, Table 8.8, there are twelve 'cells', each correspond-ing to a particular binomial observation. The expected frequencies for these cells would be given by the binomial distribution, multiplied by the sample size of 1000,

$$1000 \times (0{\cdot}71)^{16} = 4{\cdot}2, \quad 1000 \times 16(0{\cdot}29)(0{\cdot}71)^{15} = 27{\cdot}2, \quad \text{etc.,} \ldots$$

and the observed and expected frequencies for each cell are set out in Table 8.9.

The statistic used to compare observed and expected frequencies is to calculate the quantity

(observed frequency $-$ expected frequency)^2/expected frequency,

for each cell of the data, and then add over all the cells. From Table 8.9 we have

$$\sum_{\text{cells}} \frac{(\text{obs} - \text{exp})^2}{\text{exp}} = 1{\cdot}15 + 0{\cdot}02 + \ldots + 0$$

$$= 14{\cdot}36.$$

This statistic is usually denoted χ^2, and clearly large values indicate that the theory used to derive the expected frequencies – in this case the binomial distribution – is unlikely to hold. In order to determine the significance of various values of χ^2, we need the distribution of the statistic. A general result says that if the null hypothesis is true, the distribution is approxi-mately the χ^2-distribution with degrees of freedom equal to

$$\begin{pmatrix} \text{number of cells over which the} \\ \text{statistic is calculated} \end{pmatrix} - 1$$

$$- \begin{pmatrix} \text{number of parameters estimated when} \\ \text{fitting expected frequencies} \end{pmatrix} \quad (8.29)$$

The unity subtracted is due to the restriction that the total number of observations is the same for both observed and expected frequencies. For our example, the parameters of the binomial distribution were given, not estimated, and so the degrees of freedom are $12 - 1 - 0 = 11$. We there-fore refer our observed result of 14·36 to tables of the χ^2-distribution for 11 d.f., and we conclude that the data are in good agreement with the theory that the observations are binomial with $n = 16$, $\theta = 0{\cdot}29$.

It is important to recognize that only *large* values of χ^2 are regarded as significant. If the χ^2 is too small, say if the above example on 11 d.f. had come to less than 4·57, this means too good agreement with the model. This may be difficult to interpret, but it usually means that the model is at fault some-where, or that the data was faked (and badly!).

This test is called the χ^2 *goodness of fit* test, and it can be used to compare

THE ANALYSIS OF DISCRETE DATA

observed frequencies with those expected under any theoretical model. For use with continuous distributions, the scale of the variable must be broken up into convenient intervals, and then the variable is treated as discrete. One of the advantages of the goodness of fit test is that it can be applied to such a wide variety of situations; however, for any specific comparison, such as the test of agreement of data with the binomial distribution given above, there are often better tests which can be employed. The reader will notice that we have calculated the χ^2 dispersion test and the χ^2 goodness of fit test upon the same data; a detailed comparison of these tests is given below.

The distribution theory of the χ^2 goodness of fit statistic will not be given here, and the reader is referred to Cramér (1946, Chapter 30). The result is in any case approximate, and the approximation is poor if there are any cells with expected frequencies of less than one, or if there are several expected frequencies less than five. These cases must be avoided by combining cells; thus in Table 8.9, it might have been better if the cells for $x = 10$ and 11 were combined, and the degrees of freedom would then be reduced by one. This necessity to combine cells for small expected frequencies very often leads to a loss of power in the tail regions, where differences are more likely to show up.

For testing agreement of a frequency distribution with the binomial or Poisson distributions, the dispersion tests of § 8.8 and § 8.9 have more power (see § 7.6) in most situations than the goodness of fit test. The exceptions are mainly when the alternative hypothesis is a distribution with a different form to the binomial (or Poisson) but the same variance; clearly the dispersion test would be of little value for such distributions. Another advantage of the dispersion test over the goodness of fit test is that it can be used for a very small number of observations, whereas a goodness of fit test needs about a 40–50 observations to use effectively.

Test of a single binomial (compare § 8.3)

For another example of the application of the general method of comparing observed and expected frequencies, we consider Example 8.5 of § 8.4.

We can set out the results for this problem as in Table 8.10, where the expected frequencies are evaluated on the null hypothesis, that the probability of children with an opalescent dentine is $\frac{1}{2}$.

Table 8.10 *Observed and expected frequencies for Example 8.5*

	Children		Total
	with condition	*without condition*	
Observed	52	60	112
Expected	56	56	112

Applying the formula $\sum (O - E)^2/E$, and summing over both cells, we have

$$\chi^2 = \frac{4^2}{56} + \frac{4^2}{56} = 0 \cdot 57,$$

and the number of degrees of freedom is 2 (one for each cell), less 1 (for forcing the totals to agree) or 1. When the final χ^2 is on one degree of freedom, the continuity correction can be used,

$$\chi^2 = \sum \frac{\{|O - E| - \frac{1}{2}\}^2}{E}$$

$$= \frac{(3\frac{1}{2})^2}{56} + \frac{(3\frac{1}{2})^2}{56} = 0 \cdot 44.$$

This test is identical with the test of § 8.3, and gives an identical result. Since the quantity (8.15) is approximately $N(0, 1)$ under the null hypothesis, its square will be approximately χ_1^2, and this is equal to the formula just used.

One important feature of the above calculations on Table 8.10 is that there are *two* cells, not one. That is, 52 children have the condition, and 60 do not; the χ^2 must be calculated over *both* cells.

The 2×2 contingency table

For a further illustration of the general method we discuss the results of the survey among young mothers, Example 8.8 and Table 8.1. Our null hypothesis is (*i*) of § 8.6, p. 188, that the true proportion of row I's is the same for both columns.

Let Table 8.2 denote the observed frequencies, then we have to calculate the frequencies expected on the null hypothesis. Of the N observations, $(a + c)$ are in column I (interviewer A) and $(b + d)$ are in column II (interviewer B). The null hypothesis states that the proportion of row I's is the same for both columns; for Table 8.1 this is that the proportion who wanted to be in hospital was the same for A and B. The best estimate of this unknown proportion is the overall proportion of row I's, $(a + b)/N$; for Table 8.1 this is 88/163. Therefore the expected number of row I's in column I on the null hypothesis is

(Total in col. I) $\times$ (proportion in row I)

$$= (a + c) \times (a + b)/N$$
$$= \text{(column total)} \times \text{(row total)/(grand total)}. \quad (8.30)$$

In fact we see that formula (8.30) can be used to calculate expected frequencies under the null hypothesis for all cells in the table, and this is done in Table 8.11.

We now apply the general method and sum {observed − expected}²/expected over the four cells. However, we notice that the resulting χ^2 will be on

Table 8.11 *Expected frequencies for Table 8.1 data*

	Interviewer		Total
	A	B	
Wanted to be in hospital	84 × 88/163 = 45·350	79 × 88/163 = 42·650	88
Wanted to be at home	84 × 75/163 = 38·650	79 × 75/163 = 36·350	75
Total	84	79	163

one degree of freedom, since there are four cells, but two proportions have been estimated and the grand total fixed. A continuity correction is usually applied for χ_1^2, and it amounts to reducing | observed − expected | by a half before squaring. The absolute difference | observed − expected | is identical for all cells, since the sum of (observed − expected) must be zero over any row or column. The calculations of χ^2 for Table 8.1 are therefore,

$$\chi^2 = \{| \text{ obs} - \text{exp} | - \tfrac{1}{2}\}^2 \times \sum_{\text{all cells}} \frac{1}{\text{expected}}$$

$$= \{5 \cdot 350 - 0 \cdot 5\}^2 \left\{ \frac{1}{45 \cdot 350} + \frac{1}{42 \cdot 650} + \frac{1}{38 \cdot 650} + \frac{1}{36 \cdot 350} \right\}$$

$$= 2 \cdot 33.$$

This is identical to the value obtained in Example 8.9 using formula (8.24).

The intuitive justification of the formula (8.30) given above can be made rigorous. The rigorous argument is based on the conditional distribution of table entries for a given set of marginal totals; the question of estimating an unknown proportion, introduced into the above argument, does not arise.

Exercises 8.10

1. Calculate the goodness of fit test for agreement of the data of Example 1.2 with the Poisson distribution.

2.* One way to regard the Poisson distribution is that it is the distribution of the number of *events*, such as warp breakages on a loom, which occur in a fixed time, when events occur randomly in time; see § 2.5(v). Suppose we observe n different processes for times $t_1, \ldots, t_n$ respectively, and the number of events occurring in each is $x_1, \ldots, x_n$. Show how to test the null hypothesis that the mean for the ith process is λt_i, where λ is an unknown constant.

8.11 Contingency tables

The method and calculations given in the previous section for a 2×2 contingency table are readily extended to deal with a contingency table having r rows and c columns. Consider the following example.

Example 8.16 (London, B.Sc. Spec., 1956). A survey was carried out in order to find out the groups of newspapers read by people, and how readership varies with social class. The data are given in Table 8.12.

Table 8.12 *Readership of newspaper groups by social class*

Newspaper group	Social class			Total
	Well-to-do	*Middle class*	*Working class*	
α	44	613	2487	3144
β	107	791	2187	3085
γ	91	457	921	1469
Totals	242	1861	5595	7698

□ □ □

At least three questions would be of interest concerning this data. Firstly, what evidence is provided by the data about differences in the percentage of the readership in the newspapers which fall in various social classes? Secondly, what evidence is there about the overall popularity of the newspapers? Thirdly, as a check on the survey, the combined social class divisions could be compared with that known to exist in the population. We shall limit ourselves here to a discussion of the first question – variations in social class make-up of readership.

The percentages of readership in the three social classes are shown in Table 8.13. It is immediately obvious from this table that there are fairly

Table 8.13 *Percentages of readership in social class*

Newspaper group	*Well-to-do*	*Middle class*	*Working class*
α	1·4	19·5	79·1
β	3·5	25·6	70·9
γ	6·2	31·1	62·7

large differences between the newspaper groups on social class readership, there being a tendency to have more middle- and upper-class readers as we go from α to β to γ. It seems very unlikely that such large differences could be due to chance fluctuations. However, a precise test of significance can be carried out as follows.

The null hypothesis is that the (population) percentages of readership in the three social classes are identical for all three newspaper groups. If this is so, the best estimates of the probabilities for each social class will be given by the bottom line of Table 8.12,

$$242/7698, \quad 1861/7698, \quad 5595/7698.$$

Since the total number of α group readers observed was 3144, the expected numbers for the α group for the three social classes, on the null hypothesis, are

$$3144 \times 242/7698, \quad 3144 \times 1861/7698, \quad 3144 \times 5595/7698$$

respectively, which are 98·84, 760·06 and 2285·10. The expected frequencies for the other newspaper groups can be found similarly, and the reader will now see that this is equivalent to using the rule (8.30) to find the expected frequencies in any cell.

Rule. To find the expected frequencies in any contingency table, on the null hypothesis of independence between the row and column classifications, we use the rule:

expected frequency = (row total) × (column total)/(grand total).

We can put this algebraically as follows. Let there be r rows and c columns to the table, and let the observed frequency in row i, column j, be O_{ij}. Write the row and column totals as

$$R_i = \sum_{j=1}^{c} O_{ij}, \quad i = 1, 2, \ldots, r.$$

$$C_j = \sum_{i=1}^{r} O_{ij}, \quad j = 1, 2, \ldots, c,$$

and write the grand total

$$T = \sum_{i=1}^{r} \sum_{j=1}^{c} O_{ij}.$$

Then the expected frequency in row i, column j, is

$$E_{ij} = R_i C_j / T. \tag{8.31}$$

The expected frequencies for Table 8.12 (on the null hypothesis) based on this rule are given in Table 8.14. When doing the calculations on a desk

Table 8.14 *Expected frequencies for Table 8.12 data*

Newspaper group	Well-to-do	Middle class	Working class
α	98·84	760·06	2285·10
β	96·98	745·80	2242·22
γ	46·18	355·13	1067·69

machine, some work can be saved by dividing, say, the row totals by the grand total, R_i/T, and then a simple multiplication is all that is needed to obtain the expected frequencies (8.31).

We now apply the general rule described in the previous section, and calculate

$$\sum_{i=1}^{r} \sum_{j=1}^{c} (O_{ij} - E_{ij})^2/E_{ij} \qquad (8.32)$$

which is referred to as contingency χ^2. By the rule (8.29) this quantity will be approximately distributed as χ^2 when the null hypothesis is true with degrees of freedom,

(no. of cells) $- 1 -$ (no. of row and column parameters estimated)
$$= rc - 1 - (r - 1) - (c - 1)$$
$$= rc - r - c + 1$$
$$= (r - 1)(c - 1)$$
$$= \text{(no. of rows} - 1)(\text{no. of cols} - 1). \qquad (8.33)$$

When the χ^2 is on one degree of freedom, the continuity correction described in § 8.10 is applied.

The calculation of contingency χ^2 from Tables 8.12 and 8.14 is

$$\chi^2 = \frac{(44 - 98{\cdot}84)^2}{98{\cdot}84} + \frac{(613 - 760{\cdot}06)^2}{760{\cdot}06} + \ldots + \frac{(921 - 1067{\cdot}69)^2}{1067{\cdot}67}$$

$$= 30{\cdot}4 + 28{\cdot}5 + \ldots + 20{\cdot}2 = 174{\cdot}7$$

which is on $(3 - 1)(3 - 1) = 4$ degrees of freedom.

Hint for calculations. Many squaring and division operations in the calculation of contingency χ^2 can be avoided by use of tables such as Barlow's Tables, Comrie (1965).

The result obtained above is very highly significant, since the 0·1% point for χ_4^2 is only 18·47, and the first cell alone contributes more than this. For this example therefore, there is very strong evidence indeed that there are real differences between the social class make-up of readership of the newspaper groups.

This example has been worked in detail for illustrative purposes. In practice it would be pointless to carry the calculations so far; the contribution for χ^2 from the first cell alone would be sufficient to establish significance at the 0·1% level. However, cases do arise where the meaning of the data is not so clear cut at a glance, and this method provides a useful means of assessing the evidence against the null hypothesis.

There are short-cut methods of calculating contingency χ^2, which do not involve calculating expected frequencies for each cell. These are based on the method indicated in Exercise 8.11.3; see Skory (1952) and Leslie (1951), However, there is usually a need to see where the greatest contributions to

χ^2 come from. Once a significant result is obtained, the hypothesis of independence of the row and column classifications is rejected, and it becomes necessary to examine in what way the classifications are dependent. The analysis *never* ends with a statement, 'The null hypothesis is rejected at the 5% level.' A comparison between observed and expected frequencies is a great aid in interpreting a significant result.

In order to compare the observed and expected frequencies for Example 8.46 we compare Tables 8.12 and 8.14, and this leads us to the same conclusions as we arrived at from a study of Table 8.13. There is a trend in the frequencies from $\alpha \rightarrow \beta \rightarrow \gamma$, so that the contributions to χ^2 from the β row are small, but the contributions from all entries in the α and γ rows are large, and there is no single observed frequency which has a dominant effect.

Exercises 8.11

1. In Example 8.16 there were two independent surveys carried out; the results of survey A are given in Table 8.12 and the results of survey B are given below.

(*i*)   Test whether there are differences between the social class readership of the newspapers in this new data.

(*ii*) Test whether the overall popularity figures of the two surveys agree, within random variations.

(*iii*) Test separately for each newspaper group whether there are significant differences between the two surveys in the social class readership.

(*iv*) Test whether the overall social class figures of the two surveys are significantly different.

(*v*)   Write up a thorough report on these surveys, including summary tables, etc.

Results of survey B

Newspaper group	Well-to-do	Middle class	Working class	Total
α	16	277	1593	1886
β	74	588	1401	2063
γ	57	307	448	812
	147	1172	3442	4761

2. (London, B.Sc. Gen., 1958). In a routine eyesight examination of 8-year-old Glasgow school children in 1955, the children were divided into two categories, those who wore spectacles (A), and those who did not (B). As a result of the test, visual acuity was classed as good, fair or bad. The children

wearing spectacles were tested with and without them. The results are given below.

Visual acuity of Glasgow schoolchildren (1955)

	Category					
	A, with spectacles		A, without spectacles		B	
	Boys	Girls	Boys	Girls	Boys	Girls
Good	157	175	90	81	5908	5630
Fair	322	289	232	222	1873	2010
Bad	62	50	219	211	576	612
Total	541	514	541	514	8357	8252

What conclusions can be drawn from these data, regarding (*a*) sex differences in eyesight, (*b*) the value of wearing spectacles?

3. Show that contingency χ^2 (8.32) is equal to

$$\chi^2 = T\left(\sum \varphi_i/R_i - 1\right)$$

where

$$\varphi_i = \sum_{j=1}^{r} O_{ij}/C_j.$$

8.12 A tasting experiment

Finally, as an example of the combined use of methods given in this chapter to interpret a set of data, we report an experiment designed and analysed by Gridgeman (1956) concerning the manufacture of jam. The analysis given below follows closely that given by Gridgeman.

'Sucrose is sweeter than glucose by a factor that has a personal equation and whose usual value has been variously estimated to fall in the range of about 1·2–2·0. Glucose, in the commercial form known as corn syrup or confectioner's glucose, possesses certain desirable qualities as an ingredient of jam. Apart from the matter of cost, there is then an open question as to whether it is a good thing to use some glucose in jam manufacture; in other words, how would the customer – and in particular a customer accustomed to all-sucrose jams – react to it? This question was recently posed in Ottawa, with special reference to a sugar mix of three parts sucrose and one part glucose, which mixture makes a jam that is certainly not obviously different in taste from a standard jam. It was decided to assay the difference on a

Table 8.15 *Results of Gridgeman's tasting experiment*

Jam	Strawberry								Raspberry					
Manufacturer	I		II		III		IV		I		III		IV	
Subject	S	P	S	P	S	P	S	P	S	P	S	P	S	P
1	0	0	1	0	0	0	1	1	0	1	1	1	0	1
2	1	0	0	1	0	1	0	1	1	0	0	1	1	0
3	0	0	0	0	0	0	0	1	0	1	0	0	1	1
4	1	0	0	1	1	0	0	1	1	0	0	1	0	1
5	0	0	0	0	1	1	0	0	1	1	0	0	1	1
6	0	0	0	0	0	0	1	1	1	1	0	1	0	1
7	0	0	0	0	0	0	1	1	1	0	1	0	0	1
8	0	1	0	1	1	0	1	0	1	0	0	1	1	0
9	0	1	1	0	0	1	0	1	1	0	0	1	0	1
10	0	0	0	0	0	1	0	0	0	0	1	0	0	0
11	1	0	0	0	0	0	0	0	1	1	0	1	0	0
12	0	0	1	1	0	0	1	1	1	1	0	0	0	0
13	1	1	1	1	0	0	1	0	0	0	0	0	0	0
14	1	0	1	0	1	1	0	0	1	0	0	1	0	1
15	0	1	0	0	0	1	1	0	1	0	1	0	0	0
16	1	0	1	1	1	1	1	1	0	0	0	0	0	0
17	0	1	1	0	0	1	1	0	1	0	1	1	1	1
18	0	0	0	1	0	0	1	1	0	1	1	0	0	+
19	0	0	0	0	0	1	0	0	0	0	1	1	1	1
20	0	1	0	1	1	0	0	1	0	1	0	+	0	1
21	+	+	+	+	+	+	+	+	+	+	+	+	+	+
22	0	0	1	0	1	0	0	1	1	0	0	0	1	1
23	0	0	0	0	0	0	0	0	0	0	1	1	0	0
24	1	0	0	1	0	1	1	1	0	1	0	0	1	0
25	1	1	0	1	1	0	0	0	1	1	1	0	1	0
26	1	1	1	0	0	1	1	0	0	0	0	1	1	0
27	1	0	1	0	1	0	0	0	0	0	1	0	0	0
28	1	0	1	1	1	0	1	1	0	0	0	1	1	1
29	1	1	0	0	1	1	1	1	0	0	1	1	0	+
30	+	+	0	1	1	0	0	0	0	1	0	1	1	0
31	0	0	1	1	1	1	0	1	1	0	1	1	1	1
32	1	0	0	1	1	1	1	1	0	1	0	1	1	1
33	0	0	0	0	1	1	1	1	1	0	1	0	0	0
34	0	0	0	0	0	1	0	1	1	0	1	1	1	0
35	1	1	1	1	1	0	1	0	1	0	1	0	0	0
36	0	0	0	1	0	1	0	0	1	0	1	1	1	0
37	0	+	0	+	1	+	0	+	1	0	0	0	0	1
38	0	0	0	0	1	1	0	0	1	1	1	1	0	1
39	1	0	0	1	+	1	0	0	0	1	1	1	1	0
40	0	+	0	0	0	1	1	0	1	+	1	+	1	0
41	0	0	0	1	1	0	1	1	1	0	0	0	0	0
42	0	0	0	1	0	1	0	1	0	0	0	1	0	1
43	0	1	0	0	0	1	1	0	0	0	+	+	0	0
44	0	0	0	0	1	1	0	0	0	1	0	0	0	0
45	1	1	1	1	0	1	0	0	0	0	1	1	1	1
46	0	0	0	0	0	1	1	0	0	0	0	0	0	0
47	0	1	0	1	0	1	0	0	1	0	1	0	1	0
48	0	0	0	0	1	0	0	1	1	0	0	0	1	0

Manufacturer II did not produce raspberry jam.

0 = vote for sucrose jam   S = votes for sweetness
1 = vote for glucose jam   P = votes for preference
+ = neutral vote

laboratory panel consisting of a few dozen local volunteers. The design and analysis of this trial are the subject of the present paper.

'The original plan called for four independent manufacturers (hereafter referred to by the first four Roman numerals), each to make parallel batches of jam from the same raw materials. One, the "regular" batch, was to contain sucrose only, and the other, the "glucose" batch, was to differ from the first by a weight/weight replacement of twenty-five per cent of the sucrose by glucose. This was to be done with two kinds of jam: strawberry and raspberry. Thus $4 \times 2 \times 2$ jam samples were expected. In the event one manufacturer did not prepare any raspberry jams so there were fourteen samples to deal with.'

Forty-eight subjects were recruited as tasters, and each was presented with pairs of samples of jam, each pair being a sucrose and glucose jam of either strawberry or raspberry, made by one manufacturer. The pairs of jams were presented together, but the order in which the seven pairs were presented to a subject was randomized. For each such pair of samples, each subject was asked to state which he preferred, and which he thought was sweeter. Subjects were urged to make a decision, no matter how little confidence they felt in so doing, but nevertheless one subject recorded neutral responses for all seven pairs, and eight others gave some neutral responses. The results are shown in Table 8.15.

We now need to summarize this data so that its implications are more clearly apparent. One method is to look down the columns, say for strawberry jam made by manufacturer I, and count the number of subjects who prefer the Regular jam (denoted R) to the Glucose jam (denoted G) for both sweetness and preference, etc. This is shown for strawberry jam for manufacturers I and III in Table 8.16 where N denotes a neutral vote; tables for the remaining manufacturers and for raspberry jams are left as an exercise. Two points immediately arise from Table 8.16,

(1) There is a marked difference between the marginal totals for manufacturers I and III, manufacturer I obtaining more votes for his Regular jam than manufacturer III.

(2) For manufacturer I, the sweetness and preference classifications appear independent. That is, the distribution of preference votes among R, G, N, is about the same for those who voted R or G for sweetness. This is not true for manufacturer III, and there is some suspicion of association. To test for this, the standard contingency table method of § 8.11 can be used, but the expected frequencies for the N responses will be too small, and probably it is best to ignore these, making a 2×2 table.

There are seven summary tables similar to those given in Table 8.16, and all these can be examined in this way. The conclusions we reach will relate to sweetness-preference association in the group as a whole, for particular jams.

Table 8.16 *Votes for sweetness and preference among tasters*

Strawberry I

Preference	Sweetness			Total
	R	G	N	
R	21	10	0	31
G	7	6	0	13
N	2	0	2	4
Total	30	16	2	48

Strawberry III

Preference	Sweetness			Total
	R	G	N	
R	9	11	0	20
G	16	9	1	26
N	0	1	1	2
Total	25	21	2	48

It does not relate to whether individuals in the group associate sweetness and preference.

Further information on sweetness and preference association can be obtained in the following way. Each subject compares seven pairs of jams for both sweetness and preference, and we can count the number of times each subject prefers the jam he thinks is the sweeter; this count will vary from zero to seven. Table 8.17 shows a table of such associations for the 39 subjects who gave no neutral judgments. Thus there were four subjects who did not prefer any jam they thought sweeter, etc. Now if no individual had a

Table 8.17 *Sweetness preference association in individuals*

No. of associations in pair contrasts	0	1	2	3	4	5	6	7	Total
Observed frequency	4	1	6	8	4	7	6	3	39

sweetness-preference association Table 8.17 would follow a binomial distribution, with $n = 7$ and some unknown value of θ. We can therefore test the hypothesis of no sweetness-preference association by the binomial dispersion test of § 8.9, and this is left as an exercise; see Exercise 8.12.2. The significance level is very high, giving quite strong evidence that there are some subjects who base their preferences on judgments of sweetness. It is easy to show that the frequencies observed for counts of 0 and 7 are both well above expectation.

Another table can be made by taking sweetness (or preference) votes separately and counting the number of occasions out of seven each subject votes for the glucose jam, again ignoring subjects giving neutral responses. This is given in Table 8.18.

Table 8.18 *Distribution of sweetness votes*

No. of votes for glucose jam	0	1	2	3	4	5	6	7	Total
Observed frequency	1	6	7	14	11	4	1	0	44

If the group of individuals was completely homogeneous, Table 8.18 would follow the binomial distribution and this can be tested by the binomial dispersion test. The result is non-significant, but Gridgeman says, 'This is probably due to the small size of the trial, and the small differences between contrasted jams.'

The completion of the analysis and general discussion and conclusions are left as exercises for the reader. Those who need assistance may consult Gridgeman (1956).

Exercises 8.12

1. Complete tables similar to Table 8.16 for all manufacturers and jams, and test for association in the way indicated in the text.

Present a summary of the total votes for sweetness and preference separately, in the categories R, G, N, so as to reveal more clearly (a) any differences between manufacturers and (b) any differences between jams.

2. Carry out a binomial dispersion test on Table 8.17 and on Table 8.18.

3. Write out a general discussion of the results of this experiment and mention any criticisms of the experimental design.

4. The following table is a plan (based on data given by Fleming and Baker) showing counts of Japanese Beetle larvae per square foot in the top foot of soil of an 18 × 8 ft area of a field planted with maize. The columns of the table correspond to the direction of cultivation in the field, i.e. to that of the rows of maize which were sown 4 ft apart.

(a) Having regard to the overall frequency distribution of the 144 counts, would you describe the dispersion of larvae as a random one?

(b) Do you regard a test or tests applied to this overall distribution as an exhaustive means of examining for heterogeneity in the distribution of larvae

over the area? What other line of attack would you suggest? Give a critical
discussion of the conclusions you draw from any analysis undertaken.

Number of Japanese Beetle larvae

0	4	2	1	3	4	6	6
2	1	3	2	0	3	6	6
3	1	1	3	0	1	4	7
1	3	2	1	1	2	1	7
5	2	1	1	1	1	0	2
3	3	2	1	4	0	0	2
5	6	4	3	1	1	1	8
3	5	0	1	1	3	4	6
2	1	0	2	3	3	3	4
3	2	4	1	2	1	2	4
2	3	2	3	3	1	3	4
1	1	4	4	1	2	5	2
2	3	4	3	1	5	3	2
2	4	3	1	1	1	3	0
3	5	4	0	0	1	1	3
5	3	1	2	2	1	1	7
1	4	3	3	1	2	0	1
5	5	4	2	0	3	2	2

(London, B.Sc. Gen., 1960)

5. The table given below, containing part of data collected by Parkes from
herd-book records of Duroc-Jersey pigs, shows the distribution of sex in litters
of 4, 5, 6 and 7 pigs.

Examine whether these data are consistent with the hypothesis that the
number of males within a litter of given size is a binomial variable, the sex
ratio being independent of litter size. If you are not altogether satisfied with
this hypothesis, in what direction or directions does it seem to fail?

No. of males in litter	Size of litter			
	4	5	6	7
0	1	2	3	—
1	14	20	16	21
2	23	41	53	63
3	14	35	78	117
4	1	14	53	104
5	—	4	18	46
6	—	—	—	21
7	—	—	—	2
Totals	53	116	221	374

(London, B.Sc. Gen., 1960)

CHAPTER NINE

Statistical Models and Least Squares

9.1 General points

In § 3.5 we began a discussion of the concept of a *statistical model*; this is a very important concept, and most statistical analyses begin with a stage of setting up a model. We have a set of data before us, and we define a precise statistical model, which involves unknown parameters – the model to be such that it would generate data similar to the observations before us. All questions of statistical inference are then reduced to questions concerning the unknown parameters of the model. This leads on to problems of obtaining estimates of the unknown parameters from data, and problems of estimating confidence intervals, carrying out significance tests, etc. The purpose of this chapter is to illustrate further the model building stage of an analysis, and describe a general technique by which point and interval estimates and significance tests can be made for parameters in the models. In the first part of this chapter we are concerned simply with the idea of model building, and not with questions of how we use the model once it is formulated.

As explained in § 3.5, the model is nearly always chosen after a preliminary examination of the data and it is important to incorporate checks of the model into any analysis. We never believe that the model holds *exactly*; in any case the concept of random variation is rather artificial. However, we must make the model realistic enough for our purposes, and this depends on the way in which the results of our analysis are going to be used.

Consider again the experiment on bioassay of wood preservatives, Example 1.8. From Figure 1.15 it appears plausible to try the model

$$\left. \begin{array}{l} E(y) = \alpha + \beta x \\ V(y) = \sigma^2 \end{array} \right\} \qquad (9.1)$$

where y is the log (% weight loss), and x the retention level, and where y is normally distributed. We can check whether this model is satisfactory by the methods given in Chapters 8 and 10, and if the model seems to hold, the essential information in the data can be summarized into estimates of the three parameters α, β and σ^2. This will facilitate the comparison of the results with the results of similar experiments on other wood, other preservatives, etc., or the information can be used as a guide to the retention level desired for practical application.

For a second example, consider the tobacco moisture experiment, Example 1.9. Here again model (9.1) appears to be appropriate, where y is $\sqrt{}$(scale reading) and x is the moisture content as determined by a direct chemical method, which we assume to be extremely accurate. Here, the main aim of the experiment is to calibrate the sets of apparatus, that is, to predict what actual moisture content corresponds to a scale measurement x on a new sample of tobacco, and the model (9.1) will do this easily. A second aim of the experiment is to compare the two sets of apparatus A and B, and this can be done easily in terms of comparisons of estimates of the parameters α, β and σ^2 for A and B.

Another example of a statistical model is given by equation (3.33) as a model for the sampling of baled wool problem, Examples 1.10 and 3.16. Here again, if the model is satisfactory, a large set of data is summarized neatly into estimates of a few parameters, namely μ, τ^2 and σ^2. In this problem, the statistical model can be used to design a more efficient sampling plan for the main sample. That is, the model can be used to predict what the standard error of estimates of μ would have been if different numbers of cores and bales had been sampled.

Once a statistical model has been tested and the parameters estimated, the main applications can be summarized as:

(a) *Prediction.* The model is used to predict what results would be obtained in a new situation.

(b) *Comparison.* The model is used to facilitate comparisons between two or more similar sets of data observed under different conditions.

We shall have examples and cases of other applications in later chapters.

Before a model is widely applied, it is most important to check the conditions under which it is valid. For example, suppose the model (9.1) is fitted for the bioassay of wood preservatives experiment, with a particular strain of fungus, a particular type and density of wood, controlled humidity and temperature, etc. If any of these conditions are changed, it is likely that different results would be obtained. The observed results from one experiment will be of very limited value unless we can compare them with other results on similar experiments, done under different conditions, and assure ourselves that the model has some general validity. Similarly with the tobacco moisture experiment, the data were obtained using a particular type of tobacco and controlled temperature and humidity, etc. It would be imperative to carry out experiments under different conditions to see whether the same relationship held, before applying the results widely.

Another general point is that unless there are special reasons, the model used is kept as simple as possible while still satisfactorily explaining the data.

Thus wherever possible a linear model is used, and the number of parameters kept as small as possible. Transformations such as log, square root, etc., may be used to simplify a model. However, if a transformation simplifies a model as far as expectation is concerned, it may make the picture much more complicated in terms of variances. In the bioassay of wood preservatives experiment we were fortunate in that one transformation made it possible to adopt a simpler model for both expectations and variances, viz., that the expectations fall on a straight line and the variances are constant.

Exercises 9.1
1. Sketch the curves

$$y = ae^{bx}, \quad y = ax^b,$$

and give the transformations which would make them linear.

9.2 An example
Example 9.1. Baker (1957) gives the analysis of an experiment concerned with the tensile strength of a plastic material.

'In plastics the tensile strength of a polymer is often obtained by moulding a flat sheet of the material and then cutting strips from it, which are tested on a tensometer. The temperature and pressure used in moulding the sheet affect the measured tensile strength. In the example the temperatures used are 150°C, 170°C and 190°C (coded 0, 1, and 2 respectively); the pressures used are 1 ton and 2 tons per square inch (coded 0 and 1 respectively); and two types of polymer are used (codes 0 and 1 respectively).' The results are shown in Table 9.1 and they are plotted in Figure 9.1.

Table 9.1 *Results of the experiment of Example* 9.1

Polymer x_1	0	0	0	0	0	0	1	1	1	1	1	1
Pressure x_2	0	0	0	1	1	1	0	0	0	1	1	1
Temperature x_3	0	1	2	0	1	2	0	1	2	0	1	2
Tensile strength/10	94	97	109	99	101	112	88	91	91	86	91	97
Observation no.	1	2	3	4	5	6	7	8	9	10	11	12

Let us start to build a model for this data by assuming that the variances of the observations are all equal; denote the variance σ^2. The expectation for each observation can be defined by a function of temperature, pressure and type of polymer; let us start with the simplest possible model by trying a linear combination of these variables. We can write this linear function

$$E(y) = \alpha_0 + \alpha_1 x_2 + \alpha_2 x_2 + \alpha_3 x_3 \qquad (9.2)$$

where the α's are constants, x_1 and x_2 are (0, 1) variables defining the polymer

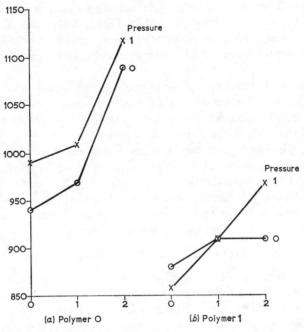

Figure 9.1 *Graphs of results of Example* 9.1

type and pressure as given in Table 9.1, and x_3 is the code denoting temperature. By (9.2) the expectation of observations 1, 6 and 12 for example would be

$$E(y_1) = \alpha_0$$
$$E(y_6) = \alpha_0 + \alpha_2 + 2\alpha_3$$
$$E(y_{12}) = \alpha_0 + \alpha_1 + \alpha_2 + 2\alpha_3, \quad \text{etc.}$$

Thus if model (9.2) holds, then apart from random variation, the pattern of the data would satisfy the following points.

(*i*) We should find that for either polymer and any temperature level, the difference caused by changing the pressure from 0 to 1 is to add a constant α_2 to the observations. Mathematically this means we have

$$E(y_4 - y_1) = E(y_5 - y_2) = E(y_6 - y_3) = E(y_{10} - y_7)$$
$$= \ldots = \alpha_2.$$

This pattern is seen to hold approximately for polymer 0, see Figure 9.1(*a*), but Figure 9.1(*b*) shows consistent discrepancies from this model. At temperature 0, the observation for pressure 0 is the higher, at temperature 1 the observations for both pressures are equal, while at temperature 2 the observation for pressure 1 is the higher. There is definite evidence here of an *interaction* between temperature and pressure.

(*ii*) We should find that the effect of changing polymer is merely to add a constant (a_1) to the observations. Thus Figure 9.1(*a*) and Figure 9.1(*b*) should have nearly identical patterns of points. This is clearly not the case and the differences appear to be systematic, and not accountable by random variation.

(*iii*) We should find that for a given polymer and pressure level, the observations for the three temperatures should fall on a straight line, the slope of this line being the same for both polymers and both pressure levels. This is clearly not the case. For polymer 0 the response to temperature is certainly not linear, whereas for polymer 1 a linear response may hold within the limits of random variation, but with a slope depending on pressure.

The model (9.2) must therefore be abandoned, and since the patterns of points for the two polymers are so different, we could treat them separately and search for a model for each. [A joint model would be preferred if a simple one existed.] The above discussion leads us to suggest

$$E(y) = \alpha_0 + \alpha_2 x_2 + \alpha_3 x_3 + \alpha_4 x_3^2 \qquad (9.3)$$

as a model for polymer 0. An examination of whether (9.3) satisfactorily explains the data would need a much more careful discussion than we have attempted so far. There are four unknown parameters in (9.3), α_0, α_2, α_3, α_4, and since there are six observations we need some means of calculating good or best estimates from the data. The remaining two degrees of freedom can be used to estimate σ^2, or if σ^2 is known or estimated from other data the two degrees of freedom can be used to check the adequacy of the model. When σ^2 is not known it is more difficult to check the model; one way is to calculate the *residuals*

$$\text{residual} = \text{observation} - \text{expectation}$$

and see if there is a pattern in them. This is rather subjective, but often fruitful; in the present example we are fitting four parameters to only six observations, and this procedure would be rather dangerous. A second method when σ^2 is not known is to add an extra term to the model, for example $\alpha_5 x_2 x_3$ could be added to (9.3), and then we examine whether this extra term has improved the fit significantly. There is little point in attempting either method with Example 9.1 as there are too few degrees of freedom remaining after the model is fitted. However, the reader will see that there is need for a theory here to provide estimates of unknown parameters, tests of significance, etc. One method of very general and widespread application is provided by the method of least squares; this is discussed in the next section.

We now return to Example 9.1 and discuss a model for polymer 1. The simplest possible model is

$$E(y) = \alpha'_0 + \alpha'_2 x_2 + \alpha'_3 x_3 \qquad (9.4)$$

where the primes indicate that these parameters are for polymer 1. Model (9.4) will clearly not be satisfactory because of the discussion under (*iii*) above. In order to allow the slope of the linear relationship with temperature to vary with pressure, we try the modified model

$$E(y) = \alpha'_0 + \alpha'_2 x_2 + \alpha'_3 x_3 + \alpha'_4 x_2 x_3. \qquad (9.5)$$

Now that we have formulated a model, we need precise methods of fitting parameters, and testing models, etc., but in this section we have been concerned merely with the model building phase of an analysis.

Exercises 9.2

1. For polymer 1, Example 9.1, draw straight lines by eye through the three points for each pressure. Find the slopes of these lines, and equate them to the slopes given by (9.5). By rough methods of this type, obtain estimates of α'_0, α'_3 and α'_4 in (9.5). Calculate the residuals from your fitted model.

How would you obtain rough estimates of the parameters for (9.3), polymer 0? Is there any way of attaching standard errors to such rough estimates?

9.3 Least squares

Once a statistical model has been set up, the parameters in it must be estimated from the data. Sometimes good estimates are intuitively obvious, but this is not always so. Further, problems of testing hypotheses about the parameters arise. There is need for a general theory by which we can derive estimates with good properties, and also obtain tests of hypotheses. Such a theory is provided by the mathematical theory of least squares; this theory is of very general application and a great deal of statistical theory is based upon it. We shall not discuss any of the mathematical theory here, but merely sketch the method and some of the important results.

Consider the following problem:

Example 9.2. A surveyor measures angles whose true values are λ and μ, and he then makes an independent determination of the combined angle $(\lambda + \mu)$. His measurements are y_1, y_2 and y_3 respectively, and they can be assumed to have no bias but are subject to random errors of zero mean and variance σ^2. Obtain estimates of λ and μ. (London, B.Sc. General, 1959.)   □ □ □

The observations y_1, y_2 and y_3 have expectations λ, μ, $(\lambda + \mu)$ respectively. The method of least squares begins by forming the sum of squares of each observation from its expectation,

$$S = (y_1 - \lambda)^2 + (y_2 - \mu)^2 + (y_3 - \lambda - \mu)^2. \qquad (9.6)$$

The estimates of $\hat{\lambda}$ and $\hat{\mu}$ of λ and μ are now found which minimize this sum of squares; in this book $\hat{\lambda}$ is used to denote the least squares estimate of λ, and $\tilde{\lambda}$ is used to denote any other estimate of λ. The minimization is done by differentiating S with respect to λ and μ, and putting the resulting expressions equal to zero, and we have

$$\frac{dS}{d\lambda} = -2(y_1 - \lambda) - 2(y_3 - \lambda - \mu)$$

$$\frac{dS}{d\mu} = -2(y_2 - \mu) - 2(y_3 - \lambda - \mu).$$

By equating these to zero and solving we obtain estimates

$$\hat{\lambda} = (2y_1 - y_2 + y_3)/3,$$

and $\qquad \hat{\mu} = (2y_2 - y_1 + y_3)/3.$

We obtain directly from the estimates that

$$E(\hat{\lambda}) = \lambda E(\hat{\mu}) = \mu$$

and $\qquad V(\hat{\lambda}) = V(\hat{\mu}) = 2\sigma^2/3.$

The minimized value of the sum of squares is

$$S_{\min} = (y_1 + y_2 - y_3)^2/3$$

and since $E(S_{\min}) = \sigma^2$, $S_{\min}$ provides an estimate of σ^2. If the y_1's are independently and normally distributed, the quantity $(y_1 + y_2 - y_3)$ is normally distributed, and $(S_{\min}/\sigma^2)$ has a χ_1^2-distribution.

Further examples of least squares are given in succeeding chapters of this book. The method has an appealing set of optimum properties. The main results are as follows:

If the observations have constant variance, and their expectations are linear functions of the unknown parameters, such as (9.3), (9.4), (9.5), then

(*i*) Least squares estimates are unbiased.

(*ii*) Least squares estimates are linear functions of the observations for any model which is linear in the unknown parameters.

(*iii*) Among all linear functions of the observations which provide unbiased estimates of the unknown parameters, least squares estimates have the smallest variance.

(*iv*) The minimized sum of squares is called the *residual sum of squares*, and this is an unbiased estimate of $k\sigma^2$, where k is the appropriate number of degrees of freedom.

(*v*) If the observations are normally distributed, the distribution of the residual sum of squares is proportional to a χ^2-distribution.

(*vi*) If the observations are normally distributed, the estimates are also normally distributed. Confidence intervals and significance tests can therefore be obtained by standard techniques.

The method can be applied when the expectations of the observations are not linear functions of the unknown parameters, but slightly less appealing optimum properties hold. The method is readily adapted to cope with observations which are not homogeneous in their variance.

Exercises 9.3

1. The diagram below represents part of an oil refinery. Oil flows in through A, and out through either B or C. There are three meters, numbered 1, 2 and 3 as shown, and all give readings of flow which are subject to normally distributed error with zero mean and variance σ^2. The observed readings are Y_1, Y_2 and Y_3, while the true (unknown) readings are θ_1, θ_2 and θ_3. Obtain least squares estimates of θ_1, θ_2 and θ_3, and give their variances. [The estimates should satisfy the restriction $\theta_2 + \theta_3 = \theta_1$.]

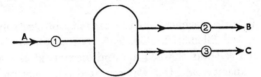

(London, B.Sc. General, 1965)

CHAPTER TEN

Linear Regression

10.1 Introduction

In § 9.1 we discussed two examples in which the *linear regression model* (9.1) is appropriate. This chapter is based directly on the theory of least squares, discussed briefly in § 9.3, and here we develop the methods of analysing data for which the model (9.1) is appropriate. The methods are *not* applicable to estimate *all* linear relationships, and the following examples make this clear.

Example 10.1. In an experiment on the capacity of electrolytic cells, four cells were taken, and filled with 0·50, 0·55, 0·60 and 0·65 ml of electrolyte respectively. Each cell was charged up, and discharged at constant current to estimate the capacity, and this was repeated ten times on each cell. The results were as follows:

Quantity of electrolyte	0·50	0·55	0·60	0·65 ml
Mean capacity	131·27	133·00	134·55	137·01

The apparent linear relationship between capacity and quantity of electrolyte (in this range) was expected, and it was required to estimate this relationship. □ □ □

In Example 10.1 the quantity of electrolyte is under the control of the experimenter and is not subject to random variation; it is said to be a *nonstochastic variable*. The capacity measurement on the other hand, fluctuates haphazardly, and can be classified as a *stochastic variable*.

Example 10.2. Madansky (1959) records measurements of yield strength and Brinell hardness of artillery shells. A linear relationship is expected between the variables, but both measurements are subject to error. The model (9.1) is not appropriate for this example, and the methods of this chapter do not apply. See Exercise 10.1.2, and § 10.8. □ □ □

In some restricted situations where both variables are stochastic, the methods of this chapter can still be used. Clearly we can do this if the error in one variable is very much less than the error in the other variable, and the tobacco moisture experiment, Example 1.9, is one example of this. The error in the direct chemical method of estimating moisture is known to be negligible, whereas the scale readings on the apparatuses are subject to error which can be estimated from the data as having a standard deviation of about 0·2; see below. The situations in which linear regression can be used are discussed further in § 10.8. Here we assume that there is one stochastic variable y and

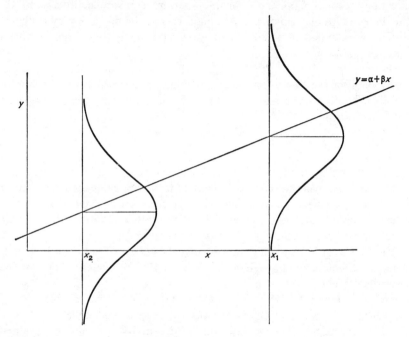

Figure 10.1 *Linear regression model*

one non-stochastic variable x, and that the expectation of y is linearly dependent on x, with constant variance

$$E(y \mid x) = \alpha + \beta x$$
$$V(y \mid x) = \sigma^2. \tag{10.1}$$

The variable y is called the *response variable*, and x is called the *explanatory variable*. The straight line $y = \alpha + \beta x$ is called the *regression line of y on x*, and the parameter β is called the *regression coefficient*. If we were able to take a very large number of observations on y for a given value of x and calculate the expectation, and then repeat this for other values of x, then the model (10.1) assumes that these expectations all lie on the straight line $y = \alpha + \beta x$;

see Figure 10.1. The parameter β is the slope of this line, and is the change in the expectation of y for a unit increase in x.

In (10.1) we are also assuming that the distribution of y is such that only its expectation changes with x. Thus for values of x, say x_1 and x_2, the distributions of y are identical except for a translation of $\beta(x_1 - x_2)$, given by the difference in their expectations.

If this model fits our data, problems of interpretation of the data can be resolved to problems of making inferences about the parameters α, β and σ involved in (10.1). In order to be able to estimate confidence intervals and carry out tests of significance on these parameters, we usually assume that the distribution of y is normal. Although this model appears very restrictive it can be extended in a number of ways to cover a very large number of practical situations.

One major difficulty with this model is that in some problems it is not clear which of the two variables is the response and which the explanatory variable. It often has a meaning to put the model in both directions, that is (10.1), and also

$$E(x) = \alpha' + \beta'y$$
$$V(x) = \sigma'^2. \tag{10.2}$$

Model (10.1) gives the regression of y on x, whereas (10.2) is the regression of x on y, and it is easy to see that in general these two regression lines are different, for they represent averages over the population of possible results in different directions; see Exercise 10.1.1. Some rules to help us decide which of the two regression lines is relevant are given in § 10.8.

The first step in analysing any set of data such as Examples 10.1 or 1.9 is to plot the data on a graph, as in Figure 1.17. From such a graph we can immediately check

(a) whether a linear relationship between $E(y)$ and x is reasonable,

(b) whether the assumption of constant variance is reasonable.

Further, a rough regression line can be drawn by eye, and approximate values of α and β calculated from this line, which provide a check on more complex methods of estimation. As discussed briefly in § 1.5, there is need for a more precise analysis in order to provide estimates of precision and also to obtain confidence intervals, significance tests, etc.

Exercises 10.1

1. Carry out the following calculations on the data given in Example 1.6, Table 1.10, for the heights of fathers and daughters. Assign to each 2-inch group interval the average height, and then calculate:

(i) the average height of daughters whose fathers have height 59, 63, 67, 71, 75 in;

(*ii*) the average height of fathers whose daughters have height 57, 61, 65, 69 in.

Plot both sets of results on the same graph. Why are the two sets of points not identical? How is this related to the difference between models (10.1) and (10.2)?

2. Formulate a model for Example 10.2.

10.2 Least squares estimates

Suppose that we have data (x_1, y_1), (x_2, y_2), . . . , (x_n, y_n), and that y is the stochastic variable and x the non-stochastic variable. We shall find it convenient to redefine model (10.1) in the form

$$E(y \mid x) = \alpha + \beta(x - \bar{x}) \atop V(y \mid x) = \sigma^2 \Bigg\} \tag{10.3}$$

where
$$\bar{x} = \left(\sum_1^n x_i\right) \Big/ n.$$

In order to obtain estimates of α and β we use the method of *least squares*. We form the sum of squared differences between the observed values of y and the expected values given by the regression line, and then minimize this sum for choice of estimates $\hat{\alpha}$ and $\hat{\beta}$ of α and β.

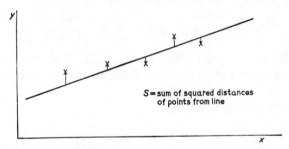

Figure 10.2 *Least squares estimates*

Thus we form

$$S = \sum_{i=1}^n \{y_i - \alpha - \beta(x_i - \bar{x})\}^2 \tag{10.4}$$

and find the values of $\hat{\alpha}$ and $\hat{\beta}$ for α and β which minimize this sum. We could do this minimization by the method indicated in § 9.3, that is, by differentiating with respect to α and β, and equating the resulting expressions to zero. However, there is an alternative method of doing this minimization in this case. The sum (10.4) can be rewritten as shown overleaf and we can easily find the values $\hat{\alpha}$ and $\hat{\beta}$ for α and β which minimize this sum.

$$S = \sum_{1}^{n} (y_i - \bar{y})^2 + n(\alpha - \bar{y})^2 + \sum (x_i - \bar{x})^2 \left\{ \beta - \frac{\sum y_i(x_i - \bar{x})}{\sum (x_1 - \bar{x})^2} \right\}^2$$

$$- \frac{\left\{ \sum y_i(x_i - \bar{x}) \right\}^2}{\sum (x_i - \bar{x})^2}$$

From this it follows that if

$$\hat{\alpha} = \bar{y} \tag{10.5}$$

and

$$\hat{\beta} = \frac{\sum y_i(x_i - \bar{x})}{\sum (x_i - \bar{x})^2} \tag{10.6}$$

then S is minimized, and has the value

$$S_{\min} = \sum (y_i - \bar{y})^2 - \frac{\left\{ \sum y_i(x_i - \bar{x}) \right\}^2}{\sum (x_i - \bar{x})^2} \tag{10.7}$$

Now by definition of S, $S_{\min}$ is the sum of squared deviations from the estimated regression line of y on x, and it will be convenient to denote it by Dev. SS(y, x). The first term on the right of (10.7) is the total corrected sum of squares for y which we shall denote CS(y, y),

$$CS(y, y) = \sum_{1}^{n} (y_i - \bar{y})^2 = \sum y_i^2 - \left(\sum y_i \right)^2 \Big/ n.$$

Therefore, the second term on the right of (10.7) {CS(y, y) − Dev. SS(y, x)}, represents the amount of variation in y attributable to the regression alone, and we shall denote this Reg. SS(y, x).

We notice in passing here that

$$\sum_{1}^{n} y_i(x_i - \bar{x}) = \sum_{1}^{n} (y_i - \bar{y})(x_i - \bar{x})$$

$$= \sum_{1}^{n} y_i x_i - \left(\sum_{1}^{n} y_i \right)\left(\sum_{1}^{n} x_i \right) \Big/ n, \tag{10.8}$$

and we denote this quantity CS(y, x). The last form, (10.8), is the one usually used in computation. In this notation we have from (10.7)

$$\text{Reg. SS}(y, x) = \{CS(y, x)\}^2 / CS(x, x).$$

Numerical example

Example 10.3. We shall illustrate the calculations for linear regression on the tobacco moisture content experiments, Example 1.9. A linear relationship is expected between $y = \sqrt{}$(scale reading), and x_j, the direct chemical determination of moisture content. The transformed data are set out in Table 10.1 for both machines A and B. We shall fit the model (10.3) separately for each apparatus, with x as the explanatory variable.

Table 10.1 *Data for Example 1.9 transformed by $y = \sqrt{}$(scale reading)*

x	11·0	11·1	7·2	8·3	12·4	14·7	5·1	21·7
y_A	3·46	3·48	2·74	2·83	4·00	4·95	2·24	6·92
y_B	3·18	3·67	2·92	3·10	4·10	4·86	2·21	6·91

x	20·2	19·1	25·0	10·2	13·3	23·6	12·0
y_A	6·57	6·18	8·30	3·44	4·47	7·59	3·87
y_B	6·83	6·19	8·05	3·46	4·18	7·43	3·85

This example will be used throughout the chapter to illustrate the numerical calculations needed. Full conclusions for this example are given as an exercise at the end of the chapter, Exercise 10.8.2.   □ □ □

The main calculations involve evaluating $CS(x, x)$, $CS(y, y)$ and $CS(y, x)$. This last calculation is carried out using (10.8), that is by subtracting $(n\bar{y}\bar{x})$ from the uncorrected sum of products $\sum y_i x_i$. The calculations are set out in Table 10.2. The equation for the regression line is found by inserting the estimates $\hat{\alpha}$ and $\hat{\beta}$ into (10.3).

Table 10.2 *Calculation of linear regression*

Machine A. Sums of squares and sums of products

	y_A	$y_A x$	x
Total	71·04	—	214·90
Mean	4·74	—	14·33
USS	387·65	1182·213	3611·43
Corr.	336·45	1017·766	3078·80
CSS	51·20	164·447	532·63

$$\beta_A = \frac{164·447}{532·63} = 0·3087$$
$$\text{Reg. } SS(y_A, x) = 50·77$$
$$\text{Dev. } SS(y_A, x) = 0·43$$

Regression line: $y_A = 0·3087x + 0·31$

Machine B. Sums of squares and sums of products

	y_B	$y_B x$	x
Total	70·94	—	214·90
Mean	4·73	—	14·33
USS	384·02	1175·850	3611·43
Corr.	335·50	1016·334	3078·80
CSS	48·52	159·516	532·63

$$\beta_B = \frac{159·516}{532·63} = 0·2995$$
$$\text{Reg. } SS(y_B, x) = 47·77$$
$$\text{Dev. } SS(y_B, x) = 0·75$$

Regression line: $y_B = 0·2995x + 0·44$

The reader should work through at least one of these sets of calculations.

Exercises 10.2

1. Check the numerical results given in Table 10.2 for at least one machine.

2. Ten batches of raw material being used for a certain process are selected, and a preliminary measurement x is made of the purity of the batch. After processing, the yield y of each batch is measured. The results are given below. You may assume that the measurement x is made with negligible error, but that the yield measurements are subject to appreciable random variation. Plot the data, and fit the model (10.3) by obtaining least squares estimates of α and β. Give the regression line, and the sum of squares due to regression.

x	4·0	5·6	2·4	5·0	3·3	5·7	2·6	5·3	4·5	4·8
y	9·2	10·3	8·9	10·1	9·3	10·2	9·3	10·8	9·9	9·3

3. Differentiate (10.4) with respect to α and β, and solve the equations $\partial S/\partial \alpha = 0$, $\partial S/\partial \beta = 0$. Check that the estimates (10.5) and (10.6) are obtained by this method.

10.3 Properties of $\hat{\alpha}$ and $\hat{\beta}$

We shall derive some properties of the estimates $\hat{\alpha}$ and $\hat{\beta}$, based on the assumptions stated, that only y is a stochastic variable, and the x's can be regarded as fixed or controlled quantities.

Firstly, the estimates $\hat{\alpha}$ and $\hat{\beta}$ are unbiased, for

$$E(\hat{\alpha}) = E(\bar{y}) = \frac{1}{n}E\left(\sum y_i\right) = \frac{1}{n}\sum E(y_i)$$

$$= \frac{1}{n}\sum \{\alpha + \beta(x_i - \bar{x})\}$$

$$= \alpha.$$

Also we have, using the generalization of rule E3, § 3.1,

$$E(\hat{\beta}) = \frac{\sum (x_i - \bar{x})\, E(y_i)}{\sum (x_i - \bar{x})^2} = \frac{\sum (x_i - \bar{x})\{\alpha + \beta(x_i - \bar{x})\}}{\sum (x_i - \bar{x})^2}$$

$$= \beta.$$

Further,
$$V(\hat{\alpha}) = V(\bar{y}) = \sigma^2/n, \tag{10.9}$$

and if the y_i are independent, we use rule V2, § 3.2, and equation (3.20) to obtain

$$V(\hat{\beta}) = V\left\{\frac{\sum y_i(x_i - \bar{x})}{\sum (x_i - \bar{x})^2}\right\} = \frac{1}{\left\{\sum (x_i - \bar{x})^2\right\}^2}\sum (x_i - \bar{x})^2\, V(y_i)$$

$$= \frac{\sigma^2}{\sum (x_i - \bar{x})^2}. \tag{10.10}$$

The student should be able to check for himself that $\hat{\alpha}$ and $\hat{\beta}$ are uncorrelated; see Exercise 10.3.2.

It can also be shown that if the y_i are independent,

$$E\{\text{Dev. SS}(y, x)\} = (n - 2)\sigma^2$$

so that an estimator of σ^2 is

$$s^2 = \frac{1}{n - 2} \text{ Dev. SS}(y, x), \tag{10.11}$$

and this is called the *mean square for deviations from regression.*

Now the sum of squares due to regression is

$$\frac{\{\text{CS}(y, x)\}^2}{\text{CS}(x, x)} = \hat{\beta}^2 \, \text{CS}(x, x),$$

and since x is not a stochastic variable the expectation of this expression is

$$\text{CS}(x, x) \, E(\hat{\beta}^2) = \text{CS}(x, x) \left\{ \frac{\sigma^2}{\text{CS}(x, x)} + \beta^2 \right\}$$
$$= \sigma^2 + \beta^2 \, \text{CS}(x, x).$$

These properties can be summed up in Table 10.3 where the mean squares are the CSS divided by the degrees of freedom. The CSS column of this table adds up exactly, since the three terms are the terms in equation (10.7). Table 10.3 shows how the total variation in y, as represented by the CSS, is divided into two parts, one attributable to a straight-line regression, and the other attributable to deviations from this straight line. The degrees of freedom are explained below. The mean square for deviations is the estimate of σ^2 given in (10.11).

Table 10.3 *Analysis of variance table for linear regression*

Source	CSS	d.f.	E(*Mean square*)
Due to regression	$\{\text{CS}(y, x)\}^2/\text{CS}(x, x)$	1	$\sigma^2 + \beta^2 \, \text{CS}(x, x)$
Deviations	Dev. SS(y, x)	$n - 2$	σ^2
Total	CS(y, y)	$n - 1$	

All these properties hold without assuming normality. If the distribution of y is normal, then $\hat{\alpha}$ and $\hat{\beta}$, being linear combinations of normal variables, are themselves normal. Thus significance tests and confidence intervals for α and β can be carried out using the t-distribution, with the variance formulae (10.9) and (10.10), and using the variance estimate (10.11), on $(n - 2)$ d.f.

[An equivalent test of the hypothesis $\beta = 0$ can be obtained from Table 10.3 by an F-test of the ratio M.S.(due to regr.)/M.S.(deviation), on $(1, n - 2)$ d.f.; see Chapter 12.] For convenience we summarize results on testing for α and β below.

The degrees of freedom in Table 10.3 can be explained intuitively or mathematically. There are $(n - 1)$ degrees of freedom altogether between the n observations $y_1, \ldots, y_n$. A straight line is fixed by the average $(\bar{y}, \bar{x})$ and one more point, so that the sum of squares due to regression must be on one degree of freedom. Mathematically, this is seen to be true since

$$\mathrm{CS}(y, x) = \sum_{i=1}^{n} y_i(x_i - \bar{x})$$

is a linear function of the y's and hence this quantity is normally distributed if the y_i are independently and normally distributed. Thus the distribution of $\{\mathrm{CS}(y, x)\}^2$ is proportional to χ_1^2 when $\beta = 0$. Having fixed $\bar{y}$ and $\hat{\beta}$, there must be $(n - 2)$ degrees of freedom remaining for the sum of squares due to deviations from regression.

Significance test and confidence intervals for β

Suppose our hypothesis is that $\beta = \beta_0$, then we calculate

$$t = \frac{(\hat{\beta} - \beta_0)}{s/\sqrt{\{\mathrm{CS}(x, x)\}}} \tag{10.12}$$

where s is the square root of (10.11), and then refer to tables of the t-distribution on $(n - 2)$ d.f. A test of the hypothesis $\beta_0 = 0$ is referred to as a test of significance of the regression, since this tests whether there is any evidence that the expected value of y does vary with x. For confidence intervals at the $100(1 - \theta)\%$ level we use

$$\hat{\beta} \pm t_{(n-2)}(\theta/2) \cdot \frac{s}{\sqrt{\{\mathrm{CS}(x, x)\}}}.$$

Significance test and confidence intervals for α

To test the hypothesis $H_0 : \alpha = \alpha_0$, calculate

$$t = \frac{\hat{\alpha} - \alpha_0}{s/\sqrt{n}} \tag{10.13}$$

and refer to tables of the t-distribution on $(n - 2)$ d.f. For confidence intervals use

$$\hat{\alpha} \pm t_{(n-2)}(\theta/2) \cdot s/\sqrt{n}.$$

Example 10.4. *Application to Example* 10.3. For machine A we obtain

$$S_A = 0 \cdot 43/13 = 0 \cdot 033,$$

as an estimate of σ_A, and an estimate of $V(\hat{\beta})$ is obtained by inserting this in (10.10),

$$\mathrm{Est}\ V(\hat{\beta}) = 0 \cdot 033/532 \cdot 63 = 0 \cdot 000062.$$

A test of significance of the hypothesis $B_A = 0$ is obtained by calculating

$$t = \frac{0.3087}{\sqrt{\{0.000062\}}} = 39.2$$

which we refer to t-tables on $n - 2 = 13$ degrees of freedom. This is very highly significant, as we would expect from Figure 1.17; such a significance test is quite superfluous for this example since it is obvious that the expected value of y depends strongly on x. The 95% confidence intervals for β_A are

$$0.3087 \pm 2.16\sqrt{(0.000062)} = (0.2917, 0.3257),$$

where 2.16 is the 95% point of t on 13 d.f.

For machine B we have similar results.

$$\hat{\sigma}_B^2 = 0.75/13 = 0.058$$

$$\text{Est } V(\hat{\beta}_B) = 0.058/532.63 = 0.00011.$$

A test of significance of the hypothesis $\beta_B = 0$ is

$$t = \frac{0.2995}{\sqrt{\{0.00011\}}} = 28.6,$$

which is referred to t-tables on 13 d.f., and is also very highly significant. The 95% confidence intervals for β_B are

$$0.2995 \pm 2.16\sqrt{(0.00011)} = (0.2768, 0.3222).$$

Analysis of variance tables are set out below.

Table 10.4 *Analysis of variance for Example* 10.3 *linear regression*

Source	Machine A			Machine B		
	CSS	d.f.	M.S.	CSS	d.f.	M.S.
Due to regression	50.77	1	50.77	47.77	1	47.77
Deviations	0.43	13	0.033	0.75	13	0.058
Total	51.20	14		48.52	14	

The meaning and uses of the analysis of variance table will become more clear in subsequent chapters, but there are four points with regard to a table such as those in Table 10.4.

(*i*) The CSS and degree of freedom add up and have a direct interpretation. The CSS for deviations, for example, separates out a sum of squares due to deviations of the observations from the regression line. The amounts of CSS due to regression and due to deviations are of interest in their own right, as summaries of the data.

(*ii*) From Table 10.3 we see that E(M.S. due to regression) is always greater than E(M.S. Deviations) if $\beta \neq 0$. If $\beta = 0$ then both mean squares are estimates of σ^2, and it can be shown that they are statistically independent.

(*iii*) An overall test of significance of the regression can be carried out using the method of § 6.3. We calculate

$$F = (\text{M.S. due to regression})/(\text{M.S. Deviations}),$$

and refer to F-tables, regarding large values as significant. This is equivalent to the t-test outlined above.

(*iv*) The ANOVA table can also be regarded as a convenient way of setting out the calculations necessary to obtain the M.S. due to deviations, which is an estimate of σ^2. ☐☐☐

Exercises 10.3

1. Write down the analysis of variance table for Exercise 10.2.2, and obtain 99% confidence intervals for σ^2. Test the significance of the regression, that is, test that β differs significantly from zero. Also obtain 95% confidence intervals for α and β.

2. Assume the model (10.3) and show that the least squares estimate $\hat{\beta}$ of β satisfies

$$C(y_i, \hat{\beta}) = \sigma^2(x_i - \bar{x})/\sum (x_i - \bar{x})^2.$$

Hence show that

$$C(\hat{\alpha}, \hat{\beta}) = 0.$$

3. Prove (10.11) as follows. Write

$$y_i = \alpha + \beta(x_i - \bar{x}) + \varepsilon_i,$$

where the ε_i are independent, with $E(\varepsilon_i) = 0$, $V(\varepsilon_i) = \sigma^2$. Show that

$$y_i - \bar{y} - \hat{\beta}(x_i - \bar{x}) = \varepsilon_i - \bar{\varepsilon}. - \tilde{\varepsilon}(x_i - \bar{x})$$

$$= z_i, \text{ say,}$$

where

$$\bar{\varepsilon}. = \sum_{i=1}^{n} \varepsilon_i/n,$$

and

$$\tilde{\varepsilon} = \sum \varepsilon_i(x_i - \bar{x})/\sum (x_i - \bar{x})^2.$$

Also show that $E(z_i) = 0$, and

$$V(z_i) = \left\{ 1 - \frac{1}{n} - \frac{(x_i - \bar{x})^2}{\sum (x_i - \bar{x})^2} \right\} \sigma^2,$$

and hence show that

$$E\{\text{Dev. SS}(y, x)\} = E\left(\sum z_i^2 \right) = \sum V(z_i) = (n - 2)\sigma^2.$$

Can you see a way of shortening this proof using (10.7) and $E(\hat{\beta}^2)$?

10.4 Predictions from regressions

Example 10.5. Suppose in Example 10.1 we wished to carry out a further large set of observations on cells containing 0·70 ml of electrolyte. What mean capacity is predicted for such cells by the data? Obtain 95% confidence intervals for this quantity.   □ □ □

Now for any given value of x, say x', we predict the mean value of y as

$$\hat{E}(y) = \hat{\alpha} + \hat{\beta}(x' - \bar{x}) \qquad (10.14)$$

where $\bar{x}$ refers to the data on which $\hat{\alpha}$ and $\hat{\beta}$ are estimated. Taking variances, we have

$$V\{\hat{E}(y)\} = V(\hat{\alpha}) + (x' - \bar{x})^2 \, V(\hat{\beta})$$

since $C(\hat{\alpha}, \hat{\beta}) = 0$, and

$$V\{\hat{E}(y)\} = \sigma^2 \left\{ \frac{1}{n} + \frac{(x' - \bar{x})^2}{CS(x, x)} \right\}. \qquad (10.15)$$

By using expressions (10.14) and (10.15) we can obtain point and interval estimates for the expectation of y at x'. Any such estimates are subject to the assumption that the regression is still linear. For Example 10.4, x' is outside the range of x originally used, and there is some doubt in this example as to whether linear regression holds or not at 0·70 ml of electrolyte.

A similar but distinct situation arises as illustrated in Example 10.6.

Example 10.6. After a period of time, one further pair of observations was taken on Example 10.3, machine A, to check that the machine was still working correctly, giving results (y', x'). Is this pair of results consistent with the previously estimated regression?   □ □ □

For a value of x, say x', the expectation of y is predicted as in (10.14) with a variance (of the prediction) given by (10.15). Now the difference we wish to test is $\{y' - \hat{E}(y)\}$, which has a variance,

$$V\{y' - \hat{E}(y)\} = V(y') + V\{\hat{E}(y)\}$$

$$= \sigma^2 \left\{ 1 + \frac{1}{n} + \frac{(x' - \bar{x})^2}{CS(x, x)} \right\}. \qquad (10.16)$$

Thus a test of significance is given by comparing

$$t = \frac{y' - \hat{E}(y)}{s\sqrt{\left\{ 1 + \dfrac{1}{n} + \dfrac{(x' - \bar{x})^2}{CS(x, x)} \right\}}}$$

with the t-distribution on ν d.f., where ν is the d.f. of the estimate s^2 of σ^2.

There is an important difference between (10.16) and (10.15). The variance (10.15) is the variance of a predicted mean, and (10.16) is the variance of the difference between an observation and a predicted mean. In (10.16) the unity within the brackets is usually the dominant term except when x' is distant from $\bar{x}$.

To summarize, the variance (10.15) applies when we are referring to a *predicted mean,* and the variance (10.16) applies when we are referring to a predicted observation.

Exercises 10.4

1. Carry out the calculations for Example 10.5.

10.5 Comparison of two regression lines

Suppose we have two independent sets of data, as in Example 10.3, yielding two independent estimates of slope $\hat{\beta}_1$, $\hat{\beta}_2$, then it may be of interest to test the hypothesis $\beta_1 = \beta_2$.

Suppose there are n_1, n_2 observations in the two groups, yielding sums of squares from regression Dev. $SS_1(y, x)$, Dev. $SS_2(y, x)$, etc. Then if the population variance σ^2 is the same for the two groups,

$$V(\hat{\beta}_1 - \hat{\beta}_2) = \sigma^2\left\{\frac{1}{CS_1(x, x)} + \frac{1}{CS_2(x, x)}\right\}. \tag{10.17}$$

A combined estimate of σ^2 is obtained as in § 6.2, giving

$$\hat{\sigma}^2 = \{\text{Dev. } SS_1(y, x) + \text{Dev. } SS_2(y, x)\}/\{(n_1 - 2) + (n_2 - 2)\}.$$

A test significance that $\beta_1 = \beta_2$ is therefore provided by comparing

$$t = \frac{\hat{\beta}_1 - \hat{\beta}_2}{\sqrt{[V(\hat{\beta}_1 - \hat{\beta}_2)]}}$$

with the t-distribution on $(n_1 + n_2 - 4)$ d.f.

Example 10.7. For the data of Example 10.3, analysed in § 10.4, we proceed as follows, using the results of Tables 10.2 and 10.4.

A test of significance that the population variances are equal is given by calculating

$$F = \frac{\hat{\sigma}_B^2}{\hat{\sigma}_A^2} = \frac{0 \cdot 058}{0 \cdot 033} = 1 \cdot 76,$$

which is less than the 20% point of $F(13, 13)$. We may therefore justifiably obtain a common estimate of variance

$$\hat{\sigma}^2 = (0 \cdot 43 + 0 \cdot 75)/26 = 0 \cdot 0454.$$

From this we obtain

$$V(\hat{\beta}_1 - \hat{\beta}_2) = 0 \cdot 0454\left(\frac{1}{532 \cdot 63} + \frac{1}{532 \cdot 63}\right) = 0 \cdot 000170.$$

A test of significance that $\beta_1 = \beta_2$ is therefore

$$t = \frac{0 \cdot 3087 - 0 \cdot 2995}{\sqrt{(0 \cdot 000170)}} = 0 \cdot 71$$

which is nowhere near being significant. The data would appear to be in good agreement with the hypothesis that there is a common slope, and the next problem which arises is to obtain point and interval estimates for this common slope. The theory for this is indicated briefly below. □ □ □

When we assume that two regression lines have a common slope this does not imply that they are necessarily identical, but that they are parallel. To obtain an estimate of a common β from two sets of data we can proceed by least squares, as in § 10.2. We form the sum of squares similar to (10.4), but containing two separate summations, one summation for each set of data, and the common β appears in both summations. The least squares estimate of β is obtained by differentiating with respect to β, as before. The common estimate of slope is

$$\hat{\beta} = \frac{CS_1(y, x) + CS_2(y, x)}{CS_1(x, x) + CS_2(x, x)} \tag{10.18}$$

and we obtain directly that the variance of this estimate is

$$\sigma^2 / \{ CS_1(x, x) + CS_2(x, x) \}$$

From the data of Example 10.3, the common estimate of slope is

$$\hat{\beta} = \frac{164 \cdot 447 + 159 \cdot 516}{2 \times 532 \cdot 63} = 0 \cdot 3041;$$

in this example the x-values are identical and therefore $CS_1(x, x)$ and $CS_2(x, x)$ are equal.

Exercises 10.5

1. Obtain 95% confidence intervals for the common estimate of slope of Example 10.3 data. [The point estimate is $0 \cdot 3041$, obtained above.]

2. Show how to obtain 95% confidence intervals for the difference between two slopes β_1 and β_2.

3. In Exercise 10.2.2, the original experiment involved selecting twenty batches of raw material, dividing them at random into two sets of ten, and processing one set by the standard procedure and the other set by a modified procedure. Results for purity x, and yield y, for the ten batches treated by the modified procedure are given below; results for the standard procedure are as in Exercise 10.2.2. By examining the slopes of the regression lines, test

for interaction between the process difference and the purity of the raw material. If there is no evidence of a difference in slopes, fit two parallel lines to the two sets of data.

Results for modified process

x	3·6	5·2	2·1	4·0	2·6	3·1	4·3	4·5	2·9	4·1
y	9·6	10·6	9·1	10·4	9·7	9·8	10·0	10·4	9·1	10·6

4. Suggest a formula for testing the hypothesis $\beta_1 = \beta_2$ when the two sets of observations y have different but unknown variances σ_1^2, and σ_2^2.

5. *Suppose you have two independent unbiased estimates $\hat{\beta}_1$ and $\hat{\beta}_2$ of a common β, with variances $\sigma^2/CS_1(x, x)$ and $\sigma^2/CS_2(x, x)$ respectively. Show that the linear combination of these

$$\bar{\beta} = \alpha\hat{\beta}_1 + (1 - \alpha)\hat{\beta}_2, \quad 0 \leqslant \alpha \leqslant 1$$

which has minimum variance is (10.18).

Compare with the result of Exercise 6.2.3.

10.6 Equally spaced x-values

If the x-values in regression data are equally spaced, then the amount of calculation involved in a regression analysis can be very much reduced.

If we make a transformation of the data

$$X = ax + b$$

then we have

$$CS(y, x) = CS(y, X)/a,$$
$$CS(x, x) = CS(X, X)/a^2,$$
$$\beta(y, x) = a\beta(y, X),$$

while from (10.7)

$$\text{Dev. } SS(y, x) = \text{Dev. } SS(y, X).$$

Suppose we have data in the following form

x	5·6	5·7	5·8	5·9	6·0	6·1
y	y_0	y_1	y_2	y_3	y_4	y_5
X	-5	-3	-1	$+1$	$+3$	$+5$

We transform the scale of x so that in the new scale, $\sum X_i = 0$, thus the correction term for $CS(x, x)$ and the product correction term are both zero.

Thus we have $a = 20$ in this example, and

$$\beta(y, x) = 20 \frac{(-5y_0 - 3y_1 \ldots + 5y_5)}{(5^2 + 3^2 + 1^2 + 1^2 + 3^2 + 5^2)} \qquad (10.19)$$

and also

$$\text{SS due to regr.} = (-5y_0 - 3y_1 - \ldots + 5y_5)^2/70. \qquad (10.20)$$

Exercises 10.6
1. Carry out an analysis of Example 10.1 using the methods given in this section.

10.7 Use of residuals

The *residuals* of any set of results are the differences between the observations and their expectations, fitting parameters where necessary. For linear regression the residuals are

$$r_i = y_i - \bar{y} - \hat{\beta}(x_i - \bar{x}) \qquad (10.21)$$

and from Exercise 10.3.3 they have expectation zero and variance

$$V(r_i) = \left\{ 1 - \frac{1}{n} - \frac{(x_i - \bar{x})^2}{\sum (x_i - \bar{x})^2} \right\} \sigma^2. \qquad (10.22)$$

These residuals are not independent, since the parameters α and β are estimated from the same data on which the residuals are being calculated. Exercise 10.7.2 shows that the covariance of two residuals is

$$C(r_i, r_j) = - \left\{ \frac{1}{n} + \frac{(x_i - \bar{x})(x_j - \bar{x})}{\sum (x_k - \bar{x})^2} \right\} \sigma^2, \quad i \neq j. \qquad (10.23)$$

When n is very small the correlation between residuals can therefore be appreciable, and could lead to a pattern appearing in them. However, for most practical problems this correlation between residuals can be ignored.

Example 10.8. The residuals for machine A, Example 10.3, are given by using the regression equation in Table 10.2 and we have

$$r_i = y_i - 0.31 - 0.3087 x_i.$$

The fifteen residuals, with associated x-values, are:

x	11·0	11·1	7·2	8·3	12·4	14·7	5·1
r_i	−0·25	−0·26	0·21	−0·04	−0·14	+0·10	+0·36

x	21·7	20·2	19·1	25·0	10·2	13·3	23·6	12·0
r_i	−0·09	+0·02	−0·03	+0·27	−0·02	+0·05	−0·01	−0·14

These residuals should add to zero, but do not quite do so because of round-ing-off errors.

The residuals can be examined for evidence of departures from assumption, and we now discuss very briefly three ways in which the residuals can be studied.

1. *Looking for patterns*

One way in which residuals can be used is to look for patterns in them, to seek for evidence of non-randomness, or of departure from the model. Some ways of looking for patterns include the following.

(*i*) We can look for a pattern in the residuals in the order in which the observations were taken. Any pattern is evidence that successive observations are not independent, or have a trend in time. If a trend exists it may be complicated, and have periodic or seasonal components. In Example 10.8, there is no evidence of any such pattern.

(*ii*) The residuals can be ordered by the x-values and then examined for a pattern. A pattern in the residuals when ordered in this way indicates a possible departure from linearity of the regression. For Example 10.8 the re-ordered residuals, multiplied by 100, are as follows:

$$36, \quad 21, \quad -4, \quad -2, \quad -25, \quad -26, \quad -14, \quad -14,$$
$$+5, \quad +10, \quad -3, \quad +2, \quad -9, \quad -1, \quad +27,$$

In this order, there is some tendency for series of positive and negative values, and also we see very large positive residuals at either end, and large negative residuals in the centre. This picture can also be seen on a graph of the original observations, if the scale is large enough, and the fitted regression line is drawn in; see Figure 1.17. There is therefore some slight evidence of non-linearity in the regression.

(*iii*) The residuals can be plotted against the fitted values, $\{\hat{\alpha} + \hat{\beta}(x_i - \bar{x})\}$, to see if there is any evidence of variations in the variance of the observations as the expectation varies. Such a variation should also be evident from (*ii*).

(*iv*) Sometimes successive observations are not independent, as for example when there is a subjective tendency for observers to try and make each observation like the previous one. To detect such dependence the residuals are set out in the order in which the observations were taken, and successive residuals in this order are plotted against each other, that is r_{i+1} versus r_i.

2. *Looking for non-normality*

The residuals can be studied for evidence of departure from normality in the underlying distribution. The difficulty here is that even if the distribution of the y's for given x is normal, the distribution of the residuals is *not* quite normal; some negative kurtosis is introduced. The reason for this is that some parameters have been estimated from the data, and hence there are linear combinations of the residuals which add to zero. However, if n is not too small, we may treat the residuals like n independent observations, and ignore the kurtosis.

One obvious way to study the normality of the residuals is to plot the empirical cumulative distribution of the residuals on normal probability paper; see Exercise 2.6.2. However, there is a nearly equivalent method, which avoids the need for drawing special graph paper.

Suppose we take n observations from a $N(0, 1)$ population and order them, then let the ordered observations be

$$x_{(1)} \leqslant x_{(2)} \leqslant \ldots \leqslant x_{(n)}.$$

[The ith ordered value in such a set is called the ith *order statistic*.]

We can repeat this an indefinite number of times, as in the sampling experiment of § 4.3, and find the average values of each order statistic. The expectations of normal order statistics have been tabulated exactly, but for our purposes we can use the approximation

$$\Phi\{E(x_{(i)})\} \simeq i/(n + 1). \tag{10.24}$$

In words, equation (10.24) says that the expected value of the ith order statistic from a sample of n standard normal variables is such that the normal probability corresponding to it is approximately $i/(n + 1)$. If $n = 15$, we have, for example

$$\Phi\{E(x_{(1)})\} = 1/16 = 0 \cdot 0625$$
$$E(x_{(1)}) \simeq -1 \cdot 53.$$

The full set of approximate expected values for $n = 15$ is

$$-1 \cdot 53, \quad -1 \cdot 15, \quad -0 \cdot 89, \quad -0 \cdot 67, \quad -0 \cdot 49, \quad -0 \cdot 32,$$
$$-0 \cdot 16, \quad 0, \quad 0 \cdot 16, \quad 0 \cdot 32, \quad \ldots, \quad \ldots, \quad 1 \cdot 53. \tag{10.25}$$

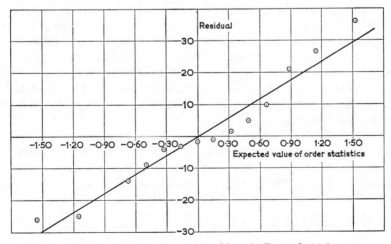

Figure 10.3 *Normal plot for residuals of machine A, Example* 10.3

It is now clear that if we plot the ordered residuals against these values just obtained, (10.25), we expect (approximately) a straight line if the underlying distribution is normal. Figure 10.3 gives such a graph for residuals in Example 10.8 and a straight line has been drawn by eye through the points. This graph is tolerably straight, indicating a reasonable fit by a normal distribution. Evidence of non-normality would be shown by a graph in which the points weave around in a continuous curve, and are *not* haphazardly placed about the line.

3. *Looking for outliers*

It is frequently found in practice that sets of data presented for analysis contain 'wild shots' or 'outliers', and a very important application of residuals is to spot these observations. The simplest way to do this is just to look for the largest residuals in size (positive or negative). Sometimes one or two residuals are so much larger than any others in magnitude, that it is intuitively clear that the observations are suspect. Another method of analysing residuals, called the Full Normal Plot, FUNOP, has been suggested by Anscombe and Tukey (1963); see also Tukey (1962) and the book by Barnett and Lewis (1978).

When suspect observations are found, it is more difficult to decide what to do. If possible laboratory records, etc., are checked to see if there is something unusual about the conditions for the suspect observations, which would give a reason for rejecting them. If this cannot be done, it is probably best to analyse the data with and without the outliers. Care must be taken before rejecting any observation outright on the basis of internal statistical evidence only.

Discussion

For the particular case of simple linear regression, much of what can be gained from a study of residuals can also be seen in a good graph of y against x, except for studying non-normality. A study of residuals is of greater value when a more complex statistical model is used, such as, for example, the models considered in subsequent chapters of this book. Today, calculations for statistical analysis are often done on a computer, so that residuals are very quickly and easily generated, which makes possible simple and vital checks of assumptions.

The methods described in this section leave much open to subjective judgment, and if readers *always* make analyses of residuals, they will rapidly gain experience in interpreting results. For further details and discussion of material related to this section, see Anscombe and Tukey (1962), Tukey (1962), Chernoff and Lieberman (1954), and references therein.

Exercises 10.7

1. Analyse the residuals for machine B, Example 10.3, and compare your results with those for machine A.

2. Use the model
$$y_i = \alpha + \beta(x_i - \bar{x}) + \varepsilon_i,$$
where ε_i are independent $N(0, 1)$, to express the residuals (10.21) as a sum of ε's. Hence or otherwise derive (10.23).

10.8 Discussion of models

In connection with any discussion of regression models, it is necessary to distinguish between two types of random variation, measurement error, and inherent random variation. For an example of the second, consider measurements of height of a sample of adult men sampled randomly from a certain group, and suppose these measurements to be accurate. The variation between these measurements is inherent in the nature of the population being sampled. Measurement errors, on the other hand, can be reduced by repeated independent measurements, etc. We can now separate off three cases.

(1) The random variation in x is negligible, while the random variation in y may be of either type. In this situation we carry out regression of y on x.

(2) The random variation in x contains negligible measurement error. The population being sampled must be well defined, and the scatter diagram of y and x shows just one cluster of points. An example of this situation is the set of measurements on heights of fathers and daughters, Example 1.6. If y contains measurement error, regression of y on x can be used, but if there is negligible measurement error in either variable, as in Example 1.6, both regressions may be valid, and the following rules can be used as a guide to which one is appropriate.

(i) If one variable is to be used as a predictor of another and both are random, we take the independent variable the one to be used as a predictor, for example, Example 10.3, the tobacco moisture experiment.

(ii) If there is a causal relationship between the two variables, and this is thought to be in a particular direction, we take regression of the final variable on the initial variable. For example, we would take regression of the strength of human hairs on their diameters.

(3) If both y and x contain measurement error, the linear regression model is not appropriate. For an illustration we discuss Example 10.2. Let Y and X be the true yield strength and Brinell hardness of the shells, and y and x the observed measurements. We write

$$y = Y + u$$
$$x = X + v, \qquad\qquad (10.26)$$

where u and v are measurement errors. Then it may be reasonable to take the model

$$Y = \alpha + \beta X. \tag{10.27}$$

This looks very similar to (10.1), but it is not, as the variables Y and X cannot be observed. The errors u and v in (10.26) may or may not be independent. Different methods of analysis are appropriate for analysing this type of model, and a description of several methods, with references, is given by Madansky (1959).

A full discussion of regression type models will not be given here, and we refer the reader to Madansky. When using linear or multiple regression in practice, it is important to examine the model critically, to ensure that it is really appropriate to the situation.

Exercises 10.8

1. In Example 10.2.2 and Exercise 10.5.3 we have examined the results on purity and yield of a process. Continue this analysis by fitting a common slope to the two sets of data, and estimate the difference in intercepts of the parallel lines. Also obtain the variance of this difference, and test whether the difference is significantly different from zero.

2. Write out full conclusions to the tobacco moisture experiment, Examples 10.3, 10.4, 10.7. Include graphs, confidence statements, etc., in your report.

3.* Wehfritz (1962) describes the calibration of a flow meter used in the petroleum industry. Known volumes, z_i, for $i = 1, 2, \ldots, n$, of a fluid are passed through a meter, and the measurements y_i of the meter noted. The observations y_i contain errors from a number of sources. It is expected that a linear relationship holds between y and z. When a meter is used in practice, a reading Y_o is obtained on an unknown volume of fluid; show how to estimate the volume of fluid which has been passed and also obtain approximate 95% confidence intervals for this volume.

CHAPTER ELEVEN

Multiple Regression

11.1 Introduction

The theory discussed in the previous chapter is readily extended to cover the case of regression on any number of variables. The theory of least squares provides estimates of unknown parameters, tests of significance, etc., and the methods are relatively simple provided the model can be expressed in the form

$$E(y) = \alpha + \beta_1(x_1 - \bar{x}) + \beta_2(x_2 - \bar{x}_2) + \ldots + \beta_k(x_k - \bar{x}_k) \Big\}$$
$$V(y) = \sigma^2 \qquad (11.1)$$

where all the y's are independent, and where $\bar{x}_1$, $\bar{x}_2$, etc., are the means of the x's. The quantities $\beta_1, \ldots, \beta_k$, are called *partial regression coefficients*, and they each measure the variation in y due directly to the variation in the respective x_i, the other variables being fixed. The essential features of this model are:

(*i*) Only y is a random variable. The x's are assumed to be either under the control of the experimenter, or else to be measurable with negligible error.

(*ii*) The expectation of y is a linear function of the unknown parameters β_j.

(*iii*) The observations y_i, $i = 1, 2, \ldots, n$, are uncorrelated and have constant variance.

For the usual tests of significance to be valid it is necessary to add the assumption that the distribution of the y's is normal; in view of (*iii*) above this makes the y's independent.

Consider the following illustrative examples.

Example 11.1. Gibson and Jowett (1957) discussed some data on the operating efficiency of blast furnaces as measured by a variable y, the output of pig iron per ton of coke. The work of smelting is done by the carbon in the coke, so that variations in y are bound to be affected by variations in the percentage of available carbon in coke, denoted x_1. Variations in y are also attributable to variations in the hearth temperature of the furnace; this variable cannot be measured directly but it is reflected in the percentage of silicon in the pig iron coming out of the furnace, which is denoted x_2. Although the variables x_1 and x_2 are not under the control of the experimenter, for various reasons, they may be considered to be measured with small error compared with the random variation on y. A model of the form (11.1), with $k = 2$, may therefore be fitted.

The partial regression coefficient β_1 is the amount of change in the average value of y (the output of pig iron per ton of coke), due to a unit increase in x_1 (the percentage of available carbon in coke), when x_2 is fixed; equation (11.1) states that this change in $E(y)$ per unit increase in x_1 is the same for all x_1 and x_2. A similar interpretation holds for β_2, and if there are changes in both x_1 and x_2, equation (11.1) states that the change in $E(y)$ is simply the addition of the two changes for x_1 and x_2 separately.   □ □ □

Example 11.2. In a certain chemical process, three variables measured were y, the percentage of calcium in the metal phase of the calcium chloride/ sodium chloride/calcium/sodium melt, x_1, the melt temperature, and x_2, the percentage of calcium in the salt phase of the melt. We shall assume that x_1 and x_2 can be measured with negligible error, and so are not random variables, but that y contains variation which may be taken to be random. The data given in Table 11.1 is reproduced by kind permission of I.C.I. Ltd, Runcorn, Cheshire. In the region in which data is available, it is expected that the expectation of y is a quadratic function of x_1 and x_2. That is, we expect the model (11.2)

$$\left. \begin{array}{l} E(y) = a' + \beta_1 x_1 + \beta_2 x_2 + \beta_3 x_1^2 + \beta_4 x_1 x_2 + \beta_5 x_2^2 \\ V(y) = \sigma^2 \end{array} \right\}. \tag{11.2}$$

Now for every observed value of y, the x's are assumed to be known constants. We therefore have, for example, for the first observation in Table 11.1,

$$E(y_1) = \alpha' + \beta_1 547 + \beta_2 63{\cdot}7 + \beta_3 (547)^2 + \beta_4 (547 \times 63{\cdot}7) + \beta_5 (63{\cdot}7)^2$$

etc., and this is a linear function of the unknown parameters β_j. Therefore we conclude that since the x_i's in (11.1) are not random variables, it is immaterial to the setting up of the statistical model what relationships may exist between them, and we can throw (11.2) in the form (1.1) by writing

$$x_1^2 = x_3, \quad x_1 x_2 = x_4, \quad x_2^2 = x_5.$$

If the model had included terms such as β_j^2, or $\exp(\beta_j x_j)$, it would no longer be a linear function of the β_j's and would not come within the scope of this chapter.   □ □ □

The examples given above are illustrative, and the absolute validity of the model (11.1) can be disputed in both cases, for different reasons; these points will be discussed in § 11.4. We shall proceed first to describe the method by

Table 11.1 *Chemical process data*

Calcium (%) y	Temperature (°C) X_1	Calcium chloride (%) X_2
3·02	547	63·7
3·22	550	62·8
2·98	550	65·1
3·90	556	65·6
3·38	572	64·3
2·74	574	62·1
3·13	574	63·0
3·12	575	61·7
2·91	575	62·3
2·72	575	62·6
2·99	575	62·9
2·42	575	63·2
2·90	576	62·6
2·36	579	62·4
2·90	580	62·0
2·34	580	62·2
2·92	580	62·9
2·67	591	58·6
3·28	602	61·5
3·01	602	61·9
3·01	602	62·2
3·59	605	63·3
2·21	608	58·0
2·00	608	59·4
1·92	608	59·8
3·77	609	63·4
4·18	610	64·2
2·09	695	58·4

which data following the model (11.1) is analysed, and we shall use Example 11.2 data as an illustration.

The ultimate aim of a multiple regression analysis may be (*i*) to obtain an equation from which y may be predicted from $x_1, \ldots, x_k$, (*ii*) to find an optimum set of x-values, as say in Example 11.2, (*iii*) to provide a means of comparison with the results of similar experiments, or (*iv*) to understand which variables affect y. In order to satisfy these aims, it is necessary to provide methods for point and interval estimation, and significance tests of the β_j's and the theory of least squares is employed.

Let the observations be denoted $(y_1, x_{11}, \ldots, x_{k1}), \ldots, (y_n, x_{1n}, \ldots, x_{kn})$ then the sum of squares of the y's from their expectations is

$$S = \sum_{i=1}^{n} \{y_i - \alpha - \beta_1(x_{1i} - \bar{x}_1) - \ldots - \beta_k(x_{ki} - \bar{x}_k)\}^2. \qquad (11.3)$$

The least squares estimates of the β_j's are the values which minimize this sum of squares, and these are the solutions of the equations

$$\frac{\partial S}{\partial \alpha} = \frac{\partial S}{\partial \beta_1} = \frac{\partial S}{\partial \beta_2} = \ldots = \frac{\partial S}{\partial \beta_k} = 0.$$

These equations are called the *normal equations* of least squares. By differentiating S with respect to α, summing over the observations, and equating to zero, we obtain the least squares estimate $\hat{\alpha} = \bar{y}$, as in simple linear regression. The remaining k equations can be set out as follows, in the notation of § 10.2.

$$\left. \begin{array}{l} \hat{\beta}_1 CS(x_1, x_1) + \hat{\beta}_2 CS(x_1, x_2) + \ldots + \hat{\beta}_k CS(x_1, x_k) = CS(y, x_1) \\ \hat{\beta}_1 CS(x_2, x_1) + \hat{\beta}_2 CS(x_2, x_2) + \ldots + \hat{\beta}_k CS(x_2, x_k) = CS(y, x_2) \\ \cdot \quad \cdot \quad \cdot \quad \cdot \quad \cdot \quad \cdot \quad \cdot \quad \cdot \quad \cdot \quad \cdot \quad \cdot \\ \hat{\beta}_1 CS(x_k, x_1) + \hat{\beta}_2 CS(x_k, x_2) + \ldots + \hat{\beta}_k CS(x_k, x_k) = CS(y, x_k) \end{array} \right\} \quad (11.4)$$

The terms on the right-hand side of the equations are

$$CS(y, x_j) = \sum_{i=1}^{n} y_i(x_{ji} - \bar{x}_j) = \sum_{i=1}^{n} (y_i - \bar{y})(x_{ji} - \bar{x}_j)$$

and arise either by inserting $\hat{\alpha} = \bar{y}$ in the equations, or else by noticing that the coefficients of α in the equations for

$$\frac{\partial S}{\partial \beta_j} = 0$$

are in any case zero.

In general (11.4) is a set of k simultaneous linear equations so that the least squares estimates of the β_j's are linear functions of the y's. Therefore if the y's are assumed to be normally distributed, the $\hat{\beta}_j$'s are also normally distributed, and standard errors and confidence intervals, or significance tests can be obtained readily, by methods similar to those used in Chapter 10. However, in general the $\hat{\beta}_j$'s are not independent, and the partial regression coefficient β_s for any variable x_s can fluctuate wildly according as other variables are or are not included in the regression equation. This makes the results of the analysis difficult to interpret. For a suitable choice of the x_{ij}'s at which observations are made, some or all of the cross product terms $CS(x_j, x_k)$, for $j \neq k$ will vanish. This makes the equations easier to solve, and the results much easier to interpret since some or all of the $\hat{\beta}$'s are independent.

We shall not discuss the general theory any further in this book, but instead consider the case $k = 2$ in detail. The solution of equations (11.4) for larger k is often a task for a computer, and standard programmes are usually available.

Exercises 11.1

1. Suppose y_i, $i = 1, 2, \ldots, n$, are independently and normally distributed random variables, with

$$E(y) = \alpha + \beta_1 \, e^{\beta_2 x}$$
$$V(y) = \sigma^2,$$

where $x_1, \ldots, x_n$ are non-stochastic variables and α, β_1 and β_2 are unknown. The theory of least squares can still be applied in such cases, but the equations for the estimates are often difficult to solve, and the estimates lose some of their optimal properties. Find the equations for the least squares estimates of α, β_1 and β_2 in this case.

11.2 Theory for two explanatory variables only

For two explanatory variables the expectation of y is defined by a plane in the space (y, x_1, x_2) as shown below. If the observations are plotted in this

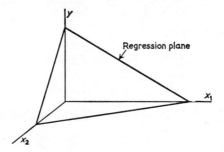

space, the sum of squares (11.3) is the sum of squared deviations of the observations from the plane measured in the y direction. The least squares estimates $\hat{\beta}_1$, $\hat{\beta}_2$ define the plane for which this sum of squares is a minimum.

For $k = 2$, the solutions of equations (11.4) are

$$\left. \begin{aligned} \hat{\beta}_1 &= \{CS(x_2, x_2) . CS(y, x_1) - CS(x_1, x_2) . CS(y, x_2)\}/\Delta \\ \hat{\beta}_2 &= \{CS(x_1, x_1) . CS(y, x_2) - CS(x_1, x_2) . CS(y, x_1)\}/\Delta \end{aligned} \right\} \tag{11.5}$$

where $\Delta = \{CS(x_1, x_1) . CS(x_2, x_2) - \{CS(x_1, x_2)\}^2$.

We can now check directly that the estimates are unbiased,

$$E(\hat{\beta}_1) = \beta_1, \quad E(\hat{\beta}_2) = \beta_2, \tag{11.6}$$

and have variances

$$V(\hat{\beta}_1) = \sigma^2 \, CS(x_2, x_2)/\Delta, \tag{11.7}$$
$$V(\hat{\beta}_2) = \sigma^2 \, CS(x_1, x_1)/\Delta,$$

and

$$C(\hat{\beta}_1, \hat{\beta}_2) = - \sigma^2 \, CS(x_1, x_2)/\Delta. \tag{11.8}$$

Clearly the least squares estimate of α is $\hat{\alpha} = \bar{y}$, and this has expectation α and variance

$$V(\hat{\alpha}) = \sigma^2/n. \tag{11.9}$$

Those acquainted with matrix algebra will notice that the matrix of variances and covariances of the $\hat{\beta}_j$'s is defined by σ^2 times the inverse of the matrix of corrected sums of squares and products of the x_j's. This is readily shown to be true for any value of k.

The variance σ^2 which appears in (11.7) to (11.9) will in general be unknown, but the theory of least squares states that it can be estimated from the minimized value of the sum of squares, which is the sum of squared deviations from the regression, and we now obtain a formula for this quantity.

Let us write for the fitted values,

$$\hat{Y}_j = \hat{\alpha} + \hat{\beta}_1(x_{1j} - \bar{x}_1) + \hat{\beta}_2(x_{2j} - \bar{x}_2),$$

then the normal equations obtained from

$$\frac{\partial S}{\partial \beta_1} = 0 \quad \text{and} \quad \frac{\partial S}{\partial \beta_2} = 0$$

can be written respectively,

$$\sum (y_j - \hat{Y}_j)(x_{1j} - \bar{x}_1) = 0, \tag{11.10}$$

and

$$\sum (y_j - \hat{Y}_j)(x_{2j} - \bar{x}_2) = 0. \tag{11.11}$$

By multiplying (11.10) by $\hat{\beta}_1$, (11.11) by $\hat{\beta}_2$ and adding, we obtain

$$\sum (y_j - \hat{Y}_j)(\hat{Y}_j - \bar{y}) = 0. \tag{11.12}$$

Put

$$y_i - \bar{y} = y_i - \hat{Y}_i + \hat{Y}_i - \bar{y},$$

then by squaring and summing over $i = 1, 2, \ldots n$, we have

$$\sum_i (y_i - \bar{y})^2 = \sum_i (y_i - \hat{Y}_i)^2 + \sum_i (\hat{Y}_i - \bar{y})^2 + 2 \sum_i (y_i - \hat{Y}_i)(\hat{Y}_i - \bar{y})$$

$$= \sum_i (y_i - \hat{Y}_i)^2 + \sum_i (\hat{Y}_i - \bar{y})^2, \tag{11.13}$$

by using (11.12). Equation (11.13) states that the total corrected sum of squares for y can be split up into the two terms on the right-hand side; these are firstly the minimized sum of squared deviations from regression, and secondly the corrected sum of squares of the fitted values, which is referred to as the sum of squares due to the regression. The fact that these sums of squares add is an algebraic property, and is independent of statistical assumptions. This is a very important point, and is used in the analysis, see below.

A simple expression can be obtained for the corrected sum of squares due to regression. We have

$$\sum_i (\hat{Y}_i - \bar{y})^2 = \sum_i \{\hat{\beta}_1(x_{1k} - \bar{x}) + \hat{\beta}_2(x_{2i} - \bar{x}_2)\}^2$$

$$= \hat{\beta}_1^2 \, \mathrm{CS}(x_1, x_1) + 2\hat{\beta}_1\hat{\beta}_2 \, \mathrm{CS}(x_1, x_2) + \hat{\beta}_2^2 \, \mathrm{CS}(x_2, x_2). \qquad (11.14)$$

Now if we take the normal equations (11.4), multiply the first and second by $\hat{\beta}_1$ and $\hat{\beta}_2$ respectively, and add, and we obtain

$$\hat{\beta}_1^2 \, \mathrm{CS}(x_1, x_1) + 2\hat{\beta}_1\hat{\beta}_2 \, \mathrm{CS}(x_1, x_2) + \hat{\beta}_2^2 \, \mathrm{CS}(x_2, x_2)$$

$$= \hat{\beta}_1 \, \mathrm{CS}(y, x_1) + \hat{\beta}_2 \, \mathrm{CS}(y, x_2).$$

Therefore from (11.14) the sum of squares due to regression is

$$\mathrm{Reg.\ SS}(y, x_1, x_2) = \sum_i (\hat{Y}_i - \bar{y})^2$$

$$= \hat{\beta}_1 \, \mathrm{CS}(y, x_1) + \hat{\beta}_2 \, \mathrm{CS}(y, x_2). \qquad (11.15)$$

These results can be presented neatly in the analysis of variance table below. The sum of squares for deviations in this table is usually obtained by subtraction from the other two quantities, although it could of course be obtained from the residuals, by squaring and summing. The residuals are defined

$$\mathrm{Residual} = y_i - \hat{Y}_i = y_i - \bar{y} - \hat{\beta}_1(x_{1i} - \bar{x}) - \hat{\beta}_2(x_{2i} - \bar{x}_2). \qquad (11.16)$$

Table 11.2 *Analysis of variance for bivariate regression*

Source	CSS	d.f.	E(Mean square)
SS due to regression	$\hat{\beta}_1 \, \mathrm{CS}(y, x_1) + \hat{\beta}_2 \, \mathrm{CS}(y, x_2)$	2	σ_r^2
SS deviations	$\sum (y_i - \hat{Y}_i)^2$	$n - 3$	σ^2
	$\sum (y_i - \bar{y})^2$	$n - 1$	

The degrees of freedom in Table 11.2 add up and have a direct interpretation, as in simple linear regression. The total line has $(n - 1)$ degrees of freedom between the observations $y_1, \ldots, y_n$. The sum of squares due to regression is a sum of squares on two degrees of freedom, since there are two constants, β_1, β_2 required in addition to the general mean, in order to fit the values $\hat{Y}_i$. The sum of squares for deviations is on $(n - 3)$ degrees of freedom since there are n contributions but three linear constraints due to fitting $\hat{\beta}_1, \hat{\beta}_2$ and the mean $\bar{y}$.

The mean square for deviations from regression is

$$\sum_i (y_i - \hat{Y}_i)^2/(n - 3)$$

and by a property of the least squares method given in (*iv*) of § 9.3, this has an expectation of σ^2; see also Exercise 11.2.3. The expectation of the mean square due to regression is derived in Exercise 11.2.2, and this is

$$E(\text{Mean square due to regr.}) = \sigma^2 + \{\beta_1^2 \, CS(x_1, x_1)$$
$$+ \, 2\beta_1\beta_2 \, CS(x_1, x_2) + \beta_2^2 \, CS(x_2, x_2)\}/2 \quad (11.17)$$

which is greater than the expectation of the mean square for deviations unless β_1 and β_2 are both zero, when the expectations are equal.

If the observations are independently and normally distributed, the mean square due to regression and the mean square for deviations are independent with a distribution proportional to χ^2 on the appropriate degrees of freedom. A significance test of the null hypothesis that there is no regression and β_1 and β_2 are both zero, is provided by an F-test of the ratio

$$\frac{\text{Mean square due to regression}}{\text{Mean square for deviations}}, \quad (11.18)$$

on $(2, \, n - 3)$ degrees of freedom, rejecting the null hypothesis for large values of F.

We can now summarize the properties of Table 11.2, the analysis of variance table:

(*i*) The sums of squares and degrees of freedom add up and have direct interpretation.

(*ii*) The mean squares have an equal expectation only if the null hypothesis is true. Otherwise, the expectation of the mean square due to regression is the greater.

(*iii*) An overall test of the significance of the regression is carried out by using the F-test (11.18).

Confidence intervals and significance tests for β_1, β_2 and α can be carried out using formulae (11.6), (11.7) and (11.9), with the mean square for deviations providing an estimate of σ^2. The methods used are parallel to those given in § 10.3 for simple linear regression, and they are illustrated in the next section.

Exercises 11.2

1. Check formulae (11.6), (11.7) and (11.18) directly from expressions (11.5).

2. Use the relationship

$$E(\hat{\beta}_1^2) = V(\hat{\beta}_1) + E^2(\hat{\beta}_1)$$

to write down $E(\hat{\beta}_1^2)$ and $E(\hat{\beta}_2^2)$. Hence obtain the expectation of the sum of squares due to regression from (11.14).

3. Write the model for bivariate regression

$$y_i = \alpha + \beta_1(x_{1i} - \bar{x}_1) + \beta_2(x_{2i} - \bar{x}_2) + \varepsilon_i$$

where the ε_i are independently distributed as $N(0, \sigma^2)$. By inserting this model, or otherwise, obtain the expectation of the mean square for deviations.

11.3 Analysis of example 11.2

In this section we illustrate the method of analysis by using Example 11.2 data, and fitting the model

$$E(y) = \alpha + \beta_1(x_1 - \bar{x}_1) + \beta_2(x_2 - \bar{x}_2) \left.\right\} \tag{11.19}$$
$$V(y) = \sigma^2$$

where $\alpha' = \alpha - \beta_1\bar{x}_1 - \beta_2\bar{x}_2$; see § 10.2. From the statement of the example, we expect this model to show up as being inadequate.

In simple linear regression the first step in any analysis is to plot y against x, and this enables us to check that the model is reasonable, and also provides a rough check on the accuracy of the results. Although multivariate regressions can be plotted using the methods of Anderson (1954) referred to in § 1.4, the graphs are much more difficult to interpret.

Figure 11.1(a) and (b) show scatter diagrams of y and x_1, and y and x_2 respectively, for Example 11.2, in each case ignoring the other variable. It seems reasonable to assume a linear relationship of y and x_2, but if the extreme point on Figure 11.1(a) is excluded, there is no evidence of any relationship between y and x_1.

Figure 11.2 shows a scatter diagram of y and x_2, with x_1 represented in four intervals by the length of the tail, as shown. If a rough regression line is drawn by eye through this data, points with long tails tend to be above this line, and the points with short tails or no tail beneath it. In this way we can see that if the points were plotted out in three dimensions, there is fairly clear evidence of a relationship between the three variables, although we cannot tell whether it is linear or quadratic.

The next step is to calculate the analysis of variance table, Table 11.2, and carry out the overall F-test, and tests and confidence intervals for β_1 and β_2. The quantities we need are as follows:

$$\bar{y} = 2\cdot917 \qquad\qquad \bar{x}_1 = 586\cdot8929 \qquad\qquad \bar{x}_2 = 62\cdot2179$$
$$\text{CS}(y, y) = 8\cdot3443 \qquad \text{CS}(y, x_1) = -118\cdot8486 \qquad \text{CS}(y, x_2) = 20\cdot2514$$
$$\text{CS}(x_1, x_1) = 21928\cdot6786 \quad \text{CS}(x_1, x_2) = -910\cdot3465 \quad \text{CS}(x_2, x_2) = 96\cdot1411$$

By solving (11.5) we obtain

$$\hat{\beta}_1 = 0\cdot005478 \quad \hat{\beta}_2 = 0\cdot2625$$

and the analysis of variance table is shown in Table 11.3. The F-ratio for the overall significance of the regression is $15\cdot5$ on $(2, 25)$ degrees of freedom; this is highly significant, since the $0\cdot1\%$ point is only $9\cdot22$.

We can now ask whether either of x_1 and x_2 separately contribute a

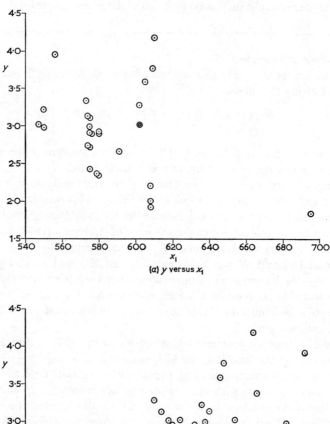

Figure 11.1 *Scatter diagrams for Example 11.2 data*

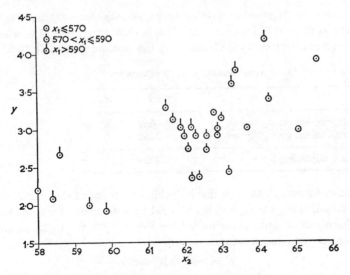

Figure 11.2 *Bivariate scatter diagram for Example* 11.2 *data*

significant amount to the overall regression. From (11.7) and (11.8), using the mean square for deviations to estimate σ^2, we have

$$V(\hat{\beta}_1) = 0{\cdot}000011 \quad V(\hat{\beta}_2) = 0{\cdot}002522$$
$$C(\hat{\beta}_1, \hat{\beta}_2) = 0{\cdot}000105$$

and following the method in § 10.3, we calculate

$$t = \frac{\hat{\beta}_1}{\surd\{V(\hat{\beta}_1)\}} = \frac{0{\cdot}005478}{\surd\{0{\cdot}000011\}} = \frac{0{\cdot}005478}{0{\cdot}003317} = 1{\cdot}65 \qquad (11.20)$$

and

$$t = \frac{\hat{\beta}_2}{\surd\{V(\hat{\beta}_2)\}} = \frac{0{\cdot}2625}{\surd\{0{\cdot}002522\}} = \frac{0{\cdot}2625}{0{\cdot}050220} = 5{\cdot}23 \qquad (11.21)$$

and we refer these values to the t-tables on 25 degrees of freedom, to find that $\hat{\beta}_2$ is highly significant (beyond the 0·1% level) but $\hat{\beta}_1$ is not significant, only reaching the 10% level.

Table 11.3 ANOVA *table for Example* 11.2 *data*

Source	CSS	d.f.	M.S.	F-ratio
Due to regression	4·6652	2	2·23	15·5
Deviations	3·6791	25	0·15	
Total	8·3443	27		

The precise meaning of the t-tests just carried out is shown more clearly by an equivalent way of computing the tests. Suppose we calculate linear regression of y on x_2, ignoring x_1, then we can calculate by subtraction the extra amount in corrected sums of squares accounted for by taking regression on x_1 and x_2 instead of just y and x_2. This is shown in Table 11.4, where the sum of squares for y on x_2 is obtained by the method given in § 10.2.

Table 11.4 ANOVA *for contribution of* x_1 *to regression*

Source	CSS	d.f.
Due to regression on x_1 and x_2	4·6652	2
Due to regression on x_2 (ignoring x_1)	4·2658	1
Due to x_1, adjusting for x_2	0·3994	1

The sum of squares in the last line in Table 11.4 is called the sum of squares due to x_1, adjusting for x_2, and is obtained by subtraction from the first two lines. For a test of significance we use the F-test, with the mean squares for deviations in the denominator,

$$F = 0·40/0·15 = 2·7$$

on (1, 25) degrees of freedom, and the result is non-significant. Now the square of t on v degrees of freedom is equal to F on (1, v) degrees of freedom, so that this F-test is exactly equivalent to the t-test (11.20) and it tests whether the extra variation accounted for by doing regression of y on x_1 and x_2 rather than just y on x_2, is significant.

The ANOVA table for testing the contribution of x_2 to the regression is shown in Table 11.5. The F-ratio is $4·02/0·15 = 26·8$, which is very highly significant. This test is equivalent to (11.21).

Table 11.5 ANOVA *for contribution of* x_2 *to regression*

Source	CSS	d.f.
Due to regression on x_1 and x_2	4·6652	2
Due to regression on x_1 (ignoring x_2)	0·6441	1
Due to x_2, adjusting for x_1	4·0211	1

We conclude from these significance tests that the addition of a linear term in x_1 when a term in x_2 is already included, does not contribute significantly to the regression, but that the partial regression coefficient of y on x_2 is very highly significant, being beyond the 0·1% level.

If an independent estimate of σ^2 had been available, it would have been possible to test the mean square for deviations against this. For example, if at every point (x_1, x_2) two observations were taken, the squares of differences between these pairs of observations would be estimates of σ^2. A significant result for the test

(mean square deviations)/(mean square for independent estimate of σ^2)

would indicate that the model is not satisfactory. One great advantage of requiring experiments to have replicated observations is that a precise test of the model is provided.

If, as in Example 11.2, there is no independent estimate of σ^2, a rough test of the model is provided by plotting the residuals out in the (x_1, x_2) plane, and then looking for a pattern. Figure 11.3 shows the residuals for Example 11.2, after having fitted (11.19) and lines have been drawn on the figure corresponding to the approximate position of zero residuals. There is very clearly a trough of negative residuals, some being very large, and

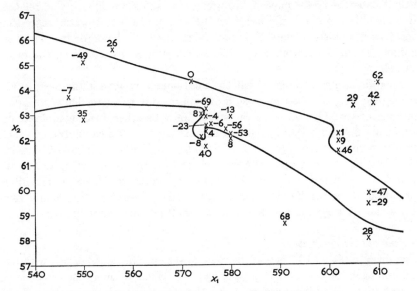

Figure 11.3 *Residuals for Example* 11.2 *using model* (11.19)

around this are a number of large positive residuals. This indicates some curvature of the regression plane, and the next step in the analysis would be to include the extra terms given in model (11.2), and test each of the β_j's for significance, plot a new set of residuals, etc. We shall not discuss these calculations any further, but the corrected sum of squares for fitting (11.2) is 6·50 on 5 degrees of freedom, which is a considerable improvement on the value 4·67 given in Table 11.3 for model (11.19).

Summary of method of analysis
Step 1. Plot the data and select a likely model.
Step 2. Calculate the estimates of the regression coefficients.
Step 3. Calculate the residuals, and carry out the analysis indicated in § 10.7.
         This step may lead to the selection of a new model, or to the detection of 'outliers', and so necessitate redoing step 2.

Step 4. Calculate tests of significance and confidence intervals for the regression coefficients.

If the calculations are carried out on a computer, steps 2 and 4 and some of step 3 can be carried out in one run on the machine.

Interpretation of significance tests

If in a regression on two independent variables, one of the partial regression coefficients is not significant, at about the 5% level, we could consider omitting it from the final equation. However, if x_1 and x_2 are strongly correlated, it is possible to get a significant overall F-test, and yet *both* partial regression coefficients non-significant! This merely means that if one of the variables is included in the model, very little is contributed by including the other as well.

There is no need to insist that the final model we give is the best, in the sense of giving the maximum sum of squares due to regression for a given number of variables. In some cases there may be other linear combinations, which are not quite the best, but which are physically meaningful, or which lead to a simpler interpretation. For example, if in (11.2) it proves possible to delete the term $\beta_4 x_1 x_2$ without serious loss, the effects of x_1 and x_2 are additive, and this greatly simplifies the interpretation. However, the interpretation of the results of multiple regression analyses involving many variables can be very complicated, and we shall not discuss it further.

Setting out the calculations

The form of Table 10.2 should be used to set out the calculations, with one column each for y, x_1, $y_1 x_1$, x_2, $y_2 x_2$, $x_1 x_2$, ... When calculations are done on a desk machine, it is important to set the results out neatly, as this reduces errors and facilitates checking.

Exercises 11.3

1. The following experiment, due to Ogilivie *et al.*, was done to determine if there is a relation between a person's ability to perceive detail at low levels of brightness, and his absolute sensitivity to light. For a given size of target, which is measured on the visual acuity scale (1/angle subtended at eye) and denoted x_1, a threshold brightness (denoted y) for seeing this target was determined [x_1 was restricted to be either 0·033 or 0·057]. In addition, the absolute light threshold (denoted x_2) was determined for each subject.

It is known that y depends on x_1, and this dependence can be taken to be linear. Use multiple regression to test whether y is significantly dependent on x_2 (for the purposes of the test, assume that any such dependence would be linear), i.e. use the model in which the expected value of y is

$$\alpha + \beta_1(x_1 - \bar{x}_1) + \beta_2(x_2 - \bar{x}_2)$$

where $\bar{x}_1$ and $\bar{x}_2$ denote the averages of x_1 and x_2 respectively.

	Subject 1	2	3	4	5	6	7	8	9
(A)	3·73	3·75	3·86	3·88	3·89	3·90	3·92	3·93	3·98
(B)	4·72	4·60	4·86	4·74	4·42	4·46	4·93	4·96	4·93
(C)	5·17	4·80	5·63	5·35	4·73	4·92	5·15	5·42	5·14

	Subject 10	11	12	13	14	15	16	17	18
(A)	3·99	4·04	4·07	4·07	4·09	4·10	4·17	4·20	4·22
(B)	4·63	5·20	5·31	5·06	5·21	4·97	4·82	5·03	5·44
(C)	5·92	5·51	5·89	5·44	6·74	5·41	5·22	5·49	6·71

	Subject 19	20	21	22	23	24	25
(A)	4·23	4·24	4·29	4·29	4·31	4·32	4·42
(B)	4·64	4·81	4·58	4·98	4·94	4·90	5·10
(C)	5·14	5·37	5·11	5·40	5·71	5·41	5·51

Rows (A) give values of x_2. Rows (B) and (C) give values of y for $x_1 = 0.033$ and $x_1 = 0.057$ respectively.

Carry out a full analysis of this data, including graphical checks, etc.) where appropriate.

The calculations can be simplified by using the following scale for x_1, represent $x_1 = 0.033$ as -1, and $x_1 = 0.057$ as $+1$, then the average for x_1 is zero and the correction for all sums of products involving x_1 is also zero. (London, Anc. Stats., 1962. See Ogilvie et al. (1960) for full description of the experiment.)

11.4 Discussion

The theory given in § 11.2 and § 11.3 readily generalizes to deal with k independent variables in the form of model (11.1). Estimates of the β,'s are obtained by solving the normal equations (11.4) and the sum of squares due to regression is

$$\hat{\beta}_1 \, \text{CS}(y, x_1) + \hat{\beta}_2 \, \text{CS}(y, x_2) + \ldots + \hat{\beta}_k \, \text{CS}(y, x_k),$$

which is on k degrees of freedom; see Table 11.2 for comparison. Significance tests for the β_j can be carried out by straight analogues of the methods given in § 11.3, but as hinted in that section difficulties arise in the interpretation of the results of such tests, and a discussion of this is beyond this volume.

One case of a generalization to k independent variables which is simple, is when all of the corrected sums of products $CS(x_i, x_j)$ for $i \neq j$, are zero. This situation arises in two dimensions, for example, if the set of x-values is a balanced array, as shown in Figure 11.4. When all the $CS(x_i, x_j)$ are zero, the solution of the normal equations (11.4) is

$$\hat{\beta}_j = CS(y, x_j)/CS(x_j, x_j), \quad j = 1, 2, \ldots, k,$$

and all the β_j's are independent and additive. The analysis can then be

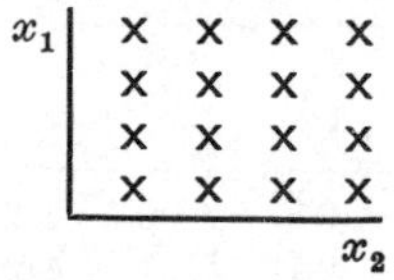

$$
\begin{array}{c|cccc}
x_1 & \times & \times & \times & \times \\
& \times & \times & \times & \times \\
& \times & \times & \times & \times \\
& \times & \times & \times & \times \\
\hline
& & & & x_2
\end{array}
$$

Figure 11.4 *Design for independent β_j's*

represented as in Table 11.6, and each term can be interpreted independently, as in Chapter 10. If therefore the placing of the x-values is under the control of the experimenters, there is a considerable advantage in arranging the cross products $CS(x_i, x_j)$ to be zero.

Table 11.6 *Analysis for independent β_j's*

Source	CSS	d.f.
Due to x_1	$\hat{\beta}_1 \, CS(y, x_1)$	1
Due to x_2	$\hat{\beta}_2 \, CS(y, x_2)$	1
.	.	.
.	.	.
.	.	.
Due to x_k	$\hat{\beta}_k \, CS(y, x_k)$	1
Deviations	(By subtraction)	$n - k - 1$
	$CS(y, y)$	$n - 1$

A difficulty which frequently arises with multiple regression is that the x-values are not under the control of the experimenter and contain a measurement error or other random variation which cannot be said to be negligible. The methods of this chapter are not strictly applicable in such a situation, and the discussion of § 10.8 is relevant here. The absolute validity of multiple regression in both Examples 11.1 and 11.2 could possibly be disputed on these grounds. A thorough investigation of the effect of this kind of departure from assumption does not seem to have been undertaken.

Exercises 11.4

1. In an investigation of the life of tools, S. M. Wu (1963) varied three factors

$$v, \text{ cutting speed (ft per min);}$$
$$f, \text{ rate of feed (in per rev);}$$
$$d, \text{ depth of cut (in).}$$

Each was used at three levels, denoted conventionally $-1, 0, 1$, which were equally spaced in terms of $x_1 = \log v$, $x_2 = \log f$, $x_3 = \log d$. Tool life, l, was measured by the number of minutes before a certain level of wear was observed; in the analysis $y = \log_e l$. An experiment was run in two blocks of six runs each in accordance with the scheme below. Previous work had suggested that a logarithmic model

$$E(y) = \beta_0 + \beta_1 x_1 + \beta_2 x_2 + \beta_3 x_3 \quad (1)$$

would be a reasonable approximation. Use the method of least squares to fit this model, supplemented by a possible block effect and give the estimated standard errors of the estimates of β_1, β_2, β_3. Outline briefly how you would examine the adequacy of the linear model (1).

Trial no.	Block no.	Coding			Log tool life
		x_1	x_2	x_3	y
1	2	-1	-1	-1	5·08
2	1	1	-1	-1	3·61
3	1	-1	1	-1	5·11
4	2	1	1	-1	3·30
5	1	-1	-1	1	5·15
6	2	1	-1	1	3·56
7	2	-1	1	1	4·79
8	1	1	1	1	2·89
9	1	0	0	0	4·19
10	1	0	0	0	4·42
11	2	0	0	0	4·26
12	2	0	0	0	4·41

[*Note.* The meaning of 'blocks' is that, for example, the experiment was done in two parts, say six runs on one day, and six on another, and that a difference between days (blocks) may arise through such factors as resetting machines, etc. Account for blocks by introducing a term $\beta_4 x_4$ into (1), with $x_4 = +1$ for block 1 and $x_4 = -1$ for block 2. Check that all the cross products $CS(x_i, x_j)$ are zero for this data, and set out the results as in Table 11.6.]

2. (This exercise is intended to prepare the reader for Chapter 12; it refers back to the material of Chapters 5 and 6.)

The data in Table 4.6 are five samples of nine observations, all taken from a normal population with expectation five and variance unity.

(i)   Combine the CSS for the five samples and make a combined estimate of variance based on the within group variation.

(ii) By considering the five group totals, calculate the sample variance of these five observations (i.e. on 4 d.f.).

(iii) What do you expect the ratio of the results obtained in (ii) and (i) to be, if the five groups of observations really are sampled from one population?

(iv) What is the effect on the ratio in (iii) if there is a difference between the group means representing some treatment difference?

(v)   Suggest a significance test for the null hypothesis that all 45 observations were drawn from the same normal population.

CHAPTER TWELVE

Analysis of Variance

12.1 The problem*

Consider the following example, which is a simplified form of a real experiment, with fictitious data.

Example 12.1. In an experiment concerning the teaching of arithmetic, 45 pupils were divided at random into five groups of nine. Groups A and B were taught in separate classes by the normal method. Groups C, D and E were taught together for a number of days and on each day C were praised publicly for their previous work, D were reproved publicly and E, while hearing the discussion of C and D, were ignored. At the end of the period all pupils took a standard test and the results are given below. What conclusions can be drawn about differences between the results of the teaching methods?

Group A (Control)	17	14	24	20	24	23	16	15	24
Group B (Control)	21	23	13	19	13	19	20	21	16
Group C (Praised)	28	30	29	24	27	30	28	28	23
Group D (Reproved)	19	28	26	26	19	24	24	23	22
Group E (Ignored)	21	14	13	19	15	15	10	18	20

In this experiment we wish to be able to say firstly, if real differences between the groups exist, and secondly, if such differences do exist, we want indications of where they are, and of their magnitude. Therefore, we wish to test the null hypothesis that there are no differences between the effects of the different teaching methods. One possible way to test this hypothesis would be to carry out all possible $^5C_2 = 10$ two-sample t-tests, and accept the null hypothesis only if none of these proved significant, but this procedure is unsatisfactory for several reasons.

Firstly, the actual significance level of the suggested procedure is in some doubt. We have

significance level = prob(at least one t-test sig. | null hypothesis true)
$$= 1 - \text{prob(no } t\text{-test significant)},$$

* Some readers may prefer to read § 12.3 before § 12.1 and § 12.2.

but only four of the ten t-tests are independent. If we proceed on the basis that they are all independent and are carried out at the nominal 5% level, then we have for the significance level of the suggested procedure

$$\text{sig. level} = 1 - (0.95)^{10} \simeq 0.401,$$

which is considerably above 5%. We could try and correct for this by operating the t-tests at a higher significance level, but it is difficult to see how to allow for the lack of independence of the t-tests in this calculation. Some of the t-tests cannot be dropped as this would leave some deviations from the null hypothesis untested. There are many difficulties in attempting to use the suggested procedure as a precise test of the null hypothesis that there is no difference between the effects of the teaching methods. If an overall test of significance can be performed, to detect whether real differences do or do not exist, then the suggested procedure may be useful as a 'follow-up' to indicate where these differences lie; see § 12.3 (*ii*) below.

A second point about the suggested procedure is that although it implies that the underlying error variance is constant for all groups, this assumption is not utilized. If the corrected sums of squares for the groups are combined on the combined degrees of freedom, then provided the population variance is really homogeneous, an improved estimate of variance is obtained. This improved estimate of variance could be used in the denominators of the t-tests, but this would increase the association between the tests. For example, if the overall estimate of variance happened to be low, most differences would tend to appear significant; thus, a common estimate of variance implies a correlation between tests.

The calculations carried out in Exercise 11.4.2 indicate the manner in which a precise test of the overall null hypothesis in Example 12.1 can be obtained.

In each group of observations in Exercise 11.4.2 the corrected sum of squares (CSS) divided by the degrees of freedom is an independent estimate of the variance σ^2. (In Exercuse 11.4.2 $\sigma^2 = 1$.) The combined CSS within groups provides a combined estimate of σ^2 based on within group variation; thus, the combined CSS is 51·9290 and the degrees of freedom are $5 \times 8 = 40$, so that we have an estimate

$$s_w^2 = 51.9290/40 = 1.298$$

based on within group variation.

Now from results given in Example 3.9 and § 4.3, the five groups totals of Exercise 11.4.2 are independent of the estimate of σ^2 just obtained, and have a variance $9\sigma^2$. Thus, provided the five groups are sampled from one normal population, the corrected sum of squares between the group means divided by nine times the degrees of freedom between groups is another estimate of σ^2 which is independent of s_w^2, and which we denote s_b^2. From Table 4.6 the group totals are respectively 47·55, 45·66, 46·17, 43·61, 39·08 and the calculations for the between groups estimate of σ^2 are as follows.

$$\begin{array}{ll}
\text{Total} \quad 222·07 & s_b^2 = \text{CSS}/(n \times \text{d.f.}) \\
\text{USS} \quad 9906·5855 & \quad = 43·5685/9 \times 4 \\
\text{Corr.} \quad 9863·0170 & \quad = 4·8409/4 \\
\text{CSS} \quad 43·5685 & \quad = 1·2102
\end{array}$$

If all the observations have been sampled from the same population, these two estimates, s_w^2 and s_b^2, should be approximately equal within the limits of random variation.

Now we notice that if every observation in one group is increased by a constant amount, then the estimate s_w^2 is unaltered, but an extra variation is imposed on the estimate s_b^2, which must on average increase it. Therefore, if the ratio s_b^2/s_w^2 is very much greater than one, this indicates that real differences between group means probably exist. The ratio s_b^2/s_w^2 is therefore a suitable test statistic to test the overall hypothesis that there are no differences between the group means; in order to obtain a precise test of significance we shall need the distribution of this ratio. Now if the null hypothesis is true, both s_b^2 and s_w^2 are estimates of variance and have distributions proportional to χ^2 on 4 and 40 degrees of freedom respectively. Therefore the ratio has an F-distribution, and significance values for the ratio can be read from F-tables, large values being significant.

One more point regarding Exercise 11.4.2 is instructive. If the observations are treated as one group we have a CSS of 56·7699 on 44 degrees of freedom, and the results would usually be presented in the following analysis of variance table.

Table 12.1 *Analysis of variance table for Exericse* 11.4.2

Source of variation	CSS	d.f.	Mean square
Between groups	4·8409	4	1·210
Within groups	51·9290	40	1·298
Total	56·7699	44	

In this table, the CSS between groups is nine times the CSS of the group means, or the CSS of the group totals divided by nine. The CSS within groups is just the combined within groups corrected sum of squares.

From the table, we notice that the CSS add up – that is, the CSS within groups plus the CSS between groups is exactly equal to the CSS obtained by treating all the data as one group. This property is established below as an algebraic identity; it implies that regardless of distribution assumptions, etc., the algebra of the analysis of variance can be used as a means of summarizing the total variation in a sample into two (or more) components, although the interpretation of these components will not always be clear. We also notice that the degrees of freedom for the between and within groups CSS add up to the number of degrees of freedom in the data as a whole.

Exercises 12.1

1. (a) If we have a set of observations or means x_A, x_B, . . . , then a linear combination $\{l_A x_A + l_B x_B + \ldots\}$ such that the coefficients add to zero, $l_A + l_B + \ldots = 0$, is called a *linear contrast* of the observations or means.

Suppose $\bar{x}_1, \ldots, \bar{x}_g$ are a set of means each of n observations, where all the observations are independently and normally distributed with expectation μ and variance σ^2.

Consider the linear contrast

$$w_1 = \sum_{i=1}^{g} \alpha_{1i} \bar{x}_i \quad \text{where} \quad \sum_{i=1}^{g} \alpha_{1i} = 0,$$

and show that $E(w_1) = 0$, and that

$$E(w_1^2) = V(w_1) = \sigma^2 \sum_{i=1}^{g} \alpha_{1i}^2 / n.$$

Then demonstrate that the distribution of $\left(n w_1^2 \middle/ \sum_{i=1}^{g} \alpha_{1i}^2 \right)$ is proportional to χ^2 on one degree of freedom.

(b) Also show that for any set of linear contrasts

$$w_j = \sum_{i=1}^{g} \alpha_{ji} \bar{x}_i \quad \text{where} \quad \sum_{i=1}^{g} \alpha_{ji} = 0$$

the condition that the w_j and w_k are uncorrelated is

$$\sum_{s=1}^{g} \alpha_{js} \alpha_{ks} = 0, \quad j \neq k.$$

Hence argue that provided this condition holds, the quantities

$$\left(n w_j^2 \middle/ \sum_{i=1}^{g} \alpha_{ji}^2 \right),$$

are independently distributed proportionally to χ^2 on one degree of freedom.

(c) Given any one linear contrast w_1 based on g means, how many other linear contrasts can be constructed which are uncorrelated with this and with each other?

12.2 Theory of one-way analysis of variance

Suppose we have g groups of observations to compare, with n observations per group,

$$
\begin{array}{ll}
\text{group 1} & x_{11} \ldots x_{1n} \\
\text{group 2} & x_{21} \ldots x_{2n} \\
\quad \cdot & \quad \cdot \qquad \cdot \\
\quad \cdot & \quad \cdot \qquad \cdot \\
\text{group g} & x_{g1} \ldots x_{gn}
\end{array}
$$

and denote the group means $\bar{x}_{1.}, \bar{x}_{2.}, \ldots, \bar{x}_{g.}$, then the analysis of variance table is as follows (*cf.* Table 12.1):

Table 12.2 *One-way analysis of variance*

Source	CSS	d.f.
Between groups	$n \sum_{i=1}^{g} (\bar{x}_{i.} - \bar{x}_{..})^2$	$(g-1)$
Within groups	$\sum_{i=1}^{g} \sum_{j=1}^{n} (x_{ij} - \bar{x}_{i.})^2$	$g(n-1)$
Total	$\sum_{i=1}^{g} \sum_{j=1}^{n} (x_{ij} - \bar{x}_{..})^2$	$gn - 1$

Let us suppose that any observation x_{ij} has the following structure,

$$x_{ij} = \mu_i + \varepsilon_{ij}, \tag{12.1}$$

where the μ_i, $i = 1, 2, \ldots, g$, are the unknown means of the g groups, and the ε_{ij} are errors which we assume to be independently distributed as $N(0, \sigma^2)$. The reader should notice the three important assumptions here, homogeneity of variance throughout the groups, and independence and normality of the errors. We can now establish the following properties.

Property 1. The corrected sums of squares in Table 12.2 add up.
Proof: Put

$$x_{ij} - \bar{x}_{..} = (x_{ij} - \bar{x}_{i.}) + (\bar{x}_{i.} - \bar{x}_{..}) \tag{12.2}$$

and then square and add to obtain

$$\sum_i \sum_j (x_{ij} - \bar{x}_{..})^2 = \sum \sum (x_{ij} - \bar{x}_{i.})^2 + n \sum (\bar{x}_{i.} - \bar{x}_{..})^2. \tag{12.3}$$

The product term obtained on squaring the right-hand side of 12.2 vanishes on summing over j. The terms of (12.3) are the corrected sums of squares in Table (12.2) and the property is proved.

Property 2. The expected value of the mean square within groups in Table 12.2 is σ^2.

Proof: The result follows because the mean square within groups is just the combined within groups estimate of σ^2. For any group i, we have

$$E\left\{\frac{1}{n-1}\sum_j (x_{ij} - \bar{x}_{i.})^2\right\} = \sigma^2$$

from Example 3.11. By summing over groups and dividing by g we obtain the result,

$$E\left\{\frac{1}{g(n-1)}\sum_i \sum_j (x_{ij} - \bar{x}_{i.})^2\right\} = \sigma^2. \tag{12.4}$$

Property 3. The expected value of the mean square between groups is

$$\sigma^2 + \frac{n}{(g-1)}\sum_i (\mu_i - \bar{\mu}_.)^2$$

where $\bar{\mu}_.$ is the average of the μ_i's.

Proof: From the model (12.1) we have

$$\bar{x}_{i.} = \mu_i + \bar{\varepsilon}_{i.}$$
$$\bar{x}_{..} = \bar{\mu}_. + \bar{\varepsilon}_{..},$$

in obvious notation, so that the corrected sum of squares between groups is

$$n\sum_i (\bar{x}_{i.} - \bar{x}_{..})^2 = n\sum_i (\mu_i - \bar{\mu}_. + \bar{\varepsilon}_{i.} - \bar{\varepsilon}_{..})^2$$

$$= n\sum_i (\mu_i - \bar{\mu}_.)^2 + n\sum_i (\bar{\varepsilon}_{i.} - \bar{\varepsilon}_{..})^2$$

$$+ 2n\sum_i (\mu_i - \mu_.)(\bar{\varepsilon}_{i.} - \bar{\varepsilon}_{..}), \tag{12.5}$$

and the mean square is obtained by dividing by $(g-1)$. When we take expectations, the product term in (12.5) vanishes, since it is a linear combination of ε_{ij}'s. By the model, the terms $\bar{\varepsilon}_{i.}$ are independent with expectation zero and variance σ^2/n, since they are means of n terms, so that by applying Example 3.11 again we have

$$E\left\{\frac{1}{(g-1)}\sum_i (\bar{\varepsilon}_{i.} - \bar{\varepsilon}_{..})^2\right\} = V(\bar{\varepsilon}_{i.}) = \sigma^2/n$$

and

$$E\left\{\frac{n}{(g-1)}\sum_i (\bar{\varepsilon}_{i.} - \bar{\varepsilon}_{..})^2\right\} = \sigma^2.$$

Therefore we obtain the result

$$E\left\{\frac{n}{(g-1)}\sum_i (\bar{x}_{i.} - \bar{x}_{..})^2\right\} = \sigma^2 + \frac{n}{(g-1)}\sum_i (\mu_i - \bar{\mu}_.)^2. \tag{12.6}$$

These three properties hold if we assume the structure (12.1) and assume that the ε's are uncorrelated. Therefore even without assuming normality, a rough test of significance is provided by looking at the mean squares between and within groups; only if the mean square between groups is greater than the mean square within groups are there any grounds for disputing the null hypothesis. For example, in Table 12.1 the mean square between groups is *less* than the mean square within groups, so that regardless of distributional assumptions, the data can be said to be in good agreement with the theory that they come from one population in respect of the variation between the group means.

To obtain a precise test of significance we must assume normality, and the corrected sum of squares within groups then has a distribution proportional to χ^2 on $g(n-1)$ degrees of freedom, regardless of the values of the μ_i. The corrected sum of squares between groups will also have a distribution proportional to χ^2 provided the null hypothesis is true, so that significance levels for the ratio

$$\frac{\text{M.S. between groups}}{\text{M.S. within groups}}$$

are found by looking up standard F-tables for a one-sided test. The properties of the analysis of variance table can now be summarized as in (*i*), (*ii*) and (*iii*) of § 11.2.

Least squares approach

The discussion given above has not been based on the theory of least squares, but instead rests to some extent on certain intuitive ideas. However, we can reach the same results using the theory of least squares.

We estimate the μ_i in (12.1) so as to minimize

$$S = \sum_i \sum_j (x_{ij} - \mu_i)^2$$

and this leads to estimates $\hat{\mu}_i = \bar{x}_{i\cdot}$. The minimized sum of squares is therefore

$$S_{\min} = \sum_i \sum_j (x_{ij} - \bar{x}_{i\cdot})^2 \tag{12.7}$$

and by the general theory of least squares this quantity has a distribution $\sigma^2\chi^2$ on $ng - g = (n-1)g$ degrees of freedom, since there are g parameters fitted.

If the null hypothesis is true, and all the μ_i are equal, say to μ, then the least squares estimate of μ is $\bar{x}_{\cdot\cdot}$, and

$$S_{\min} = \sum \sum (x_{ij} - \bar{x}_{\cdot\cdot})^2 \tag{12.8}$$

which has a distribution $\sigma^2\chi^2$ on $(ng - 1)$ degrees of freedom. We can now fill in Table 12.2 and find by subtraction that the corrected sum of squares between groups has a distribution $\sigma^2\chi^2$ on $(g-1)$ degrees of freedom.

Exercises 12.2

1. Suppose we have g groups of observations for a one-way analysis of variance, with unequal numbers of observations per group, as follows.

$$\begin{array}{ll} \text{group 1} & x_{11} \ldots x_{1,n_1} \\ \text{group 2} & x_{21} \ldots x_{2,n_2} \\ \quad \cdot & \quad \cdot \quad \cdot \\ \quad \cdot & \quad \cdot \quad \cdot \\ \quad \cdot & \quad \cdot \quad \cdot \\ \text{group } g & x_{g1} \ldots x_{g,n_g} \end{array}$$

Assume that the observations are all independent, and for the ith group, x_i is distributed $N(\mu_i, \sigma^2)$, where σ^2 is unknown. The variances of the g groups can be assumed identical.

Show that the CSS in the analysis of variance table below add up, and interpret the d.f. Also find the expectations of the mean squares.

Analysis of variance

Source	CSS	d.f.
Between groups	$\displaystyle\sum_{1}^{g} n_i(\bar{x}_{i.} - \bar{x}_{..})^2$	$g - 1$
Within groups	$\displaystyle\sum_{i=1}^{g}\sum_{j=1}^{n_i} (x_{ij} - \bar{x}_{i.})^2$	$\sum n_i - g$
Total	$\displaystyle\sum_{i=1}^{g}\sum_{j=1}^{n_i} (x_{ij} - \bar{x}_{..})^2$	$\sum n_i - 1$

12.3 Procedure for analysis

There are three phases in any analysis of variance. Firstly the calculation of the ANOVA table, down to the F-test of significance. Secondly some follow-up procedure is required to find out where the differences in means are, and there are several methods in common use. These follow-up procedures can be used whether or not the F-test is significant at a high enough level, since they indicate where real differences might exist. Thirdly there are checks of assumptions using residuals.

(i) Procedure for calculating the ANOVA table

The steps involved when we have g groups with n observations per group are as follows.

Step 1. Calculate the group totals and means, and the grand total.
Step 2. Calculate the correction, which is (grand total)$^2/ng$.

Step 3. Calculate the uncorrected sum of squares (USS) for the whole data.
Step 4. The CSS for the whole data is (3) — (2).
Step 5. Calculate the USS between groups, which is the sum of (group total)$^2/n$ over groups.
Step 6. The CSS between groups is (5) — (2).
Step 7. The CSS within groups is obtained by subtraction, (4) — (6). Write out the ANOVA table, and obtain mean squares except for the total line.

Example 12.2. For Example 12.1 these calculations are as follows.

Step 1. *Group*	*A*	*B*	*C*	*D*	*E*	*Grand total*
Total	177	165	247	211	145	945
Mean	19·7	18·3	27·4	23·4	16·1	

Step 2. Correction $= (945)^2/45 = 19,845$.
Step 3. $17^2 + 14^2 + \ldots + 18^2 + 20^2 = 21,041$.
Step 5. $\{177^2 + 165^2 + \ldots + 145^2\}/9 = 185,109/9 = 20,567\cdot7$.

Steps 1, 2, 3 and 5 enable us to carry out steps 4 and 6 which are usually set out as follows:

	Whole data		*Between groups*
USS	21,041	USS	20,567·7
Corr.	19,845	Corr.	19,845·0
CSS	1196	CSS	722·7

Step 7 is carried out by writing out the ANOVA table.

Table 12.3 *Analysis of variance for Example* 12.1

Source	CSS	d.f.	M.S.
Between groups	722·7	4	180·68
Within groups	473·3	40	11·83
Total	1196·0	44	

The F-test of the null hypothesis that the groups all have the same population mean is to refer

$$180\cdot68/11\cdot83 = 15\cdot3$$

to the F-tables for $(4, 40)$ degrees of freedom. From F-tables we find that $F(4, 40)$ at 1% is about 3·8 and at the 0·1% level is about 5·7, so that this result is significant at well beyond the 0·1% level, giving very strong evidence that real differences in the group means exist. This concludes the first phase of the analysis. □ □ □

(ii) Least significant difference analysis

Now that we have very good evidence that real differences between the group means exist in Example 12.1, we need to set about finding where these differences are. One method is to revert to using t-tests in the form (6.3), with the within groups mean square as an estimate of σ^2. This is, of course, what we have condemned in § 12.1, but we are on more sure ground now – we have an F-test of whether or not any differences exist. We must bear in mind that this use of t-tests will tend to give us too many significant differences.

When all the groups have the same number of observations in them, the denominators of all possible t-tests will be identical. We can therefore save a lot of calculation by evaluating the least difference in any pair of means which would be significant at a given level. We illustrate this, for Example 12.1.

The standard error of the difference of any two group means for Example 12.1 is estimated at

$$\sqrt{\left\{\sigma^2\left(\frac{1}{n} + \frac{1}{n}\right)\right\}} = \sqrt{\left\{11 \cdot 83\left(\frac{1}{9} + \frac{1}{9}\right)\right\}} = 1 \cdot 621.$$

The 5% and 1% points of the t-distribution on 40 degrees of freedom are $2 \cdot 02$ and $2 \cdot 70$ respectively. Therefore the least difference in means to be significant is

$$1 \cdot 621 \times 2 \cdot 02 = 3 \cdot 275$$

at 5%, and

$$1 \cdot 621 \times 2 \cdot 70 = 4 \cdot 378$$

at 1%. We call these least significant differences, and denote them LSD(5%, 40 d.f.), etc.

The next step is to compare the differences between the group means with these LSD's, and in this way it becomes fairly clear that we can make the following conclusions for Example 12.1:

(i)   The two control groups do not differ significantly.
(ii) Group C achieves significantly higher scores than any other group.
(iii) Group E had significantly lower results than any other group except B.
(iv) Group D had significantly higher results than the control groups, but significantly less than Group C.

Any statement of conclusions should also give confidence intervals for the differences between pairs of means which are of interest. This is done by adding and subtracting the LSD from the observed difference of two means. For example, 95% confidence intervals for the difference between the means of groups D and E are

$$(23 \cdot 4 - 16 \cdot 1) \pm 3 \cdot 275 = (4 \cdot 0, 10 \cdot 6).$$

In Example 12.1 there is another set of comparisons which are of greater interest. The control group has been replicated, and provided there are no important differences between the experimental conditions of the two control groups, and that this is borne out by the differences between the sample means being non-significant, then we can combine the group means. For example, we would require groups A and B to be groups of children of similar age, sex, intelligence, etc., and to be taught by the same teacher; if groups A and B had been significantly different we would suspect that the groups did differ in some vital respect. Assuming that it is valid to combine in Example 12.1, the average of groups A and B is 19·0 and we are now interested in comparing the means for C, D and E with this combined control mean.

The variance of a difference between the combined average and C, D or E is

$$\sigma^2(\tfrac{1}{18} + \tfrac{1}{9}) = \frac{\sigma^2}{6},$$

where σ^2 is estimated from the mean square within groups, on 40 degrees of freedom. The new LSD (5%, 40 d.f.) is therefore

$$2 \cdot 02 \times \sqrt{\left(\frac{11 \cdot 83}{6}\right)} = 2 \cdot 83$$

We now look at the differences between the means for C, D and E, and 19·0, and see whether they are greater or less than 2·83. The conclusions from this analysis are not substantially different from those made above. Confidence intervals for the difference between the combined average of the control groups and groups C, D or E are obtained as before, by simply adding and subtracting the new LSD from the observed differences.

In some experiments the comparison of new treatments or processes with a standard or control is vital, and these comparisons can be made much more precise by using more observations on the standard or control than on any of the other treatments; see Exercise 12.3.3.

Analysis by the studentized range
Another method of analysing differences between group means in the analysis of variance is to make use of the studentized range. If $y_1, \ldots, y_n$ are independent normally distributed random variables with the same mean, and with a variance σ^2, and if s^2 is an independent estimate of σ^2 on ν degrees of freedom, the studentized range is

$$\text{Range}(y_1, \ldots, y_n)/s \tag{12.9}$$

and percentage points of this are given in Biometrika tables; see also the short table in Appendix II of this book. Let $q(\alpha, n, \nu)$ denote the upper $\alpha\%$ point of the studentized range of n observations when the variance is estimated on ν degrees of freedom. For example, $q(5\%, 5, 40) = 4 \cdot 04$, which means that only once in twenty samples is the ratio (12.9) greater than 4·04, when $n = 5$ and the variance is estimated on 40 degrees of freedom.

Now the distribution of the ratio

$$\frac{\text{Range of } g \text{ group means of } n \text{ observations}}{\sqrt{(\text{mean square within groups})}/\sqrt{n}} \qquad (12.10)$$

will be the same as the distribution of (12.9) when there are no real differences between the groups. That is, if the null hypothesis is true, the range of g group means of n observations will be greater than

$$q\{5\%, g, g(n-1)\} \times \sqrt{\{\text{Mean square within groups}\}}/\sqrt{n}$$

only once in twenty times, in repeated trials under the same conditions. For our example

$$q(5\%, 5, 40) = 4.04,$$

$$\sqrt{\{\text{mean square within groups}\}}/\sqrt{n} = 1.14,$$

and any difference in group means of more than $4.04 \times 1.14 = 4.60$ can be regarded as significant at the 5% level. We therefore find that the pairs of groups (A, C), (B, C) and (B, D) are significantly different and that (A, D) and (A, E) are large enough to be suggestive of a possible difference.

Rule. Look up the appropriate percentage point of the studentized range, and multiply it by the standard error of a group mean; regard any means differing by more than this as significant at the level in question.

This method of analysing means has the property that, if the null hypothesis is true, only once in twenty times would any repetition of the experiment contain a pair of means which would be regarded as significant at the 5% level. The method is therefore not only an alternative to the LSD method of analysing differences in means, but it is also an alternative to calculating the F-test in the ANOVA table. That is, it is an alternative solution to the problem discussed in § 12.1, of providing a precise significance test that there are no differences between a set of means.

Discussion on the analysis of means phase
The following facts can be stated about the relative merits of the two procedures given above for analysing means.

If a difference in means is studied simply because it is the largest difference then the range method must be used. If a particular comparison between means is of interest on *a priori* grounds, then the LSD method is appropriate. Apart from these two conditions either method may be used.

The range method can be criticized as being over-conservative, while the LSD approach gives too many false positives. However, the 5% and 1% levels must *never* be observed strictly, and when viewed in this light there is not too much difference between the methods when the number of degrees of freedom are small.

The range method does not readily give confidence intervals, as the LSD approach does, and it is impractical to adapt the range method to deal with, for example, a control group having more observations than any other.

There are other methods of analysing means; one will be given in §12.5, and for others see Tukey (1949a), Kurtz *et al* (1965), the book by Miller (1966) and the review paper by Wetherill and O'Neill (1971).

(iii) Analysis of residuals

The normality, additivity and homogeneity of variance assumptions can all be checked by examining the residuals in an appropriate way. For one-way analysis of variance, the residuals are

$$r_{ij} = x_{ij} - \bar{x}_{i.}, \quad i = 1, \ldots, g; \quad j = 1, \ldots, n.$$

The methods of analysis are similar to those used in § 10.7 and this is left as an exercise for the reader; see Exercise 12.3.1.

(iv) Conclusions

It is most important that any statistical analysis should be finished with a careful statement of the conclusions which can be drawn from the various phases of the analysis. Such a statement should include standard errors, confidence intervals and indications of the need for more data, etc.

Exercises 12.3

1. Carry out an analysis of residuals of Example 12.1 data, using the methods described in § 10.7.

2. Carry out an analysis of Exercise 6.5.2 by treating the data as a one-way analysis of variance, with four groups of eight observations (and assuming that the variance is homogeneous).

3. An experiment is run to compare each of t treatments with a control. The structure of the observations is

$$x_{ij} = \mu_i + \varepsilon_{ij},$$

where ε_{ij} are independent $N(0, \sigma^2)$, μ_i, $i = 1, 2, \ldots, t$, are the means of the treatments under comparison, and μ_0 is the mean of the control treatment. Let m observations be taken on the control treatment and n on each of the other treatments, so that $(tn + m)$ observations are used in all.
(i) Find the variance of the estimated difference between one of the new treatments and the control treatment.
(ii) Given that N observations should be used in all, what values should m and n have to minimize the variance obtained in (i)?

S

12.4 Two-way analysis of variance

The data we have considered so far in this chapter have been ordered in one direction only, and in practice data are often classified in several ways. Consider the following example.

Example 12.3. The data given below are part of the results of an experiment by Cane and Horn (1951) which was designed to investigate the timing and correctness of responses of children to spatial perception problems. The questions were arranged in sets of eighteen, the sets being of balanced difficulty. Four groups of thirty children took part, these being 14-year-old boys, 13-year-old boys, 14-year-old girls and 13-year-old girls respectively, and each child completed six sets of eighteen questions. The design of the experiment was balanced in such a way that over each group of children, each set of eighteen questions appeared equally often in each order of presentation, 1st, 2nd, ..., 6th. Every time a set of eighteen questions was used, the order within it was randomized. This design ensures that if there were no effects due to order of presentation or of groups of children, the total scores of any group of children in any order of presentation would be expected to be the same.

The data in Table 12.4 below are the total scores for each group of thirty children for the questions given in each position.

Table 12.4 *Total scores for groups in Cane and Horn's experiment*

Group number	Order of presentation 1	2	3	4	5	6	Total
I	188	200	181	181	175	170	1095
II	170	191	171	170	178	160	1040
III	175	167	145	153	165	174	979
IV	129	125	139	124	140	129	786
Totals	662	683	636	628	658	633	3900

In this example we expect two different types of effect. Firstly there may be differences between the groups of children, and secondly the children may tend to get better (or worse) results as the test proceeds, either as they become accustomed to the type of question, or eventually get tired, etc.

Suppose that we can assume the following structure for an observation x_{ij} in the ith row and jth column,

$$E(x_{ij}) = \mu + \rho_i + \gamma_j, \tag{12.11}$$

where
$$i = 1, 2, \ldots, r \quad j = 1, 2, \ldots, c$$

and
$$\sum_1^r \rho_i = \sum_1^c \gamma_j = 0.$$

This structure assumes that $E(x_{ij})$ is made up of an overall mean, plus a component ρ_i expressing the difference between the expected result in the ith row and the overall average μ, and lastly of a similar component γ_j representing the difference between the expected result in the jth column and μ. The structure of equation (12.11) also assumes that the row and column effects are additive; that is, if there are no errors, we might observe the following results in three rows and three columns. In Table 12.5, the

Table 12.5 *Example of additive effects*

Row	Column			Row effect
	1	2	3	
1	1	−2	−5	−2
2	3	0	−3	0
3	5	2	−1	+2
Column effect	+3	0	−3	

difference between corresponding observations in any two given rows is constant along the rows, and similarly for the columns. For example, the difference between column one and two is three, whichever row we look in; see § 3.5. If we suspect that in a particular experiment the effects will not be additive, but perhaps multiplicative, then the following analysis will not apply (unless we make a transformation, such as to take logarithms if a multiplicative model is appropriate).

Table 12.6 *Two-way analysis of variance*

Source	CSS	d.f.	E(Mean square)
Rows	$c \sum_{i=1}^{r} (\bar{x}_{i.} - \bar{x}_{..})^2$	$r - 1$	$\sigma^2 + \dfrac{c}{(r-1)} \sum_{i=1}^{r} \rho_i^2$
Columns	$r \sum_{j=1}^{c} (\bar{x}_{.j} - \bar{x}_{..})^2$	$c - 1$	$\sigma^2 + \dfrac{r}{(c-1)} \sum_{j=1}^{c} \gamma_j^2$
Residual	$\sum_{j=1}^{c} \sum_{i=1}^{r} (x_{ij} - \bar{x}_{i.} - \bar{x}_{.j} + \bar{x}_{..})^2$	$(r-1)(c-1)$	σ^2
Total	$\sum_{j=1}^{c} \sum_{i=1}^{r} (x_{ij} - \bar{x}_{..})^2$	$rc - 1$	

Finally, we assume that we have additive independent errors ε_{ij}, normally distributed with $E(\varepsilon_{ij}) = 0$, $V(\varepsilon_{ij}) = \sigma^2$, so that

$$x_{ij} = \mu + \rho_i + \gamma_j + \varepsilon_{ij}, \tag{12.12}$$

with $\sum \rho_i = \sum \gamma_j = 0$, as before.

The student should check for himself that under the conditions just given, the results in Table 12.6 can be obtained, and the corrected sums of squares and degrees of freedom add up, as in Table 12.3; see Exercises 12.4.4 and 12.4.5.

Procedure for analysis

The calculation of the analysis of variance table follows the pattern outlined in § 12.3.

Step 1. Calculate the row and column totals and means, and the grand total.

Step 2. Calculate the correction, which is

$$\text{(grand total)}^2/\text{(no. of obs.)}.$$

Step 3. Calculate the USS for the whole data.

Step 4. The CSS for the whole data is (3) — (2).

Step 5. Calculate the USS between rows, which is the sum of (row total)^2/c, over rows.

Step 6. The CSS between rows is (5) — (2).

Step 7. Calculate the USS between columns, which is the sum of (column total)^2/r over columns.

Step 8. The CSS between columns is (7) — (2).

Step 9. The CSS for residual is obtained by subtraction, (4) — (6) — (8). Write out the ANOVA table and obtain the mean squares except for the total line.

This procedure leads to the analysis of variance table for Example 12.3, which is shown in Table 12.7.

Table 12.7 *Analysis of variance for Example 12.3*

Source	CSS	d.f.	M.S.	F
Between groups (rows)	9060·3	3	3020·1	30·8 (very highly sig.)
Between positions (cols.)	566·5	5	113·3	1·2 (not sig.)
Residual	1473·2	15	98·2	
Total	11,100·0	23		

We now proceed with F-tests for rows and columns independently, comparing each mean square against the mean square for residual. For rows, we have

$$F = 3020\cdot1/98\cdot2 = 30\cdot8$$

on (3, 15) degrees of freedom, which is very highly significant. For columns we have

$$F = 113 \cdot 3/98 \cdot 2 = 1 \cdot 15$$

on (5, 15) degrees of freedom, which is not significant. We can therefore conclude this phase of the analysis by saying that the data give very strong evidence that the groups of children achieve different average scores, but there is no evidence of any effect due to the position of a test in the sequence of six.

We now proceed with the second phase of the analysis separately for both rows and columns, using the LSD method or the range method; the LSD's for rows and columns will be different, since the numbers of rows and columns differ. This phase of the analysis is left as an exercise for the reader and we shall give an alternative method of analysis in the next section.

The third phase of the analysis, which is checking of the assumptions, also proceeds in the manner indicated previously. The residuals for a two-way layout are

$$r_{ij} = x_{ij} - \bar{x}_{i.} - \bar{x}_{.j} + \bar{x}_{..}. \tag{12.13}$$

The final step in the analysis is to make a detailed report of the conclusions, including confidence intervals, etc. A report for Example 12.3 is given in the next section, as an illustration.

Exercises 12.4

1. Obtain the residuals (12.13), for Example 12.3 data, and examine them for outliers and for any consistent patterns. Plot the residuals out so as to test for departures from normality, using the method given in § 10.7(2).

2. Continue the previous exercise by plotting the residuals against the fitted values, where

$$\text{fitted value} = \bar{x}_{i.} + \bar{x}_{.j} - \bar{x}_{..}.$$

From this plot, judge whether there is any evidence that the variance of an observation depends on its expectation.

3. Examine the difference between groups in Example 12.3 by the LSD method, considering the following questions;
(i)   Do boys (and separately, girls) improve their performance with increase in age?
(ii) Do boys have higher performance than girls of the same age?
(iii) Do girls improve more between the ages of 13 and 14 than boys do?

4. Prove that the sums of squares in Table 12.6 add up, as follows. Write

$$x_{ij} - \bar{x}_{..} = (x_{ij} - \bar{x}_{i.} - \bar{x}_{.j} + \bar{x}_{..}) + (\bar{x}_{i.} - \bar{x}_{..}) + (\bar{x}_{.j} - \bar{x}_{..})$$

and then square and add over i and j.

5. Insert the model (12.12) into the corrected sums of squares in Table 12.6, and hence obtain the expectations of the mean squares. [Follow the method of proof given in § 12.2.]

12.5 Linear contrasts

In this section we introduce another method of analysing a set of means in the second phase of an analysis of variance, and the method is based on the results of Exercise 12.1.1.

Frequently we find in the analysis of a set of means, that particular contrasts are of interest. In Example 12.1, we are particularly interested in making inferences about the differences between the mean of groups A and B, and the other groups. These differences can be represented by the contrasts $(\bar{x}_A + \bar{x}_B - 2\bar{x}_C)$, $(\bar{x}_A + \bar{x}_B - 2\bar{x}_D)$, etc. We can set these contrasts out as follows:

Table 12.8 *Contrasts for Example* 12.1

	$\bar{x}_A$	$\bar{x}_B$	$\bar{x}_C$	$\bar{x}_D$	$\bar{x}_E$
w_1	$+1$	$+1$	-2	0	0
w_2	$+1$	$+1$	0	-2	0
w_3	$+1$	$+1$	0	0	-2
w_4	$+1$	-1	0	0	0

The last contrast in Table 12.1 is the difference between the means of groups A and B.

For a second example, a set of interesting contrasts between the groups of children in Example 12.3 are as follows:

Table 12.9 *Contrasts for Example* 12.3

Group	Age and sex			
	(14, b) 1	(13, b) 2	(14, g) 3	(13, g) 4
w_1	1	1	-1	-1
w_2	1	-1	1	-1
w_3	1	-1	-1	1

The first contrast here gives the average difference between boys and girls, and the second contrast gives the average difference due to improvement between the ages 13 years and 14 years. The last contrast gives the difference between the amount girls improve between 13 years and 14 years, and the amount boys improve between the same ages; this is called the sex–age interaction.

Thus in general we see that if we have a set of g means each based on n observations, there will frequently be linear contrasts

$$w_j = \sum_{i=1}^{g} \alpha_{ji}\bar{x}_i, \quad \text{where} \quad \sum_{i=1}^{g} \alpha_{ji} = 0, \tag{12.14}$$

which are of interest. Now in the analysis of variance table, the CSS for g means is on $(g-1)$ degrees of freedom. We can split the CSS up into $(g-1)$ terms, each having one degree of freedom, and the method is indicated in Exercise 12.1.1. We set up $(g-1)$ linear contrasts (12.14), such that

$$\sum_{i=1}^{g} \alpha_{ji}\alpha_{ki} = 0, \quad j \neq k \tag{12.15}$$

and then the $(g-1)$ terms $\left(nw_j^2 \middle/ \sum_{i=1}^{g} \alpha_{ji}^2 \right)$ are corrected sums of squares

on one degree of freedom, which add up to the total corrected sums of squares between groups given in the analysis of variance table. [We have not proved that the CSS add up, and this is difficult without using matrix algebra. However, it is easy to demonstrate it numerically on an example.]

In doing the calculations on a desk machine, it will often be more accurate to use group totals rather than group means. If we denote the group totals T_i, $i = 1, 2, \ldots, g$, we therefore use the formulae

$$W_j = \sum_{i=1}^{g} \alpha_{ji}T_i$$

$$\text{CSS} = W_j^2 \middle/ \left(n . \sum_{i=1}^{g} \alpha_{ji}^2 \right). \tag{12.16}$$

Example 12.4. For Example 12.3 a set of contrasts is given in Table 12.9, and this set obeys the condition (12.15). The group totals are respectively 1095, 1040, 979 and 786. Therefore we have

$$w_1 = 1095 + 1040 - 979 - 786 = 370$$

and the sum of squares with one degree of freedom is

$$\frac{nw_1^2}{\sum \alpha_{ji}^2} = (370)^2/6 \times 4 = 5704 \cdot 16.$$

We also have

$$w_2 = 1095 - 1040 + 979 - 786 = 248$$
$$\text{CSS} = (248)^2/24 = 2562 \cdot 67,$$

and

$$w_3 = 1095 - 1040 - 979 + 786 = -138$$
$$\text{CSS} = (-138)^2/24 = 793 \cdot 50.$$

These results can be set out as in Table 12.10, which shows that these three sums of squares add exactly to the total CSS between groups, as given by Table 12.7.

Table 12.10 *Individual degrees of freedom for Example* 12.3

Source	CSS	d.f.
Sex	5704·16	1
Age	2562·67	1
Sex × age	793·50	1
Between groups	9060·33	3

Under the assumptions, the three CSS on one degree of freedom are statistically independent, and under the null hypothesis will have an expectation of σ^2. They can be tested against the mean square for residual, using the F-test, large values being significant.

For Example 12.3 the tests of significance are:

$$\text{Sex} \qquad F = 5704\cdot16/98\cdot2 = 58\cdot1$$
$$\text{Age} \qquad F = 2562\cdot67/98\cdot2 = 26\cdot1$$
$$\text{Sex} \times \text{age} \quad F = 793\cdot50/98\cdot2 = 8\cdot1.$$

All of these are on (1, 15) degrees of freedom, so that all three are very highly significant. Even the smallest of the three, the sex × age interaction, is very nearly significant at the 1% level. It is also of interest that more than 60% of the CSS between groups is attributable to the difference between sexes.

This procedure is equivalent to doing a particular set of t-tests on contrasts of the means, and is open to objection on various grounds raised in § 12.1. We must be particularly careful about any linear contrasts chosen after the data are seen. However, if the linear contrasts are chosen *a priori*, as in Tables 12.8 and 12.9, it is entirely legitimate to look at the degrees of freedom separately, and this results in a much more powerful test than the overall F-test between groups.

Sometimes the contrasts which interest us are not orthogonal, that is, they do not obey the restriction (12.15). One example of this is given by Table 12.8, with reference to Example 12.1. The linear contrasts between the mean for the control groups and C, D and E separately, do not obey (12.15) and so are not independent. The procedure can still be used, but the effect of this is that the CSS do not add up to the CSS between groups, and we cannot set them out as in Table 12.10.

There is no need to split all the degrees of freedom up when using this procedure. The importance of the method lies in estimating and testing meaningful contrasts, and there is no point in creating extra contrasts which have no meaning, merely to make the degrees of freedom and corrected sums of squares add up.

We return to Example 12.3 for a further illustration. We might expect a gradual rise (or fall) in the scores as the test proceeds. In such circumstances we would be interested in trying linear regression on the order of presentation. The regression coefficient is, however, a linear combination of the observations, and if the method of this section is used we obtain the sum of squares due to linear regression as a single degree of freedom. The group totals, with a suitable scale for the order of presentation, are as follows:

Order	1	2	3	4	5	6
Scale	−5	−3	−1	1	3	5
Total over groups	662	683	636	628	658	633

The linear contrast which is proportional to the regression coefficient is

$$w = -5 \times 622 - 3 \times 683 - \ldots +5 \times 633 = 228.$$

When (12.16) is applied we have

$$\text{CSS} = (228)^2/\{4 \times (5^2 + 3^2 + 1^2 + 1^2 + 3^2 + 5^2) = 185 \cdot 66,$$

and this is the corrected sum of squares due to regression of the position totals on the position number. We present the results as in Table 12.11.

Table 12.11 *Taking out linear regression CSS*

Source	CSS	d.f.	M.S.
Due to regression	185·7	1	185·7
Deviations	380·8	4	95·2
Between positions	566·5	5	

The deviations line here is filled in by subtraction, although we could create extra linear contrasts to account for it, if we wished. Both sums of squares are tested against the mean square for residual in Table 12.7 by the F-test. The sum of squares due to regression is clearly not significant; in any case it is obvious from the position totals that there is no evidence of linear regression. The mean square for deviations is less than the mean square for residual, and so is not significant either.

We have now completed a number of analyses on Example 12.3 and we conclude this section with a report of the conclusions on this experiment, as an example of the kind of statement which should *always* be given with any analysis.

Conclusion for Example 12.3

In order to test the assumptions underlying the analysis of variance of this data, values were fitted to each cell by estimating position and group effects, assuming an additive model. A plot of residuals against normal scores (Exercise 12.4.1) gave a remarkably straight line, so that there is good agreement with the normal distribution. This is what we would expect from the central limit theorem (see § 2.6(iv)), since each entry is a sum of thirty scores. A plot of residuals against fitted values (Exercise 12.4.2) showed no evidence of heterogeneity of the variance.

The analysis of variance table shows that there is no evidence of any effect due to the position of a test in the sequence of six, the F-ratio for this being only just over one, on (5, 15) degrees of freedom. A further analysis by linear regression on position also showed no effect.

The sum of squares between groups was very highly significant (well beyond the 0·1% level). Over 60% of the variation between groups can be attributed to sex differences, and about a further 30% to age differences. The group means are as follows:

	13 *yrs*	14 *yrs*	*Difference*
Girls	131·0	163·2	32·2
Boys	173·3	182·5	9·2
Difference	42·3	19·3	

The estimated standard error of a group mean is 4·04 and the estimated standard error of a difference of two means is 5·72. Boys achieve significantly higher scores than girls of the same age and there is a consistent improvement with age. However, girls greatly improve their scores between 13 years and 14 years, the difference being 32·2, whereas the improvement of boys between the same ages is only 9·2, which is not significant, not quite reaching the 10% level. The difference between the age improvements of the sexes, $32·2 - 9·2 = 23·0$, is called the sex–age interaction; this has a standard error of 8·1 and is significant at close to the 1% level. Thus although girls' performance is lower than boys', the data give very strong evidence that girls improve their performance much more between the ages of 13 and 14 than boys do. This suggests the need for data on responses by both sexes over a greater age range for a fuller understanding of the situation.

This conclusion relates, of course, to the part of the results we have considered. For full details of the experiment, its design and conclusions, see the original paper.

Exercises 12.5
1. Define some meaningful contrasts for Exercise 12.3.2, and then calculate and test these contrasts for significance.

12.6 Randomized blocks

Although we stated at the close of § 1.1 that some important implications for experimental design arise out of statistical theory, we have made very little explicit reference to this. The presentation of two-way analysis of variance in § 12.4 enables us to introduce here an important type of experimental design called a *randomized blocks design*. We begin by returning to the paired comparisons experiment on chlorination of sewage, Example 6.6.

The main source of uncontrolled variation in Example 6.6, other than sampling and measurement errors in the determination of the coliform density of each batch, arose from variations between the coliform density of the selected batches of sewage. By selecting batches of sewage in pairs, each pair being selected close together in time, pairs of batches were obtained which were as alike as possible, but such that there was considerable variation between the pairs. The two treatments A and B were then randomly assigned to the two batches in each pair; the function of randomization was to guard against unconscious selection biases, and any systematic differences between the first and second batches in the pairs. The treatments A and B were then compared using the differences within each pair; these differences are not affected by differences between the averages of the pairs, however large these might be. (For example, the addition of an arbitrary amount to both results for any one pair does not alter the differences between A and B for this or any other pair.) The effect of the paired comparisons design is therefore that it should eliminate some of the uncontrolled variation from the experimental comparisons.

If more than two treatments are to be compared, the paired comparisons design can be extended in an obvious way. For a comparison between, say, five treatments, we select our experimental units in such a way that we have groups of five units, called *blocks*, such that the units within each block are as alike as practical considerations permit. The five treatments must then be randomly assigned to the five units within each block. Consider the following example.

Example 12.5. Cochran and Cox (1957, § 4.23) report an experiment in which there were five different levels of application of potash to a cotton crop. The five levels of potash used were 36, 54, 72, 108 and 144 lb of potash per acre, but for our present purpose we may simply think of them as five different treatments, and label them T_1, T_2, . . . , T_5. One of the purposes of the experiment was to examine the effect of the treatments on the fibre strength of the cotton, and a procedure for obtaining a measure of fibre strength in arbitrary units, was laid down.   □ □ □

The experiment of Example 12.5 was discussed by Cox (1958, pp. 26–30), and he notes that there are several sources of uncontrolled variation in the observations, including variation due to the fertility etc. of the land on which the crop was grown. However, the effect on comparisons between treatments of variation due to the fertility of the land can be reduced by use of the randomized block design. Blocks of land are chosen which comprise five plots each and for which the uncontrolled variation from plot to plot within blocks is approximately minimized; this is usually achieved by arranging the plots within each block so that the blocks are roughly square. The five treatments are then randomly assigned to the five plots in each block.

For example, suppose that in a field there is an expected or known fertility difference in one direction across a field, and that strips of land in a direction at right angles with this are expected to be of homogeneous fertility. These strips can be used as blocks, and they are divided into five plots each, and the treatments randomized within each strip. In Figure 12.1 this is illustrated

Fertility gradient	Block 1	T_5 7·46	T_4 7·17	T_2 8·14	T_1 7·62	T_3 7·76
	Block 2	T_3 7·73	T_1 8·00	T_2 8·15	T_5 7·68	T_4 7·57
	Block 3	T_1 7·93	T_4 7·80	T_3 7·74	T_5 7·21	T_2 7·87

Figure 12.1 *A possible experimental arrangement of Example* 12.5

for Example 12.5. Each block contains $T_1, \ldots, T_5$ once, and the order has been randomized for each block. [The actual experimental arrangement used in Example 12.5 is not described by Cochran and Cox, and Figure 12.1 is obtained by randomizing their results; see Cox (1958, p. 28).] Thus in Figure 12.1, each treatment occurs the same number of times in each block (once), and comparisons between treatments are not affected by differences between blocks, however large these might be.

In general, if we are to compare k treatments by this method, we must find blocks of k plots such that each block is as homogeneous as possible with respect to the sources of uncontrolled variation; any source of uncontrolled variation which is identified with blocks is eliminated from the experimental comparisons. By randomizing the allocation of treatments to plots within each block, we guard against biases from any other sources of uncontrolled variation which arise out of the allocation of the treatments to the plots.

The standard analysis of a randomized block experiment is straightforward. The blocks are written out as, say, rows, and the treatments ordered and put as columns in a two-way analysis of variance; the analysis then follows the pattern described in § 12.4. For Example 12.5 the rearranged results are shown in Table 12.12. The analysis of variance table is shown in

Table 12.12 *Results of Example* 12.5

Block	T_1	T_2	T_3	T_4	T_5
1	7·62	8·14	7·76	7·17	7·46
2	8·00	8·15	7·73	7·57	7·68
3	7·93	7·87	7·74	7·80	7·21

Table 12.13 and this should be compared with Table 12.6. Differences between treatment means can be analysed by any of the methods described

Table 12.13 ANOVA *table for Example* 12.5

Source	d.f.
Block (rows)	2
Treatments (cols)	4
Residual	8
Total	14

above. There will usually be little interest in testing the blocks sum of squares for significance, or analysing differences between block means, except as a guide to the design of further experiments of the same type. The blocks were formed merely to exclude some of the uncontrolled variation from the treatment comparisons.

The value of the randomized block design can be shown more clearly by comparing it with an alternative, the *completely randomized design*. Suppose we have the field layout into 15 plots as in Figure 12.1, then a completely randomized design would be obtained by randomizing the allocation of T_1, $T_2, \ldots, T_5$, to the 15 plots so that each treatment occurred three times. One possible layout is shown in Figure 12.2. To analyse the completely random-

T_1	T_4	T_3	T_2	T_4
T_3	T_5	T_5	T_2	T_1
T_5	T_3	T_4	T_2	T_1

Figure 12.2 *A completely randomized design for Example* 12.5

ized design, we merely group the plot results by treatments, and we then have a one-way analysis of variance, as in Table 12.14.

By comparing Table 12.13 and Table 12.14, we see that in the randomized block design two degrees of freedom for blocks have been separated from the

Table 12.14 ANOVA *table for Figure* 12.2

Source	d.f.
Treatments	4
Residual	10
Total	14

residual. If these two degrees of freedom represent relatively large sources of variation, the randomized block design yields a more precise comparison of treatment means than the completely randomized design. On the other hand, if in Table 12.13 the blocks mean square is equal to (or less than) the residual sum of squares, all we have done is lose two degrees of freedom in the residual line, and this in effect reduces the precision of the treatment comparisons.

We shall not continue this discussion. For further reading on experimental design, Cox (1958) gives an excellent yet simple account, with references.

Exercises 12.6

1. Analyse the results given in Figure 12.1 for Example 12.5. Plot the treatment means against the amount of potash per acre, and write a report of the conclusions of your analysis.

2. (Cambridge Diploma, 1954, modified.) A field plan is given below of an agricultural experiment on oats, arranged in three randomized blocks of six equal plots each. The two-figure code denotes, 0, 1 or 2 units of nitrogenous fertilizer applied, and the second digit 0 or 1 unit of a phosphatic fertilizer. The other figure in the square is the yield of grain, calculated as cwt per acre.

Block A		Block B		Block C	
11	20	10	21	20	00
23·3	21·9	15·3	19·5	14·8	11·1
01	00	20	00	21	10
17·4	20·0	16·5	12·8	16·3	13·3
21	10	11	01	11	01
21·5	15·7	18·4	16·4	15·5	16·9

Carry out an analysis of this data by treating the blocks as rows and the treatments 00, 01, 10, 11, 20, 21 as columns in a 3 × 6 two-way analysis of variance. Separate single degrees of freedom for the column effects as follows, and give a physical meaning to each linear contrast.

Contrast	Treatment					
	00	01	10	11	20	21
1	−1	1	−1	1	−1	1
2	−1	−1	0	0	1	1
3	−1	−1	2	2	−1	−1
4	1	−1	0	0	−1	1
5	1	−1	−2	2	1	−1

Prepare summary tables to present the most important features of the data, and give a report on your conclusions.

12.7 Components of variance

Up to now we have been considering the analysis of variance in what we call the 'fixed effects' or 'model 1' situation. There is another type of analysis of variance, which we call the 'components of variance' or 'model 2' type; in components of variance although some of the numerical calculations are formally the same as for fixed effects, the questions being asked and the inferences to be made are quite different. In both Examples 12.1 and 12.3 the underlying questions being asked were comparisons between a set of means, these representing some particular chosen treatments. Example 1.10 data – see also Example 3.16 – is superficially very similar to Example 12.1, but in Example 1.10 the questions being asked of the data are quite different. In Example 1.10 we know that the average clean content of the bales varies from bale to bale, and also that the clean content varies within any given bale. Let us assume that we have the following model.

Let the average clean content of a bale be m, and suppose that considering the distribution of m among bales, we have a normal distribution with mean μ and variance σ_b^2. Further, let us assume that within each bale the average clean content varies (normally) about its overall average value m, with a variance σ^2. Thus the jth observation on the ith bale, x_{ij}, has the following structure,

$$x_{ij} = m_i + \varepsilon_{ij}, \tag{12.17}$$

where the m_i are distributed independently $N(\mu, \sigma_b^2)$, and the ε_{ij} are distributed independently $N(0, \sigma^2)$. With these assumptions, our problem is to estimate the three unknown parameters μ, σ^2 and σ_b^2 which describe the set-up. That is, whereas in Examples 12.1 and 12.2 our problem was to estimate and test differences between certain fixed effects, our main problem in Example 1.10 is to estimate the components of variance, σ_b^2 and σ^2.

Before proceeding we notice that we have made four important assumptions.

(a) That the within bale variation is homogeneous for all bales.

(b) That the particular bales selected are a random sample from an infinite population.

(c) That the observations are independent.

(d) That the population distributions are normal.

Assumptions (a), (c) and (d) are common to analysis of variance. Assumption (b) can be relaxed if necessary, although it leads to a more complicated analysis. In Example 1.10, assumption (b) can be regarded as a reasonable approximation.

The first phase of the analysis follows that for Example 12.1 to the analysis of variance table. Suppose we have g groups and n observations per group, then we have Table 12.15.

Table 12.15 *Analysis of variance for one-way random effects model*

Source	CSS	d.f.	M.S.	E(M.S.)
Between groups	$n \sum_{i=1}^{g} (\bar{x}_{i.} - \bar{x}_{..})^2$	$g - 1$	s_b^2	$\sigma^2 + n\sigma_b^2$
Within groups	$\sum_{j=1}^{n} \sum_{i=1}^{g} (x_{ij} - \bar{x}_{i.})^2$	$g(n - 1)$	s_w^2	σ^2
Total	$\sum_{j=1}^{n} \sum_{i=1}^{g} (x_{ij} - \bar{x}_{..})^2$	$gn - 1$		

Let us assume the set-up (12.17), then we have two sampling operations involved, see § 3.5, and we can take expectations of the mean squares over each population. Following the theory given in § 12.1 we have

$$E_\varepsilon(s_w^2) = \sigma^2$$

$$E_\varepsilon(s_b^2) = \sigma^2 + \frac{n}{g - 1} \sum (m_i - m_.)^2$$

where E_ε denote expectation over the within groups variation. By taking expectation of s_b^2 over the between group variation, we have

$$E_m E_\varepsilon(s_b^2) = \sigma^2 + n\sigma_b^2$$

Therefore the mean square between groups s_b^2 contains contributions from both the variation of the population of bale means, σ_b^2, and from the variation within bales about the bale means, σ^2. Whereas we can estimate σ^2 by the mean square within groups, s_w^2, an unbiased estimate of σ_b^2 is provided by

$$\tilde{\sigma}_b^2 = \left(\frac{s_b^2 - s_w^2}{n} \right). \tag{12.18}$$

The estimate of μ is $\bar{x}_{..}$, and the variance of this estimate is

$$E_m E_\varepsilon(\bar{x}_{..} - \mu)^2.$$

to obtain this we write

$$\bar{x}.. - \mu = (\bar{x}.. - m.) + (m. - \mu),$$

then by squaring and taking expectations we have

$$E_m E_e(\bar{x}.. - \mu)^2 = \frac{\sigma^2}{ng} + \frac{\sigma_b^2}{g} = \frac{1}{ng}(\sigma^2 + n\sigma_b^2). \quad (12.19)$$

Therefore an estimate of this variance is given by s_b^2/ng, on $(g-1)$ degrees of freedom, and confidence intervals for μ can be obtained using this estimate.

The usual F-ratio test of s_b^2/s_w^2 on $\{(g-1), g(n-1)\}$ degrees of freedom will test null hypothesis that $\sigma_b^2 = 0$, but much greater interest usually centres on point and interval estimation than on significance tests. Confidence intervals for σ^2 can be obtained using the χ^2-distribution, as explained in § 5.4. Unfortunately the estimate $\tilde{\sigma}_b^2$ is not distributed as χ^2, and confidence intervals for σ_b^2 cannot be obtained exactly. However, the distribution of s_b^2 is χ^2, and following the theory of § 6.3 we can obtain confidence intervals for the ratio $(\sigma^2 + n\sigma_b^2)/\sigma^2 = 1 + n\sigma_b^2/\sigma^2$. Using this, we have confidence intervals for the ratio σ_b^2/σ^2.

If we put $s_1^2 = s_w^2$, $s_2^2 = s_b^2$, then (6.10) gives a confidence interval for $(1 + n\sigma_b^2/\sigma^2)$ as

$$\frac{s_b^2}{s_w^2 F_{1-\alpha/2}\{g-1, g(n-1)\}}, \quad \frac{s_b^2}{s_w^2}F_{1-\alpha/2}\{g(n-1), g-1\}. \quad (12.20)$$

We then subtract unity and divide by n to get confidence intervals for the ratio σ_b^2/σ^2.

Example 12.6. *Analysis of Example* 1.10. The first phase of the analysis to the analysis of variance table follows the outline given in § 12.3. Subsequently, instead of analysing differences between means, we obtain point and interval estimates for σ^2, σ_b^2 and μ.

Table 12.16 *Analysis of variance of Example* 1.10 *data*

Source	CSS	d.f.	M.S.	F
Between bales	65·9628	6	10·99	1·76
Within bales	131·4726	21	6·26	
Total	197·4354	27		

The 5% point for $F(6, 21)$ is 2·57, so that the hypothesis $\sigma_b^2 = 0$ cannot be rejected. Point and interval estimates for the components of variance are as follows.

$$\tilde{\sigma}^2 = 6·26 \quad \tilde{\sigma}_b^2 = (10·99 - 6·26)/4 = 1·18.$$

The $2\frac{1}{2}\%$ and $97\frac{1}{2}\%$ points of χ^2_{21} are $10\cdot28$ and $35\cdot48$, so that 95% confidence intervals for σ^2 are, from (5.15),

$$\left\{\frac{131\cdot4726}{35\cdot48}, \frac{131\cdot4726}{10\cdot28}\right\} = (3\cdot70, 12\cdot76).$$

If we insert into formula (12.20) $s_b^2 = 10\cdot99$, $s_w^2 = 6\cdot26$, $F_{97\frac{1}{2}\%}(6, 21) = 3\cdot09$, $F_{97\frac{1}{2}\%}(21, 6) = 5\cdot18$, we have $(0\cdot568, 8\cdot111)$. We now subtract unity and divide by four to get $(-0\cdot11, 2\cdot03)$ which is a 95% confidence interval for the ratio σ_b^2/σ^2.

The negative value for the lower bound in the 95% confidence interval for σ_b^2/σ^2 needs some explanation. It simply reflects the fact that the F-ratio test for $\sigma_b^2 = 0$ was not significant, so that the data are compatible with this being true. Clearly we cannot have $\sigma_b^2 < 0$, but it is disputed whether more information is given by quoting the negative value or not. [For example suppose we had arrived at 95% C.I. for the ratio σ_b^2/σ^2 to be $-2\cdot50$, $+2\cdot02$.] In many practical problems this situation will not arise.   □ □ □

The need for estimates of components of variance arises mostly in two situations. One case is when the estimates are needed in order to design an efficient sampling plan, or in other ways are required to guide future planning; Example 12.7 illustrates this with reference to Example 1.10. Another case is when the estimates are required primarily to explain an existing set of data. Example 12.8 illustrates this, although some elements of future planning are also involved in it.

Example 12.7. The data given in Example 1.10 were only a pilot sample, and a much larger sample was taken subsequently in order to obtain a more precise estimate of μ. Suppose a given amount of money (or time) is available for this larger sample, then we can spend this by taking a large number of cores from a few bales, or vice versa, etc. We can use the estimates of the components of variance from the pilot sample to make an efficient design for the larger sample.

Our estimate of μ is $\bar{x}_{..}$, and from (12.19) we have for G groups with N observations in each,

$$V(\bar{x}_{..}) = (\sigma^2 + N\sigma_b^2)/NG. \tag{12.21}$$

Suppose that it costs l units to sample a bale, and one unit to sample a core from a bale, and the final sample is to cost no more than k units, then we require

$$Gl + NG = k. \tag{12.22}$$

For this given cost, N and G can be chosen to minimize (12.21). Using (12.22) we have

$$V(\bar{x}_{..}) = \frac{\sigma^2}{k - lG} + \frac{\sigma_b^2}{G}$$

$$\frac{dV}{dG} = +\frac{l\sigma^2}{(k - lG)^2} - \frac{\sigma_b^2}{G^2}.$$

The optimum choice for G and N are therefore

$$G = \frac{k\sqrt{(\rho/l)}}{1 + \sqrt{(\rho l)}}, \quad N = \sqrt{(l/\rho)},$$

where $\rho = \sigma_b^2/\sigma^2$. By using the value of ρ estimated from the pilot sample, values of G and N can be estimated so as to minimize approximately the variance of $\hat{\mu}$, subject to (12.22). ▢▢▢

A detailed numerical illustration of this application of components of variance is given by Cameron (1951).

Example 12.8. Desmond (1954) describes in detail some problems arising in the operation of a quality control scheme on car voltage regulators. The regulators were required to operate within the range 15·8 to 16·4 volts, and the final production operation consisted of setting the regulators to work within this range. The regulators were then submitted to a short run under load, and inspected at one of a number of testing stations to confirm that the setting had been carried out correctly. Any defectives were fed back into the production line for setting in the usual way and Desmond says, 'Records show that about eighteen per cent of all regulators required this rectification under good conditions, and this figure sometimes rose above fifty per cent.'

It is clear that there are two sources of variation in final measurements of operation voltage. There is the natural variation between regulators coming off the production line, and there is a measurement error. As a first stage in dealing with the problem, some data were collected to estimate these components of variance. The data given below is an extract of the data given by Desmond. Eight regulators were chosen randomly from the production going through one setting station, and passed through all four testing stations. The results are given in Table 12.17.

Table 12.17 *Extract of results reported by Desmond*

| Regulator no. | Testing station | | | |
	1	2	3	4
1	16·5	16·5	16·6	16·6
2	15·8	16·7	16·2	16·3
3	16·2	16·5	15·8	16·1
4	16·3	16·5	16·3	16·6
5	16·2	16·1	16·3	16·5
6	16·9	17·0	17·0	17·0
7	16·0	16·2	16·0	16·0
8	16·0	16·0	16·1	16·0

It is reasonable to assume as a model that the measurement x_{ij} on regulator i by testing station j has the structure

$$x_{ij} = m_i + \gamma_j + \varepsilon_{ij},$$

where $\sum \gamma_j = 0$, m_i is $N(\mu, \sigma_b^2)$, and ε_{ij} is $N(0, \sigma^2)$. Thus regulators coming for setting are assumed to have true measurements which are distributed normally about a mean μ with variance σ_b^2, and the error of measurement is assumed to be normal with variance σ^2. Estimates of σ^2 and σ_b^2 are obtained by a components of variance analysis of variance, and this is left as an exercise for the reader; see Exercise 12.7.1.

This use of components of variance is primarily to explain an existing situation, so that appropriate action can be taken. Further details of this experiment are given by Desmond (1954). □ □ □

Exercises 12.7
1. Carry out an analysis of the data of Example 12.8 estimating confidence intervals etc. where appropriate. Incorporate checks of the model using residuals. [*Note*. This is a two-way analysis of variance, in which the testing stations are considered fixed effects and the regulators random effects.]

12.8 Departures from assumptions
The discussion here of the effect of departures from assumption on inferences made in the analysis of variance will be very brief, as there are some excellent accounts available; see Chapter 10 of Scheffé (1959), and Eisenhart (1947), and references.

The assumptions made are as follows:

Additivity. The observations are assumed to follow a model such as

$$x_{ij} = \mu + \rho_i + \gamma_j + \varepsilon_{ij}$$

for a two-way analysis of variance, see (12.11), where the ε_{ij} are errors having some distribution with expectation zero:

Homogeneity of variance. It is assumed that $V(\varepsilon_{ij}) = \sigma^2$ for all observations. One example where this assumption is obviously false is the Rockwell hardness measurements experiment, *Example* 3.15; in that example Table 3.2 shows clearly that the variances of observations differ.

Independence. The errors ε_{ij} are assumed to be independent (see Eisenhart (1947) on this point). It is clear from § 5.7 that lack of independence can have serious consequences. Sometimes it is possible to design experiments so that independence can be safely assumed for physical reasons.

Normality. It is necessary to assume the ε_{ij} are normally distributed to justify the common tests of significance etc. The expectation properties of the analysis of variance table hold without this assumption, and a fixed effects analysis is not critically dependent on normality. (It is clear from § 5.7 that for a random effects model, which therefore concerns inferences about variances, normality of the between group random variables will be quite important; see Atiqullah, 1963.)

All of these assumptions can be checked by an appropriate analysis of residuals. A method of separating one degree of freedom for non-additivity is given by Tukey (1949b), and an excellent method of testing for normality is given by Shapiro and Wilk (1965).

It is not easy to state briefly the effects of various departures from these assumptions, but Chapter 10 of Scheffé contains some simple tables which enable the reader to form a judgment about this. In general, independence and additivity are the most important.

Transformations. If for a particular set of data the assumptions fail through non-additivity, heterogeneity of the variance, or non-normality, it may help to transform the observations. Consider the following example.

Example 12.9. (London, B.Sc. Gen., 1964.) An experiment was carried out to examine the effectiveness of three new preparations for curing a minor skin complaint. This complaint, which only occurs during cold weather, sometimes clears up spontaneously and the following experimental design was therefore used.

In each of four towns 160 patients suffering from the complaint during a cold spell were divided into 4 groups of 40 each. The patients in one group received preparation A, those in the second and third received preparations B and C respectively and those in the fourth group received preparation D, which was thought to have no curative properties. In each group the number of patients in whom the complaint disappeared during the experimental period of one week was noted. Thus in town 1, 23 out of the 40 on preparation A were cured, 17 out of the 40 on B were cured, and so on.

Table showing the number cured out of each group of 40

Town	Preparation			
	A	B	C	D
1	23	17	23	26
2	22	16	30	17
3	16	24	21	21
4	31	25	24	24

Analyse these data and test for differences between preparations, for differences between towns and for interaction between towns and preparations.

□ □ □

In Example 12.9, each of the sixteen observations is the number of patients out of 40 in whom the complaint disappeared. We therefore expect them to have binomial distributions with $n = 40$, and different probabilities θ_{ij} for each cell, where i denotes the row and j the column. The expectation and variance of observations r_{ij} for each cell are therefore $E(r_{ij}) = 40.\theta_{ij}$, and $V(r_{ij}) = 40.\theta_{ij}(1 - \theta_{ij})$ respectively, and if the expectations of the cells differ, so do the variances. In this example therefore we have heterogeneous variances and a non-normal distribution. Now the binomial probabilities in Example 12.9 are not too extreme, so that the binomial distributions will not be very skew and lack of normality is probably unimportant here. Also, some of the probabilities in Example 12.9 appear to be near 0·5 and the most extreme proportion observed is $31/40$, and the variance for $\theta = 0·5$ is $40 \times 0·5 \times 0·5 = 10$, and for $\theta = 31/40$ is $40 \times 31/40 \times 9/40 \simeq 7$. Therefore the variances for the observations do not differ greatly and it is probably safe to analyse the data of Example 12.9 as it stands.

If the variances of the cells in Example 12.9 differed more, we should use the transformation given by (8.2), which would render all variances approximately equal at about $1/160$, and also would make the distribution more nearly normal. So far in considering Example 12.9 we have side-stepped the additivity assumption. If we analyse the data as it stands, we are assuming the probabilities have the structure

$$E(\theta_{ij}) = \mu + \rho_i + \gamma_j$$

which could lead to probabilities outside $(0, 1)$, but is reasonable if the θ_{ij}'s have a small enough range. If we use the arcsine transformation (8.2), we are assuming the probabilities to be additive in a very peculiar scale. We shall not continue the discussion of this example but we notice that we are trying to find a single transformation which will simultaneously satisfy approximately the assumptions of normality, homogeneity of variance and additivity, and in general this cannot be done. Despite the difficulties over additivity, the arcsine transformation is frequently employed when the variances of observations are very heterogeneous. Somewhat similar remarks apply to the use of the square root transformation for Poisson variables; see § 8.2.

As a further example on the use of transformations consider Example 3.15, § 3.5, where we discussed the experiment concerning Rockwell hardness measurements. The second part of Table 3.2 records nine sample variances for this experiment, each being estimated on seven degrees of freedom. An analysis of these sample variances is appropriate to study how the variability of hardness measurements is related to surface preparation, but the nine sample variances cannot be analysed as they stand because their own variances are not homogeneous; see Exercise 5.2.3. However, Exercise 6.3.5 shows that this difficulty is avoided by taking (natural) logarithms, and the variances of the transformed values are equal at $2/7$. The logarithm transformation also makes the distribution of s^2 more nearly normal, and further, it seems reasonable to assume an additive model in log σ^2, for this corresponds to multiplicative effects on the variances. This example, therefore, illustrates the situation in which one transformation does justify the three assumptions of additivity, homogeneity of variance, and normality.

A more recent contribution to the theory of transformations is Box and Cox (1964). These authors give a constructive method of deriving a suitable transformation, but a discussion of the method is beyond the scope of this volume.

Exercises 12.8

1. Carry out an analysis of Example 12.9.

2. Analyse the nine sample variances given in Table 3.2, for Example 3.15, as described above. Write down explicitly the model you are assuming in carrying out the analysis. Notice that the mean square for residual in the analysis (based on four degrees of freedom) has an expectation of 2/7 from Exercise 6.3.3, and the χ^2-test of § 5.6 can be used on this mean square. A significant result on this last test indicates either that wild observations are present, or that the distribution assumptions do not hold.

3.* Suppose that in data consisting of a large number of groups each of the same number of observations, a plot of log (group range) against log (group mean) is roughly linear with slope γ. What transformation would you suggest to stabilize the variance? [Hint: use the result in Exercise 8.2.5 in order to find the $f(x)$ giving approximately constant variance.]

Miscellaneous Exercises

1. Discuss reasons for the importance of the Normal (Gausian) distribution in statistics.

The calorific value of a stream of gas varies in a normal frequency distribution, having a mean of 150 units (Btu/cu ft) and a standard deviation of 30 units. If several streams are mixed, the instantaneous calorific value is the average of the calorific values for the separate streams, which may be assumed to fluctuate independently, all with the same mean and standard deviation. How many streams should be mixed to ensure that the calorific value is below 110 units for only 0·5% of the time?

(London, B.Sc. Gen., 1958)

2. An experiment is designed to assess the effect of washing on the breaking extension of cotton yarn. Six 18-in lengths are selected and each cut into two halves. For each pair, one half, selected at random, is washed, (W), then tested and the other half, serving as a control, (C), is tested without washing. The results, recorded in ounces were:

	Length 1	2	3	4	5	6
1st half	12·1(W)	16·2(C)	13·9(C)	12·9(W)	15·8(C)	13·1(C)
2nd half	13·2(C)	16·5(W)	13·6(W)	13·4(C)	15·8(W)	12·9(W)

Comment on the experimental design used. Obtain 95% confidence limits for the mean difference between washed and control lengths.

(London, B.Sc. Gen., 1958)

3. The quantities $x_{ij}(i = 1, 2, \ldots, r; \ j = 1, 2, \ldots, c)$ have independent normal distributions with common variance σ^2 (unknown), and expectations

$$E(x_{ij}) = \mu + \rho_i + \gamma_j,$$

where

$$\sum_{i=1}^{r} \rho_i = \sum_{j=1}^{c} \gamma_j = 0.$$

298

The residuals r_{ij} are defined by

$$r_{ij} = x_{ij} - \bar{x}_{i.} - \bar{x}_{.j} + \bar{x}_{..},$$

where

$$\bar{x}_{i.} = \frac{1}{c}\sum_{j=1}^{c} x_{ij}, \quad \bar{x}_{.j} = \frac{1}{r}\sum_{i=1}^{r} x_{ij}, \quad \text{and} \quad \bar{x}_{..} = \frac{1}{rc}\sum_{i=1}^{r}\sum_{j=1}^{c} x_{ij}.$$

Show that the r_{ij} have zero expectation, and that their variances and covariances are given by

$$V(r_{ij}) = \frac{(r-1)(c-1)}{rc}\sigma^2$$

$$C(r_{ij}, r_{ik}) = -\left(\frac{r-1}{rc}\right)\sigma^2 \quad \text{for } j \neq k,$$

$$C(r_{ij}, r_{hj,}) = -\left(\frac{c-1}{rc}\right)\sigma^2 \quad \text{for } i \neq h,$$

$$C(r_{ij}, r_{hk}) = +\frac{1}{rc}\sigma^2 \qquad \text{for } i \neq h, j \neq k,$$

Show also that

$$E\left(\sum r_{ij}^2\right) = (r-1)(c-1)\sigma^2.$$

Discuss briefly the relevance of these results to a two-way (unreplicated) analysis of variance, and say how the residuals could be used to check the assumptions usually made.

(You may assume any standard results about the joint distribution of linear combinations of random variables.)   (*London, B.Sc. Gen., 1965*)

4. (*a*) Explain carefully what is meant by a significance test. (You may discuss a particular test of significance, if you wish, instead of discussing the subject in general.) Discuss the role of significance tests in the interpretation of data.

(*b*) Suggest a test of significance for each of the following situations, and state clearly how your tests are carried out.

(*i*) Random variables $x_1, \ldots, x_n$ are known to have a Poisson distribution with mean μ. It is required to discriminate between the hypotheses $\mu = 1$ and $\mu = 2$.

(*ii*) In the routine analysis of a certain chemical, many independent determinations of the percentage of a given compound are made. From past results it is known that these independent determinations are normally distributed with a (known) variance σ_0^2. A new member of staff, Mr A., makes ten independent determinations, $x_1, \ldots, x_{10}$, on the chemical. Test the hypothesis that the variance σ^2 of the determinations made by Mr A. is equal to σ_0^2 against the alternative that $\sigma^2 > \sigma_0^2$.

(*London, B.Sc. Gen., 1966*)

5. In order to estimate two parameters θ and ϕ it is possible to make observations of three types:

(*i*)　the first type have expectation $\theta + \phi$,

(*ii*)　the second type have expectation $2\theta + \phi$,

(*iii*)　thè third type have expectation $\theta + 2\phi$.

All observations are subject to uncorrelated errors of mean zero and constant variance σ^2.

If n observations of type 1, m observations of type 2 and m observations of type 3 are made, obtain the least-square estimates $\hat{\theta}$ and $\hat{\phi}$ and prove that

$$\text{var}(\hat{\theta}) = \text{var}(\hat{\phi}) = \frac{n + 5m}{m(2n + 9m)}\sigma^2.$$

(London, B.Sc. Gen., 1958)

6. In a certain sampling problem R batches are selected randomly from a large number, and C items are drawn randomly from each of the batches selected. (You may assume that the number of items in each batch is large compared with C.)

Define components of variance, σ_b^2 and σ_w^2 between and within batches. Set up the analysis of variance table for the data, derive the expectations of the mean squares, and show how estimates of σ_b^2 and σ_w^2 can be obtained.

Also show how to calculate confidence intervals for the ratio σ_b^2/σ_w^2.

(London, B.Sc. Gen., 1966)

7. From each of two fluids, twenty samples of equal volume were taken, and chemical analyses were carried out on each sample. The analysis on ten of the samples from each fluid (chosen randomly from the twenty) estimated the amount present of a certain chemical A: the analysis on the remaining ten samples from each fluid estimated the amount present of a chemical B.

The results, expressed in milligrams, are given below:

Fluid 1		Fluid 2	
Chemical A	Chemical B	Chemical A	Chemical B
10·0	7·4	6·8	10·9
10·3	7·3	7·4	8·5
11·0	11·9	7·1	11·2
10·3	13·4	6·7	10·6
11·9	12·1	7·2	10·1
8·4	11·4	5·4	8·2
10·9	10·0	7·0	9·1
8·3	10·2	6·5	9·0
10·5	12·2	6·5	11·0
11·3	8·8	7·8	9·1

Examine these results, investigating differences in mean and in variability, and stating any assumptions you make.

Write a short summary of your conclusions.   (*London, Anc. Stats., 1960*)

8. Discuss briefly the use of non-parametric methods in the analysis of data, with particular reference to your own field of study.

In a study of the deterioration of intelligence in schizophrenia, Haywood and Moelis gave intelligence (*IQ*) tests to male schizophrenic patients in an American mental hospital. The patients who took part in the study had all been in hospital for a considerable period and each patient was clinically diagnosed as being improved or unimproved since admission to the hospital. All patients had been given *IQ* tests on admission. Twenty improved patients were selected and each was matched as far as possible with respect to age, *IQ* on admission and severity of attack on admission, with an unimproved patient.

The data obtained are given in the following table. Use a non-parametric test such as a sign test to examine whether any deterioration in *IQ* between admission and the time of the study is greater for unimproved than for improved patients.

IQ scores of 20 matched pairs of improved and unimproved schizophrenic patients

Improved patients		Pair no.	Unimproved patients	
Admission IQ	*Present IQ*		*Admission IQ*	*Present IQ*
107	97	1	106	90
104	109	2	105	105
110	118	3	102	94
98	98	4	99	92
94	104	5	95	78
102	104	6	93	93
92	68	7	94	91
93	120	8	91	95
87	98	9	78	80
80	75	10	83	81
72	80	11	71	69
72	76	12	80	77
122	139	13	104	97
61	70	14	63	66
67	91	15	83	76
111	117	16	110	116
110	124	17	112	76
89	94	18	91	138
106	116	19	103	102
92	127	20	96	90

(*London, Anc. Stats., 1964*)

9. It has been suggested that the drug coramine is useful in helping newborn babies to breathe better. To test this point 97 babies were divided randomly into two groups, 50 being used as controls and given no treatment, and 47 being given the drug. The minute-volume x, i.e. the amount of air breathed in cubic cm per minute, was recorded for each baby at (i) 3–4 hours and (ii) 6 days after birth, the drug being supplied to the treated babies after the first measurement was made. The data (after Please and Roberts) are given below. Investigate whether the drug does seem to improve breathing (i.e. increase the minute-volume). You may assume that log (minute-volume) is normally distributed.

Controls

Baby no.	(i) x	(i) $\log\left(\dfrac{x}{100}\right)$	(ii) $\log\left(\dfrac{x}{100}\right)$	Baby no.	(i) x	(i) $\log\left(\dfrac{x}{100}\right)$	(ii) $\log\left(\dfrac{x}{100}\right)$
1	600	·778	1·155	26	693	·841	1·062
2	1000	1·000	·928	27	866	·938	·721
3	547	·738	·976	28	600	·778	·954
4	693	·841	·851	29	436	·639	1·044
5	856	·932	·875	30	500	·699	·846
6	1451	1·162	·754	31	866	·938	·990
7	528	·723	·954	32	654	·816	·841
8	945	·975	·868	33	725	·810	·892
9	780	·892	·854	34	824	·916	1·063
10	517	·713	·843	35	452	·655	·946
11	340	·531	·822	36	356	·551	·761
12	750	·875	1·188	37	843	·926	·813
13	1111	1·046	·746	38	701	·846	1·046
14	622	·794	1·113	39	857	·933	1·113
15	1177	1·071	·716	40	698	·844	933
16	567	·754	·956	41	266	·425	·618
17	368	·566	·857	42	843	·926	·945
18	578	·762	1·071	43	1096	1·040	·941
19	623	·794	·952	44	1959	1·292	·989
20	674	·829	·736	45	736	·867	·982
21	810	·908	·894	46	623	·794	1·024
22	536	·729	·805	47	1155	1·063	·979
23	503	·702	·926	48	625	·796	·963
24	664	·822	·950	49	445	·648	·926
25	630	·799	·898	50	666	·823	·729

Coramine treated

Baby no.	(i) x	(i) $\log\left(\dfrac{x}{100}\right)$	(ii) $\log\left(\dfrac{x}{100}\right)$	Baby no.	(i) x	(i) $\log\left(\dfrac{x}{100}\right)$	(ii) $\log\left(\dfrac{x}{100}\right)$
1	345	·538	1·304	25	1200	1·079	1·046
2	503	·702	1·063	26	411	·614	·940
3	982	·992	1·256	27	678	·831	·841
4	811	·909	·975	28	750	·875	1·132
5	390	·591	1·155	29	945	·975	1·000
6	858	·933	·761	30	363	·560	1·165
7	1018	1·008	1·134	31	689	·838	·846
8	405	·607	1·078	32	663	·822	·860
9	640	·806	·903	33	1111	1·046	1·036
10	945	·975	1·193	34	566	·753	·859
11	522	·718	1·079	35	405	·607	1·180
12	567	·754	·959	36	756	·879	1·032
13	600	·778	·695	37	836	·922	·938
14	555	·744	·996	38	298	·474	·931
15	619	·792	1·041	39	379	·579	·829
16	685	·836	·873	40	422	·625	·871
17	436	·639	·795	41	726	·861	·959
18	520	·716	1·047	42	810	·908	1·264
19	1006	1·003	1·098	43	769	·886	·966
20	845	·927	·969	44	1062	1·026	·940
21	672	·827	·746	45	557	·746	·945
22	482	·683	·954	46	476	·678	·763
23	2311	1·364	1·062	47	857	·933	·810
24	882	·945	·989				

(*London, B.Sc. Gen., 1958*)

10. A randomized block experiment was carried out with wheat to investigate the effect on the yield of manuring with different amounts of ammonium sulphate. Four levels of manuring treatment were tested, viz.

    A unmanured control (0 lb ammonium sulphate per acre)
    B 20 lb ammonium sulphate per acre
    C 40 lb ammonium sulphate per acre
    D 60 lb ammonium sulphate per acre

The experiment was laid out in six blocks of four plots. The yields, in lb per plot, are shown below, with the relevant treatment given in brackets after each yield figure.

Analyse the data and report on your conclusions. Include in your report 95% confidence limits for any contrast or contrasts you consider to be of interest.

Yields of wheat (lb per plot)

Blocks 1	2	3	4	5	6
17·1(D)	17·5(B)	15·3(A)	13·1(A)	14·0(D)	20·5(C)
18·6(C)	19·3(D)	19·5(D)	18·0(D)	12·6(C)	16·5(B)
17·1(B)	19·7(C)	21·6(C)	15·1(C)	13·1(A)	15·4(A)
13·3(A)	14·4(A)	19·9(B)	15·9(B)	16·0(B)	18·2(D)

(London, B.Sc. Gen., 1966)

11. The data given below is from an experiment due to Fraser, involving the 'clock test'. A pointer on a clock moves in a series of discrete irregular jumps (called stimuli), and each subject is asked to observe the clock for periods of 20 minutes, and record each stimulus he sees by pressing a button. Each subject repeated the experiment three times, with the clock at different angles (vertical, 45° and horizontal), the order being randomized for each subject. The data given below are the numbers of missed stimuli out of 200 given in each 20-minute period.

Analyse the data to see what statement is justified about which angle of display gives the least number of missed stimuli. State any assumptions made in your analysis. Comment on the experimental design.

Subject	V	45°	H	Subject	V	45°	H
1	1	5	3	10	0	4	3
2	7	6	17	11	0	2	1
3	0	0	2	12	1	3	5
4	1	8	8	13	2	1	1
5	2	2	2	14	2	0	2
6	2	3	2	15	4	4	3
7	2	9	8	16	6	1	9
8	1	9	5	17	1	1	2
9	3	4	6				

(London, Anc. Stats., 1962)

12. In an investigation of the properties of a certain electrical component the voltage across the component was measured at 10-second intervals. During the first part of the experiment, from the start until 2 minutes later, the voltage increased, but exactly 2 minutes after the start an alteration was suddenly made in the electrical circuit and during the second part of the experiment, from 2 minutes to 4 minutes, the voltage decreased. The measurement of voltage at 2 minutes after the start was omitted.

It is known that during each separate part of the experiment the logarithm of the voltage was linearly related to the time from the start of the experiment.

It is not known if there is any relation between the rates of change of log voltage with time for the two parts of the experiment. The values of the logarithm of the voltage, to base 10, are given in the table below.

First part of experiment			Second part of experiment		
Time		*log voltage*	*Time*		*log voltage*
min.	*sec.*		*min.*	*sec.*	
	0	1·34	2	10	2·21
	10	1·48	2	20	2·19
	20	1·53	2	30	2·00
	30	1·68	2	40	1·96
	40	1·74	2	50	1·96
	50	1·90	3	0	1·83
1	0	1·89	3	10	1·82
1	10	1·89	3	20	1·80
1	20	2·03	3	30	1·56
1	30	2·08	3	40	1·58
1	40	2·07	3	50	1·38
1	50	2·32	4	0	1·43

Estimate the rate of change of log voltage with time for each part of the experiment separately and give standard errors of your estimates.

Test whether the rate of increase in the first part is significantly different from the rate of decrease in the second part, stating your assumptions.

(*London, B.Sc. Gen., 1964*)

13. In a genetics experiment involving two factors A and B, the progeny were classified according to the following 2×2 table:

	A	not A	Total
B	75	14	89
not B	15	10	25
Total	90	24	114

Examine the hypothesis that the odds for each factor are 3 : 1, independently of the other factor.

(*Camb. Dip., 1951*)

14. The sequence of 250 digits given below was obtained by dictating haphazardly as fast as possible onto a tape recorder.

60277	35918	47655	32900	78092	35171	86905	42680
17399	50271	84206	77335	91708	23552	13066	18790
12785	55201	89709	68831	32755	00280	97653	37682

28783	50199	27676	38950	12547	09823	46570	11178
62739	81889	75022	43478	09613	75090	22213	16508
78235	17164	33489	27106	05154	26542	65809	71834
24764	55485						

Test whether the process may be regarded as random.   (*Oxford Dip.*)

15. The table below contains the results of four radiation experiments in a series to investigate the effects of X-rays and beta-rays on mitotic rates in grasshopper neuroblasts. Embryos from the same egg were divided into three groups, one serving as a control and the other two being exposed to physically equivalent doses of X- and beta-radiation. After irradiation, approximately equal numbers of cells in each group were examined for a specified time, and the number passing through mid-mitosis noted.

By means of a suitable analysis examine what conclusions can be drawn about the effect of irradiation on the proportion of cells reaching mid-mitosis. Is there any evidence of a difference between the effect of the two types of radiation?

Experiment	Group			
	Control	X-ray	Beta	Total
1	a 12	3	4	19
	b 10	15	12	37
2	a 14	5	6	25
	b 3	10	9	22
3	a 9	5	2	16
	b 11	17	17	45
4	a 17	5	7	29
	b 2	14	13	29

[*a* denotes the number of cells reaching mid-mitosis, *b* the number not reaching mid-mitosis.]
                                                        (*Camb. Dip., 1959*)

For further exercises in the style of this book see particularly examination papers set for one- and two-year ancillary statistics and the B.Sc. General of Pure Mathematics and Statistics, both of London University. Other London University examinations which may also provide suitable exercises are the B.Sc. Special Statistics, Diploma in Statistics and M.Sc. (Statistics) papers. These may all be obtained from the London University Publications Department, 45 Gordon Square, London, W.C.1.

Notes on Calculation and Computing

A1. General

Nowadays most statistical analyses are carried out on a computer, especially when the data set involved is large. Package programs have been written for many advanced methods of analysis, but most statisticians find that they want to use one of the main programming languages, for example FORTRAN or APL, available on large computer installations, to devise algorithms of their own. Later in this Appendix we give programs in the BASIC language for the calculations appropriate to a selection of the statistical techniques discussed in this book.

Despite the fairly ready availability of computers, it is still important for statisticians to be able to perform small-scale analyses on a desk or pocket calculator, and the use of pencil and paper is not yet to be despised! In this Appendix we give first some hints on the organization and layout of simple calculations, so as to help avoid the most common sources of error.

(i) Setting out calculations

One major source of errors is lack of neatness in setting out results, the jotting down of results on scraps of paper, etc. It is best to note calculations on sectional quadrille paper, ruled in $\frac{1}{4}$-in squares. Numbers can be written down with two digits per square, with decimal points aligned with a vertical rule, and each number should be clearly labelled, as follows:

Total	472·90
Mean	23·645

A method of setting out calculations for the sample mean and variance is given below. Each sheet should have a clear identification and at least enough subsidiary headings to ensure that someone else can follow what has been done.

In some cases calculations can be arranged so as to avoid the necessity of writing down the result of intermediate steps on paper, except where these are required for checking. Transfers of numbers from machine to paper and back again are major sources of error, and are very time-consuming. The handbooks to some calculators contain illustrations of how to avoid writing down intermediate results where this can be done. One illustration is given below in the calculation of the corrected sum of squares. If the correction term is found while the uncorrected sum of squares is still stored in a memory, the correction can be subtracted on the machine, and there is no need to record the uncorrected sum of squares, except for checking; see § A2(ii), steps 3 and 4.

Pencil is to be preferred to ink in recording results; errors can then be erased rather then crossed out.

(ii) Checks

The safest method of checking is to have the whole calculation worked independently by another person, using a different method, but this is rarely possible. However, in many calculations there are simple checks on accuracy which are easily included and do not reduce speed very much; some examples are given below. It is important not to rely too much on these checks, but to perform the calculations carefully and aim at avoiding errors.

With all calculations it is vital to check that the answers are qualitatively reasonable. This guards against bad blunders, misplacing decimal points, etc. If the calculations are aimed at estimating an unknown parameter, quick rough estimates may be available by another method, which can be used as a check. For example, when estimating the standard deviation of a normal population by the corrected sum of squares formula, the range estimate can be used as a check.

(iii) Accuracy

Several types of statistical analysis require calculations to be done in stages, with the result of one stage being needed in a later one. The question of the number of significant digits to carry in the calculation naturally arises.

Unfortunately, there is no simple answer. Using too many figures increases the chance of errors when recording results, while using too few gives inaccuracy through rounding error. One important pointer to the number of significant figures to carry in a calculation is the accuracy needed in the final result. Thus, for example, if a t-test is to be performed, the tables of t (Table 3, p.326) are mostly to the accuracy of three significant digits, so the t-statistic itself should be calculated to three or, preferably four, significant digits, so as to err on the side of safety.

A2. Calculation of the sample mean and standard deviation
Ungrouped data

The sample variance of a set of values $x_1, \ldots, x_n$ is defined in Equation 1.1

on p.29, and the standard deviation is, of course, its square root. The variance can be calculated using either of the two formulae in Equation 1.1, although the second, i.e.

$$s^2 = \frac{1}{n-1} \{\sum x_i - (\sum x_i)^2/n\}$$

is almost always to be preferred. We show below the steps needed for calculating mean and standard deviation for a sample of size n. In both cases advantage can be taken of the device discussed in § 1.2 of changing the origin and scale of the numbers for arithmetical convenience; for example, if the numbers are all recorded to one decimal place and lie between 830 and 840, it may be convenient to subtract 830 from each and then multiply by 10, reversing the operations at the end of the calculation.

(i) *Method I*
1. (Optional) Rescale the n observations [say, subtract a, then divide by b].
2. Add the observations and divide by n to obtain the mean, $\bar{x}$.
3. Subtract $\bar{x}$ from each observation to obtain the difference.
4. Square the differences and add, obtaining the corrected sum of squares (CSS).
5. Divide CSS by $(n-1)$ to give s^2, the sample variance; take the square root of the variance to obtain s, the sample standard deviation.
6. Convert results back to the original scale, if necessary. For s, multiply by b; for $\bar{x}$, multiply by b, then add a.

This method suffers from two disadvantages. It is (at least when used on a calculator) inefficient in that the sample members need to be operated on twice, first to calculate the mean in step 2 and again in steps 3 and 4. More seriously, if the numbers are such that the mean cannot be calculated exactly in the number of decimal places available, step 3 will introduce a rounding error and the calculation will therefore be inaccurate.

Both these problems are avoided in Method II, the latter by the device of leaving divisions till the last possible step.

(ii) *Method II*
1. (Optional) Rescale, as in Method I.
2. Obtain the sum S and the uncorrected sum of squares USS of the n numbers.
3. Find the mean S/n and the correction factor S^2/n.
4. Subtract the correction factor from USS, to give CSS, the corrected sum of squares.
5, 6. As Method I.

Example 1. A series of radioactivity counts (CPM) were made on a solution, and the observations were as follows:

898, 860, 834, 875, 842, 839, 875, 881.

1. Subtract 800 for convenience.
2. $S = 98 + 60 + \ldots + 81 = 504$,
   $USS = 98^2 + 60^2 + \ldots + 81^2 = 35\,456$.
3. Mean $= 504/8 = 63.0$,
   Corr. $= 504 \times 504/8 = 31\,752$.
4. CSS $= 35\,456 - 31\,752 = 3704$.
5. $s^2 = 3704/7 = 529.1$, $s = \sqrt{529.1} = 23.0$.
6. No change needed for s. Mean $= 63.0 + 800 = 863$.

Note that if a calculator with several memories is used, USS need never be recorded explicitly, since the correction factor can simply be subtracted directly on the machine.

Grouped data

It is important to realize that in principle exactly the same calculations as in the last section need to be done for grouped data, the only difference being that observations in a group are all assumed to be at the mid-point of the group. Thus, in Example 1.4, on p.20, two observations are taken to be at 97.45, one at 102.45, five at 107.45 and so on. (The first group contains observations which, rounded to one decimal place, lie between 95.0 and 99.9 inclusive; the original, unrounded, values must therefore have been between 94.95 and 99.95.) In calculating the sum and sum of squares of the observations we must therefore remember to multiply where appropriate by the number of observations in the group.

(iii) Method III

1. (Optional, but usually advisable.) Rescale the group mid-points.
2. For each group, multiply the scale value by the frequency, and add, to give the sum S.
3. Obtain the uncorrected sum of squares, USS; as step 2, but multiply the square of the scale value by the frequency.
4. Continue with steps 3–6 of Method II.

Example 2. The data given in Table 1 below are 1000 observations on the breadth of the lamina of runner bean leaves, in centimetres.

1. Table 1 gives the scale chosen; the formula is:
   $$\text{Scale value} = 2 \times (\text{group mid-point} - 5.5)$$
2. $S = 325$, as in Table 1.
3. USS $= 1907$, as in Table 1.
4. Mean $= S/n = 0.325$.
   Corr. $= S^2/n = 105.6$.
   CSS $=$ USS $-$ Corr. $= 1801.4$.
   $s^2 = \text{CSS}/(n-1) = 1.803$.
   $s = \sqrt{s^2} = 1.343$.

Table 1. *Calculation involving grouped data*

Lower end of group	Frequency	Group mid-point	Scale value	(scale) × (frequency)	(scale)^2 × (frequency)
3·25	4	3·5	−4	−16	64
3·75	12	4·0	−3	−36	108
4·25	65	4·5	−2	−130	260
4·75	182	5·0	−1	−182	182
5·25	286	5·5	0	0	0
5·75	270	6·0	1	270	270
6·25	131	6·5	2	262	524
6·75	43	7·0	3	129	387
7·25	7	7·5	4	28	112
	1000			*Total* 325	*USS* 1907

Rescaling.

Mean $= 5.5 + \frac{1}{2} \times 0.325 = 5.662$ cm.

St. Dev. $= \frac{1}{2} \times 1.343 = 0.671$ cm.

□ □ □

The standard deviation is, of course, unaffected by the arbitrary choice of 5.5 for the working origin, so that this figure is not used in rescaling the standard deviation at the end of the calculation. Note also that in calculations of this sort there can be some doubt about the exact mid-points of the groups. The quoted mid-points of 3.5, 4, 4.5, . . ., are of course accurate if the initial allocation of observations to groups was itself done accurately. On the other hand, if the original observations had been recorded accurate to two decimal places, and then regrouped as shown in Table 1, then the lowest group would have contained observations whose exact values lay between 3.245 and 3.745, the mid-point being 3.495, not 3.5.

A3. Some statistical computer programs
(i) General
In this section we sketch out some programs for elementary statistical calculations in the BASIC programming language. The language is the most common one in use on microcomputers, and is generally available also in larger computer installations.

It is important to remember that different computers will (in their terminology) support different dialects of the language. In simple terms, this means that a program written using the facilities available on one computer will usually not work on another without modification. There is, however, an American Standard for what is called Minimal BASIC, and most implementations of BASIC consist of this language together with certain extra features. The programs below have been written in accordance with the American

Standard, so that they will run without modification on several different implementations of BASIC.

The additions (often called enhancements) to Minimal BASIC appearing on most popular microcomputers mostly concern the ability to handle non-numerical and graphics characters. The programs presented here have thus had largely to exclude graphical and tabular output; the output is therefore almost always in simple numerical form, which can readily be converted into a pictorial presentation on most microcomputers.

For any statistical problem, it is not possible to write one single program which will, on all computers and for all data sets, be 'the best'. Desirable characteristics of a program depend, a bit, on the sophistication of the user, and depend very much on the sort of data to be handled. A good routine for finding the variance of numbers in the millions may be rather poor when the numbers vary between 0.55 and 0.56. It follows that the programs presented in this section cannot be thought of as ideals; in particular, a balance has had to be struck between brevity and refinement. It is of course essential that the program should not give incorrect results when applied to the sort of data sets likely to occur in practice. However, we have felt it unnecessary for present purposes that the program prevent the user attempting an impossible analysis, for example, analysing a sample of size zero, as long as the program or computer system signals this by an error message.

Accordingly, the programs here have not been cluttered with REM (for 'remark') statements, or checks for plausibility of data to be input. It is hoped that the shorter programs resulting from this policy will remain easy to follow, particularly when read in conjunction with the accompanying comments and the relevant sections of the main body of this text. The exception to this policy of brevity is that the user is provided with a prompt for each piece of data required, so that it will always be clear what part of the data set needs to be entered. (Regular users of these programs may well get bored with these repetitive prompts, but straightforward amendments to the PRINT statements are all that are needed to reduce the tedium.)

(ii) Sample mean and variance

We start with some routines to calculate the sample mean and variance of a set of N numbers, typically a random sample of that size from some distribution. In Program 1A we assume that the data is ungrouped and that N, the sample size, is known in advance. Note that the program is very straightforward, making no attempt at sophistication. (For example, it will quickly come to grief if a sample size of 1 is tried, since there is then a divisor of 0 for the sample variance, or if N is not an integer; a test for these could easily be inserted after line 210.) On the other hand, the calculations are organized efficiently, and the rather common approach of first storing the data in an array and only then performing the calculations is avioded. Note also that we avoid another common but very inefficient practice, that of first calculating the sample mean $\bar{x}$ and then calculating the variance by subtracting $\bar{x}$ from each observation before squaring it.

As noted, the program avoids storing the data in an array. It is sometimes valuable, however, to keep the data for subsequent analyses, and a very simple amendment to the program does this; just declare an array of appropriate size at the start, e.g.

$$100 \text{ DIM } X(50),$$

for a sample of size up to 50, and then write X(I) for X in lines 250–270.

Program 1B is a slight variant of Program 1A, covering the case where the sample size is not known in advance. In this situation we need to use some number as a code to indicate that all sample members have been entered. In the program as presented here we assume that a value of –999 will suffice: the test appears in line 255 of Program 1B, which can of course be easily changed if –999 seems likely to appear in any sample.

Program 1C is another variant of Program 1A, covering the case of grouped data. Here the changes are very slight; notice, however, that we now need to calculate N, the sample size, as the sum of the individual frequencies.

(iii) Student's t-test
It is obviously possible to write programs to perform all the one-sample and two-sample tests discussed in Chapters 5 and 6. In Program 2 we give one example, that of the two-sample t-test, the test for the comparison of sample means when the population variances are unknown but assumed to be equal. Notice that the calculations for Example 6.2 as set out on p.148 start by calculating the mean and CSS for each sample separately. This pattern is followed in Program 2, in which the body of Program 1A has been incorporated as a subroutine, starting at line 1000. A side-effect of this labour-saving approach is that (without a simple modification) the program will not allow either sample size to be 1; this is, strictly, a disadvantage, since the theory of the test allows one sample (but not both) to be of size 1. However, few statisticians would regard it as sensible to do such a test for a sample of that size.

(iv) Linear regression
Program 3 deals with simple linear regression, following the methods in Chapter 10. Notice that in the program (line 210) the positions of the points have been stored in an array. (Line 100 indicates that 50 points can be stored, although it is trivial to increase this if required.) Storage of the data is needed only if the residuals (line 630; see also §10.7) are wanted, and can otherwise be omitted, with X(I) replaced by X and Y(I) by Y throughout. As §10.7 shows, however, residuals are of great value in checking the reasonableness of the various assumptions made in linear regression. Program 3 will, as written, print out the basic information needed for these checks, and the appropriate plots can be drawn by hand or by a routine appended to Program

3; note though that plotting commands differ considerably from one computer to another.

As with earlier programs, Program 3 has not been designed so as to trap all errors. For example, it is not sensible to use linear regression if n, the number of points, is less than three. The program does not explicitly check for this, or for other statistical misdemeanours, such as giving every point the same value of x. However, the way in which the calculations are done does mean that when the data is not sensible, the program will not pretend to give sensible answers.

(v) Analysis of variance

In Program 4 we deal with the subject of Chapter 12, the Analysis of Variance. The program presented is for a one-way analysis, dealt with in §§ 12.1 to 12.3, and will deal with groups of any size. (Since the theory of Analysis of Variance allows some groups to be of a size as low as 1, so does the program, though it would be extremely unusual to find this in practice, and checks such as the analysis of residuals in § 12.3 (*iii*) would be impossible.)

The reader will find it instructive to check through the program to see where the quantities needed for one-way analysis of variance are stored. It is clear that N(I) stores the sample size in group I (and, incidentally, from line 100 that up to 20 groups are catered for, though again this can be increased if necessary). Line 220 indicates that T(I) stores the total for group I, but we leave the investigation of the contents of S1, S2, S3 and C1, C2, C3 and C4 to the reader.

The program does not cover parts of § 12.3 other than § 12.3(*i*), the calculation of the ANOVA table. It can easily be extended (when all the sample sizes N(1), ..., N(G) are equal) to calculate LSDs and other aspects of the analysis of means. To calculate and analyse residuals the data would need to be stored in an array on input; simply incorporate X(20, 30) in line 100 (to cater for a maximum group size of 30) and refer in lines 210–230 to X(I, J) instead of X. The residual for observation X(I, J) is then simply X(I, J)–T(I)/N(I), and can be calculated at any time after line 240.

A4. Simulation

(i) General

The topic of simulation was introduced in §§ 4.2 and 4.3, in particular in relation to sampling distributions. It is in this area that some of the greatest effects of the recent developments in computing have been seen in statistics, since the easy availability of pseudo-random numbers on microcomputers enables statistics students to obtain empirical proof of results very easily. Two immediate examples will suffice. First, a fair amount of effort used to be needed to produce a diagram like Figure 4.3, illustrating the statistical independence of sample mean and sample variance when sampling from a normal distribution. Nowadays the calculations involved are readily programmed, all that is needed being to generate a sample of normal random numbers and apply Program 1A as a subroutine; however, the production of a scatter diagram requires more than Minimal BASIC can conveniently offer. As a second example, consider the discussion in § 2.6 of the importance of the normal distribution, and in particular the note that the Central Limit Theorem shows that under certain conditions such a distribution is to be expected. The proof of this theorem is quite outside the scope of this book, but a graphical demonstration of the implications will be given below.

(ii) Random numbers from particular distributions
The heart of a computer simulation is the pseudo-random number generator. We refer the reader to § 4.2 for details of how these generators work, noting here only that the generator is called in Minimal BASIC by the function RND, each mention of the function in the program giving a new member of the sequence of pseudo-random numbers. It should also be pointed out that it is worth trying the tests on pp.106 and 107 on any generator so as to satisfy oneself that the sequence is a reasonable one. (Note also that on many micro-computers the syntax of RND is not exactly as in Minimal BASIC.)

A satisfactory pseudo-random number generator produces a set of numbers that are indistinguishable from a random sample from the rectangular distribution on $(0,1)$; see § 2.6(ii). If sampling experiments are to be carried out it is necessary to devise methods for obtaining random samples from other distributions, discrete and continuous. Methods suitable for hand calculation for obtaining these are given in § 2.6; we now supplement the discussion there by indicating other methods more geared to computers.

Binomial. The method given on p.107 (known as a table look-up method) can be readily programmed, but an alternative method given in Program 5 simply simulates the model [§ 2.5(iii)] underlying the binomial distribution. The BASIC instructions are given in the form of a subroutine called from a main program by a GOSUB 5000 instruction. Appropriate values of the index N and parameter P (called θ in § 2.5) must be supplied, and the subroutine returns a value X having that binomial distribution. Patently the time needed to execute the subroutine is proportional to N.

Exponential. As Exercise 4.2.3 makes clear, use of the expression —L*LOG(RND) in a BASIC program returns a random number from the exponential distribution with parameter L^{-1} i.e. mean L.

Poisson. The table look-up method can be used. Alternatively, the facts that exponential random numbers can be obtained simply and that there is a link between the Poisson and exponential distributions (see Exercises 2.5.5 and 2.6.4) can be exploited to give Poisson random numbers. The link is, in essence, that the gap between occurrences in a Poisson process is exponentially distributed, while the number of occurrences in some fixed time has a Poisson distribution.   One can thus obtain a Poisson random number by seeing how many exponential random 'gaps' are obtained before their sum exceeds the time given. Slight manipulation to improve speed results in Program 6, called by GOSUB 6000.

Normal. Strangely, the normal distribution is one which is particularly tricky to sample from. Here we concentrate on sampling from the standard normal distribution, $N(0,1)$; if y is obtained from $N(0,1)$, then of course $x = \mu + \sigma y$ will be a random number from $N(\mu, \sigma^2)$. (See p.70 for details.) If an approximate method is acceptable, a simple way of proceeding is to appeal to the Central Limit Theorem and add, say, 12 rectangular random numbers together. For practical purposes the numbers resulting are almost always acceptable as substitutes for normal deviates (the kurtosis coefficient $\gamma_2 = -0.1$) except when the investigation is particularly concerned with the extreme tails of the distribution. However, the method is not especially fast, and the view is generally taken now that one might as well use alternative methods which are both exact and (on most computers) faster.

An example of such a method is given in Program 7, which uses the so-called rejection technique. The details of the theoretical background are too lengthy to discuss here, but in essence the technique is that a variable is sampled from a convenient distribution, A say, with a density function similar in shape to that of the required distribution, B. A proportion of the variables sampled is rejected, mostly in the region where the density of A is greater than that of B, in such a way that the distribution of accepted numbers is B. In Program 7, A is the exponential distribution, and B the right-hand half of $N(0,1)$; the program is completed in line 1130 by assigning a random sign. Again, the program is presented in subroutine form, being called by a GOSUB 1000 statement and returning in X a $N(0,1)$ random number.

(iii) Applications
We are now in a position to use simulation to demonstrate statistical methods in action. We give below two examples, and note also that the reader is now in a position to carry out the sampling experiment in §4.3 using a micro-computer.

Confidence interval. We show below in Program 8 how one can simulate the production of a sequence of confidence intervals for a normal mean, with variance known. The program is offered as an example of how a suitable simulation can illuminate awkward statistical concepts. Confidence intervals are repeatedly calculated and printed, with an indication of whether or not the (random) interval covers the true, user-chosen, population mean, and with the percentage of intervals to date covering that mean. It is this latter per-centage that should, of course, settle down to the desired confidence co-efficient. Note that in line 180 the normal percentage point is required, appropriate for a two-sided interval, so that for a 95% interval 1.96 should be entered.

Central limit theorem. The final program offered here obtains, by simulation, the sampling distribution of the sum of several (N, say) independent, rectangularly distributed random variables. As the notes on p.71 make clear, the sampling distribution could be expected to be approximately normal for large N, while of quite different shape for small N (say, $N < 6$). Lines 560–580 assign to X the sum of N rectangular random numbers, and the later lines merely compile and then display a frequency distribution and histogram of the K values of X calculated. For a satisfactory demonstration, values of K up to about 1000 and H about 20 are usually suitable. The main use of Program 9 would be to observe the different shape of the histogram for $N = 1, 2, \ldots$

PROGRAM 1A

```
100    REM    PROGRAM FOR SAMPLE MEAN AND VARIANCE
200    PRINT "INPUT SAMPLE SIZE ";
210    INPUT N
220    LET S = 0
230    LET T = 0
240    FOR I = 1 TO N
245     PRINT "ENTER SAMPLE MEMBER ";
250     INPUT X
260     LET S = S+X
270     LET T = T+X*X
280    NEXT I
300    PRINT "MEAN =";S/N
310    PRINT "VARIANCE =";(T-S*S/N)/(N-1)
320    END
```

PROGRAM 1B

```
100    REM    PROGRAM FOR SAMPLE MEAN AND VARIANCE
210    LET N = 0
220    LET S = 0
230    LET T = 0
245    PRINT "ENTER SAMPLE MEMBER ";
250    INPUT X
255    IF X = -999 THEN 300
260    LET S = S+X
270    LET T = T+X*X
280    LET N = N+1
290    GOTO 245
300    PRINT "MEAN =";S/N
310    PRINT "VARIANCE =";(T-S*S/N)/(N-1)
320    END
```

PROGRAM 1C

```
100    REM PROGRAM FOR SAMPLE MEAN AND VARIANCE
190    REM    PROGRAM FOR SAMPLE MEAN AND VARIANCE
200    PRINT "INPUT NO. OF GROUPS ";
210    INPUT G
220    LET N = 0
230    LET S = 0
240    LET T = 0
250    FOR I = 1 TO G
260     PRINT "ENTER GROUP MID-POINT ";
270     INPUT X
280     PRINT "ENTER FREQUENCY ";
290     INPUT F
300     LET N = N+F
310     LET S = S+F*X
320     LET T = T+F*X*X
330    NEXT I
340    PRINT "MEAN =";S/N
350    PRINT "VARIANCE =";(T-S*S/N)/(N-1)
360    END
```

PROGRAM 2

```
100    REM    PROGRAM FOR 2-SAMPLE T-TEST
200    PRINT "SAMPLE NUMBER 1"
210    GOSUB 1000
220    LET N1 = N
230    LET M1 = M
240    LET C1 = C
250    LET S1 = C1/(N1-1)
300    PRINT "SAMPLE NUMBER 2"
310    GOSUB 1000
320    LET N2 = N
330    LET M2 = M
340    LET C2 = C
350    LET S2 = C2/(N2-1)
360    LET F = S1/S2
370    PRINT "VARIANCE RATIO = ";F;" ON ";N1-1;"AND ";N2-1;" D.F."
400    LET D = M1-M2
410    PRINT "DIFFERENCE IN MEANS IS ";D
450    LET V = (C1+C2)/(N1+N2-2)
460    PRINT "POOLED ESTIMATE OF VARIANCE IS ";V
500    LET T = D/SQR(V*(1/N1+1/N2))
510    PRINT "T = ";T;" ON ";N1+N2-2;" D.F."
515    REM  SEE PAGE 135, EQN.(5.6) FOR THIS RESULT
517    IF N1+N2 < 3 THEN 550
520    LET T = T*SQR((N1+N2-2)/(N1+N2-4))
530    PRINT
540    PRINT "APPROX. EQUIVALENT NORMAL DEVIATE IS ";T
550    END
1000   REM    PROGRAM FOR SAMPLE MEAN AND VARIANCE
1010   PRINT "INPUT SAMPLE SIZE ";
1020   INPUT N
1030   LET S = 0
1040   LET T = 0
1050   FOR I = 1 TO N
1055    PRINT "ENTER SAMPLE MEMBER ";
1060    INPUT X
1070    LET S = S+X
1080    LET T = T+X*X
1090   NEXT I
1095   LET M = S/N
1100   PRINT "MEAN =";M
1105   LET C = T-S*S/N
1110   PRINT "VARIANCE =";C/(N-1)
1120   RETURN
```

PROGRAM 3

```
90     REM    PROGRAM FOR SIMPLE LINEAR REGRESSION
100    DIM X(50),Y(50)
120    PRINT "SAMPLE SIZE =   ";
130    INPUT N
150    LET S1 = 0
160    LET S2 = 0
170    LET S3 = 0
180    LET S4 = 0
190    LET S5 = 0
200    FOR I = 1 TO N
205      PRINT "ENTER POINT";I;"AS A PAIR  X,Y ";
210      INPUT X(I),Y(I)
220      LET S1 = S1+X(I)
230      LET S2 = S2+Y(I)
240      LET S3 = S3+X(I)*X(I)
250      LET S4 = S4+Y(I)*Y(I)
260      LET S5 = S5+X(I)*Y(I)
270    NEXT I
300    LET C1 = S3-S1*S1/N
310    LET C2 = S4-S2*S2/N
320    LET C3 = S5-S1*S2/N
330    LET C4 = C3*C3/C1
340    LET C5 = C2-C4
400    LET B = C3/C1
410    LET A = S2/N-B*S1/N
420    PRINT "REGRESSION LINE IS"
425    IF B < 0 THEN 450
430    PRINT "Y =";A;" + ";B;"*X"
440    GOTO 500
450    PRINT "Y =";A;" - ";-B;"*X"
500    REM  NOW CALCULATE ANOVA
505    PRINT
506    PRINT "ANALYSIS OF VARIANCE"
507    PRINT
510    PRINT "  SOURCE   ","DF","SS","MS"
520    PRINT "REGRESSION"," 1",C4,C4
530    PRINT " RESIDUAL ",N-2,C5,C5/(N-2)
535    PRINT
540    PRINT "  TOTAL    ",N-1,C2
580    PRINT
585    PRINT
587    PRINT "F-STATISTIC IS ";C4*(N-2)/C5;"ON 1 AND";N-2;"D.F."
590    PRINT
600    PRINT "Y-VALUES","FITTED","RESIDUALS"
610    FOR I = 1 TO N
620      LET F = A+B*X(I)
630      PRINT Y(I),F,Y(I)-F
640    NEXT I
1000   END
```

PROGRAM 4

```
90      REM    PROGRAM FOR ONE-WAY ANOVA
100     DIM N(20),T(20)
110     PRINT "NO. OF GROUPS = ";
120     INPUT G
130     LET S1 = 0
140     LET S2 = 0
147     LET S3 = 0
148     LET N1 = 0
150     FOR I = 1 TO G
160      LET T(I) = 0
170      PRINT "SAMPLE SIZE IN GROUP";I;"= ";
180      INPUT N(I)
190      LET N1 = N1+N(I)
200      FOR J = 1 TO N(I)
205       PRINT "ENTER SAMPLE MEMBER, GROUP";I;":";
210       INPUT X
220       LET T(I) = T(I)+X
230       LET S2 = S2+X*X
240      NEXT J
250      LET S1 = S1+T(I)
255      LET S3 = S3+T(I)*T(I)/N(I)
260     NEXT I
300     LET C1 = S1*S1/N1
310     LET C2 = S2-C1
320     LET C3 = S3-C1
330     LET C4 = C2-C3
400     REM ANOVA
410     PRINT
420     PRINT "ANALYSIS OF VARIANCE"
430     PRINT
450     PRINT "      SOURCE      "," CSS","DF","  MS"
460     PRINT "BETWEEN GROUPS",C3,G-1,C3/(G-1)
470     PRINT "WITHIN GROUPS",C4,N1-G,C4/(N1-G)
480     PRINT
490     PRINT "     TOTAL      ",C2,N1-1
500     PRINT
600     PRINT "GROUP","TOTAL","MEAN"
610     PRINT
620     FOR I = 1 TO G
630      PRINT I,T(I),T(I)/N(I)
640     NEXT I
1000    END
```

PROGRAM 5

```
5000    REM SUBROUTINE TO GIVE RANDOM NUMBER
5010    REM FROM THE BINOMIAL DISTRIBUTION.
5020    REM INDEX N AND PARAMETER P NEEDED.
5100    LET X = 0
5110    FOR I = 1 TO N
5120     IF RND > P THEN 5150
5130      LET X = X+1
5150    NEXT I
5200    RETURN
```

PROGRAM 6

```
6000    REM SUBROUTINE TO GIVE RANDOM NUMBER
6010    REM FROM A POISSON DISTRIBUTION.
6020    REM MEAN M TO BE SUPPLIED.
6100    LET P = EXP(M)*RND
6110    LET X = 0
6120    IF P > 1 THEN 6150
6130    RETURN
6150    LET P = P*RND
6160    LET X = X+1
6170    GOTO 6120
```

PROGRAM 7

```
1000    REM SUBROUTINE TO GIVE RANDOM NUMBER FROM
1010    REM  NORMAL DISTRIBUTION WITH MEAN 0
1020    REM  AND STANDARD DEVIATION 1.
1100    LET X = -LOG(RND)
1110    IF RND > EXP(-.5*(X-1)*(X-1)) THEN 1000
1120    LET X = X*SGN(RND-.5)
1130    RETURN
```

PROGRAM 8

```
  50    REM    PROGRAM TO SIMULATE REPEATED CONSTRUCTION
  60    REM  OF A CONFIDENCE INTERVAL FOR THE MEAN OF
  70    REM  A NORMAL DISTRIBUTION, THE VARIANCE BEING
  80    REM  KNOWN.
 100    PRINT "POPULATION MEAN =";
 110    INPUT M
 150    PRINT "AND VARIANCE =";
 160    INPUT V
 170    LET S = SQR(V)
 200    PRINT "INPUT SAMPLE SIZE :";
 210    INPUT N
 250    PRINT "MULTIPLE OF S.D. =";
 260    INPUT K
 300    LET W = S*K/SQR(N)
 310    LET Z = 0
 320    LET Y = 0
 400    LET T = 0
 410    FOR I = 1 TO N
 420      GOSUB 1000
 430      LET T = T+(X*S+M)
 440    NEXT I
 450    LET X1 = T/N
 460    LET Y = Y+1
 470    LET L = X1-W
 480    LET U = X1+W
 500    IF M < L THEN 560
 510    IF M > U THEN 560
 520    LET Z = Z+1
 530    LET P = 100*Z/Y
 540    PRINT L,U,"IN",P
 550    GOTO 600
 560    LET P = 100*Z/Y
 570    PRINT L,U,"OUT",P
 600    GOTO 400
1000    REM SUBROUTINE TO GIVE RANDOM NUMBER FROM
1010    REM  NORMAL DISTRIBUTION WITH MEAN 0
1020    REM  AND STANDARD DEVIATION 1.
1100    LET X = -LOG(RND)
1110    IF RND > EXP(-.5*(X-1)*(X-1)) THEN 1000
1120    LET X = X*SGN(RND-.5)
1130    RETURN
```

```
      PROGRAM 9

100   REM   PROGRAM TO SIMULATE THE SUM
110   REM   OF RECTANGULAR RANDOM VARIABLES.
120   DIM A(30)
150   PRINT "HOW MANY VARIABLES?";
160   INPUT N
200   PRINT "HOW MANY SAMPLES? ";
210   INPUT K
300   PRINT "HOW MANY CATEGORIES FOR PRINTING? ";
310   INPUT H
400   FOR H1 = 1 TO H
410     LET A(H1) = 0
420   NEXT H1
500   FOR K1 = 1 TO K
550     LET X = 0
560     FOR N1 = 1 TO N
570      LET X = X+RND
580     NEXT N1
600     LET H1 = INT(1+X*H/N)
610     LET A(H1) = A(H1)+1
620   NEXT K1
690   LET M = 0
700   FOR H1 = 1 TO H
710     PRINT H1;A(H1)
720     IF A(H1) < M THEN 740
730     LET M = A(H1)
740   NEXT H1
800   FOR H1 = 1 TO H
810     LET B = INT(A(H1)*30/M)
820     IF B = 0 THEN 900
850     FOR B1 = 1 TO B
860      PRINT "*";
870     NEXT B1
900     PRINT
950   NEXT H1
1000  END
```

Statistical Tables

Table 1 *Cumulative distribution function of the standard normal distribution*
(a) For x in 0·1 intervals

x	$\Phi(x)$	x	$\Phi(x)$	x	$\Phi(x)$
0·0	0·5000	1·3	0·9032	2·6	0·9953
0·1	0·5398	1·4	0·9192	2·7	0·9965
0·2	0·5793	1·5	0·9332	2·8	0·9974
1·3	0·6179	1·6	0·9452	2·9	0·9981
0·4	0·6554	1·7	0·9554	3·0	0·9987
0·5	0·6915	1·8	0·9641	3·1	0·9990
0·6	0·7257	1·9	0·9713	3·2	0·9993
0·7	0·7580	2·0	0·9772	3·3	0·9995
0·8	0·7881	2·1	0·9821	3·4	0·99966
0·9	0·8159	2·2	0·9861	3·5	0·99977
1·0	0·8413	2·3	0·9893	3·6	0·99984
1·1	0·8643	2·4	0·9918	3·7	0·99989
1·2	0·8849	2·5	0·9938	3·8	0·99993

(b) For x in 0·01 intervals

x	$\Phi(x)$	x	$\Phi(x)$	x	$\Phi(x)$
1·60	0·9452	1·87	0·9693	2·14	2·9838
1·61	0·9463	1·88	0·9699	2·15	0·9842
1·62	0·9474	1·89	0·9706	2·16	0·9846
1·63	0·9484	1·90	0·9713	2·17	0·9850
1·64	0·9495	1·91	0·9719	2·18	0·9854
1·65	0·9505	1·92	0·9726	2·19	0·9857
1·66	0·9515	1·93	0·9732	2·20	0·9861
1·67	0·9525	1·94	0·9738	2·21	0·9865
1·68	0·9535	1·95	0·9744	2·22	0·9868
1·69	0·9545	1·96	0·9750	2·23	0·9871
1·70	0·9554	1·97	0·9756	2·24	0·9875
1·71	0·9564	1·98	0·9761	2·25	0·9878
1·72	0·9573	1·99	0·9767	2·26	0·9881
1·73	0·9582	2·00	0·9772	2·27	0·9884
1·74	0·9591	2·01	0·9778	2·28	0·9887
1·75	0·9599	2·02	0·9783	2·29	0·9890
1·76	0·9608	2·03	0·9788	2·30	0·9893
1·77	0·9616	2·04	0·9793	2·31	0·9896
1·78	0·9625	2·05	0·9798	2·32	0·9898
1·79	0·9633	2·06	0·9803	2·33	0·9901
1·80	0·9641	2·07	0·9808	2·34	0·9904
1·81	0·9649	2·08	0·9812	2·35	0·9906
1·82	0·9656	2·09	0·9817	2·36	0·9909
1·83	0·9664	2·10	0·9821	2·37	0·9911
1·84	0·9671	2·11	0·9826	2·38	0·9913
1·85	0·9678	2·12	0·9830	2·39	0·9916
1·86	0·9686	2·13	0·9834	2·40	0·9918

The function tabulated is

$$\Pr(x < X) = \int_{-\infty}^{x} \frac{1}{\sqrt{(2\pi)}} e^{-\frac{1}{2}x^2}\, dx = \Phi(X)$$

Table 2 *Percentiles of the standard normal distribution*

P	x	P	x	P	x
20	0·8416	5	1·6449	2	2·0537
15	1·0364	4	1·7507	1	2·3263
10	1·2816	3	1·8808	0·5	2·5758
6	1·5548	2·5	1·9600	0·1	3·0902

This table gives one-sided percentage points,

$$P/100 = \int_{x}^{\infty} \frac{1}{\sqrt{(2\pi)}} e^{-\frac{1}{2}x^2}\, dx$$

The two-sided percentage appropriate to any x is $2P$.

Table 3 *Percentage points of the t-distribution*

	Probability in per cent					
	20	10	5	2	1	0·1
1	3·08	6·31	12·71	31·82	63·66	636·62
2	1·89	2·92	4·30	6·96	9·92	31·60
3	1·64	2·35	3·18	4·54	5·84	12·92
4	1·53	2·13	2·78	3·75	4·60	8·61
5	1·48	2·01	2·57	3·36	4·03	6·87
6	1·44	1·94	2·45	3·14	3·71	5·96
7	1·42	1·89	2·36	3·00	3·50	5·41
8	1·40	1·86	2·31	2·90	3·36	5·04
9	1·38	1·83	2·26	2·82	3·25	4·78
10	1·37	1·81	2·23	2·76	3·17	4·59
11	1·36	1·80	2·20	2·72	3·11	4·44
12	1·36	1·78	2·18	2·68	3·05	4·32
13	1·35	1·77	2·16	2·65	3·01	4·22
14	1·34	1·76	2·14	2·62	2·98	4·14
15	1·34	1·75	2·13	2·60	2·95	4·07
20	1·32	1·72	2·09	2·53	2·85	3·85
25	1·32	1·71	2·06	2·48	2·79	3·72
30	1·31	1·70	2·04	2·46	2·75	3·65
40	1·30	1·68	2·02	2·42	2·70	3·55
60	1·30	1·67	2·00	2·39	2·66	3·46
120	1·29	1·66	1·98	2·36	2·62	3·37
∞	1·28	1·64	1·96	2·33	2·58	3·29

This table gives two-sided percentage points,

$$P/100 = 2 \int_t^\infty f(x \mid v)\, dx$$

where $f(x \mid v)$ is the p.d.f. of the t-distribution (5.3).

For one-sided percentage points the percentages shown should be halved.

Table 4 *Factors for converting a range to an estimate of σ*

n	γ_n	n	γ_n	n	γ_n	n	γ_n
2	0·886	6	0·395	10	0·325	14	0·294
3	0·591	7	0·370	11	0·315	15	0·288
4	0·486	8	0·351	12	0·307	16	0·283
5	0·430	9	0·337	13	0·300	20	0·268

Table 5 *Percentage points of the χ^2-distribution*

Degrees of freedom (v)	1	5	90	95	99	99·9
1	0·0³157	0·00393	2·71	3·84	6·63	10·83
2	0·0201	0·103	4·61	5·99	9·21	13·81
3	0·115	0·352	6·25	7·81	11·34	16·27
4	0·297	0·711	7·78	9·49	13·28	18·47
5	0·554	1·15	9·24	11·07	15·09	20·52
6	0·872	1·64	10·64	12·59	16·81	22·46
7	1·24	2·17	12·02	14·07	18·48	24·32
8	1·65	2·73	13·36	15·51	20·09	26·12
9	2·09	3·33	14·68	16·92	21·67	27·88
10	2·56	3·94	15·99	18·31	23·21	29·59
11	3·05	4·57	17·28	19·68	24·73	31·26
12	3·57	5·23	18·55	21·03	26·22	32·91
14	4·66	6·57	21·06	23·68	29·14	36·12
16	5·81	7·96	23·54	26·30	32·00	39·25
18	7·01	9·39	25·99	28·87	34·81	42·31
20	8·26	10·85	28·41	31·41	37·57	45·31
22	9·54	12·34	30·81	33·92	40·29	48·27
24	10·86	13·85	33·20	36·42	42·98	51·18
26	12·20	15·38	35·56	38·89	45·64	54·05
28	13·56	16·93	37·92	41·34	48·28	56·89
30	14·95	18·49	40·26	43·77	50·89	59·70

The table gives the percentage points χ^2, where

$$P = 100 \int_0^{\chi^2} g(y \mid v)\, dy$$

where $g(y \mid v)$ is the probability density function of the χ^2-distribution.

For $v > 30$, $\sqrt{(2\chi^2)}$ is approximately normally distributed with mean $(2v - 1)$ and unit variance.

Table 6 *Percentage points of the F-distribution*

(a) 95% points

| | | Degrees of freedom of numerator (ν_1) | | | | | | | | |
		1	2	3	4	5	6	8	12	24	∞
	1	161·4	199·5	215·7	224·6	230·2	234·0	238·9	243·9	249·0	254·3
	2	18·51	19·00	19·16	19·25	19·30	19·33	19·37	19·41	19·45	19·50
	3	10·13	9·55	9·28	9·12	9·01	8·94	8·85	8·74	8·64	8·53
	4	7·71	6·94	6·59	6·39	6·26	6·16	6·04	5·91	5·77	5·63
	5	6·61	5·79	5·41	5·19	5·05	4·95	4·82	4·68	4·53	4·36
	6	5·99	5·14	4·76	4·53	4·39	4·28	4·15	4·00	3·84	3·67
	7	5·59	4·74	4·35	4·12	3·97	3·87	3·73	3·57	3·41	3·23
	8	5·32	4·46	4·07	3·84	3·69	3·58	3·44	3·28	3·12	2·93
	9	5·12	4·26	3·86	3·63	3·48	3·37	3·23	3·07	2·90	2·71
	10	4·96	4·10	3·71	3·48	3·33	3·22	3·07	2·91	2·74	2·54
	11	4·84	3·98	3·59	3·36	3·20	3·09	2·95	2·79	2·61	2·40
	12	4·75	3·89	3·49	3·26	3·11	3·00	2·85	2·69	2·51	2·30
	14	4·60	3·74	3·34	3·11	2·96	2·85	2·70	2·53	2·35	2·13
	16	4·49	3·63	3·24	3·01	2·85	2·74	2·59	2·42	2·24	2·01
	18	4·41	3·55	3·16	2·93	2·77	2·66	2·51	2·34	2·15	1·92
	20	4·35	3·49	3·10	2·87	2·71	2·60	2·45	2·28	2·08	1·84
	25	4·24	3·39	2·99	2·76	2·60	2·49	2·34	2·16	1·96	1·71
	30	4·17	3·32	2·92	2·69	2·53	2·42	2·27	2·09	1·89	1·62
	40	4·08	3·23	2·84	2·61	2·45	2·34	2·18	2·00	1·79	1·51
	60	4·00	3·15	2·76	2·53	2·37	2·25	2·10	1·92	1·70	1·39
	∞	3·84	3·00	2·60	2·37	2·21	2·10	1·94	1·75	1·52	1·00

Degrees of freedom of denominator (ν_2)

(b) 97·5% points

| | | Degrees of freedom of numerator (ν_1) | | | | | | | | |
		1	2	3	4	5	6	8	12	24	∞
	1	648	800	864	900	922	937	957	977	997	1018
	2	38·5	39·0	39·2	39·2	39·3	39·3	39·4	39·4	39·5	39·5
	3	17·4	16·0	15·4	15·1	14·9	14·7	14·5	14·3	14·1	13·9
	4	12·2	10·6	9·98	9·60	9·36	9·20	8·98	8·75	8·51	8·26
	5	10·0	8·43	7·76	7·39	7·15	6·98	6·76	6·52	6·28	6·02
	6	8·81	7·26	6·60	6·23	5·99	5·82	5·60	5·37	5·12	4·85
	7	8·07	6·54	5·89	5·52	5·29	5·12	4·90	4·67	4·42	4·14
	8	7·57	6·06	5·42	5·05	4·82	4·65	4·43	4·20	3·95	3·67
	9	7·21	5·71	5·08	4·72	4·48	4·32	4·10	3·87	3·61	3·33
	10	6·94	5·46	4·83	4·47	4·24	4·07	3·85	3·62	3·37	3·08
	11	6·72	5·26	4·63	4·28	4·04	3·88	3·66	3·43	3·17	2·88
	12	6·55	5·10	4·47	4·12	3·89	3·73	3·51	3·28	3·02	2·72
	14	6·30	4·86	4·24	3·89	3·66	3·50	3·29	3·05	2·79	2·49
	16	6·12	4·69	4·08	3·73	3·50	3·34	3·12	2·89	2·63	2·32
	18	5·98	4·56	3·95	3·61	3·38	3·22	3·01	2·77	2·50	2·19
	20	5·87	4·46	3·86	3·51	3·29	3·13	2·91	2·68	2·41	2·09
	25	5·69	4·29	3·69	3·35	3·13	2·97	2·75	2·51	2·24	1·91
	30	5·57	4·18	3·59	3·25	3·03	2·87	2·65	2·41	2·14	1·79
	40	5·42	4·05	3·46	3·13	2·90	2·74	2·53	2·29	2·01	1·64
	60	5·29	3·93	3·34	3·01	2·79	2·63	2·41	2·17	1·88	1·48
	∞	5·02	3·69	3·12	2·79	2·57	2·41	2·19	1·94	1·64	1·00

Degrees of freedom of denominator (ν_2)

(c) 99% points

| | | Degrees of freedom of numerator (v_1) | | | | | | | | |
		1	2	3	4	5	6	8	12	24	∞
Degrees of freedom of denominator (v_2)	1	4052	4999	5403	5625	5764	5859	5981	6106	6235	6366
	2	98·50	99·00	99·17	99·25	99·30	99·33	99·37	99·42	99·46	99·50
	3	34·12	30·82	29·46	28·71	28·24	27·91	27·49	27·05	26·60	26·13
	4	21·20	18·00	16·69	15·98	15·52	15·21	14·80	14·37	13·93	13·46
	5	16·26	13·27	12·06	11·39	10·97	10·67	10·29	9·89	9·47	9·02
	6	13·74	10·92	9·78	9·15	8·75	8·47	8·10	7·72	7·31	6·88
	7	12·25	9·55	8·45	7·85	7·46	7·19	6·84	6·47	6·07	5·65
	8	11·26	8·65	7·59	7·01	6·63	6·37	6·03	5·67	5·28	4·86
	9	10·56	8·02	6·99	6·42	6·06	5·80	5·47	5·11	4·73	4·31
	10	10·04	7·56	6·55	5·99	5·64	5·39	5·06	4·71	4·33	3·91
	11	9·65	7·21	6·22	5·67	5·32	5·07	4·74	4·40	4·02	3·60
	12	9·33	6·93	5·95	5·41	5·06	4·82	4·50	4·16	3·78	3·36
	14	8·86	6·51	5·56	5·04	4·69	4·46	4·14	3·80	3·43	3·00
	16	8·53	6·23	5·29	4·77	4·44	4·20	3·89	3·55	3·18	2·75
	18	8·29	6·01	5·09	4·58	4·25	4·01	3·71	3·37	3·00	2·57
	20	8·10	5·85	4·94	4·43	4·10	3·87	3·56	3·23	2·86	2·42
	25	7·77	5·57	4·68	4·18	3·86	3·63	3·32	2·99	2·62	2·17
	30	7·56	5·39	4·51	4·02	3·70	3·47	3·17	2·84	2·47	2·01
	40	7·31	5·18	4·31	3·83	3·51	3·29	2·99	2·66	2·29	1·80
	60	7·08	4·98	4·13	3·65	3·34	3·12	2·82	2·50	2·12	⊥·40
	∞	6·63	4·60	3·78	3·32	3·02	2·80	2·51	2·18	1·79	1·00

The table gives for various degrees of freedom, v_1, v_2, the values of F such that

$$P = 100 \int_0^F h(z \mid v_1, v_2)\, dz$$

where $h(z \mid v_1, v_2)$ is the probability density function of the F-ratio (6·6).

Table 7 *Percentage points of the studentized range q* ($\alpha\%$, n, v)
(a) 5% points

v	n 2	3	4	5	6	7	8	10	12
5	3·64	4·60	5·22	5·67	6·03	6·33	6·58	6·99	7·32	8·21
6	3·46	4·34	4·90	5·30	5·63	5·90	6·12	6·49	6·79	7·59
7	3·34	4·16	4·68	5·06	5·36	5·61	5·82	6·16	6·43	7·17
8	3·26	4·04	4·53	4·89	5·17	5·40	5·60	5·92	6·18	6·87
9	3·20	3·95	4·41	4·76	5·02	5·24	5·43	5·74	5·98	6·64
10	3·15	3·88	4·33	4·65	4·91	5·12	5·30	5·60	5·83	6·47
11	3·11	3·82	4·26	4·57	4·82	5·03	5·20	5·49	5·71	6·33
12	3·08	3·77	4·20	4·51	4·75	4·95	5·12	5·39	5·61	6·21
13	3·06	3·73	4·15	4·45	4·69	4·88	5·05	5·32	5·53	6·11
14	3·03	3·70	4·11	4·41	4·64	4·83	4·99	5·25	5·46	6·03
15	3·01	3·67	4·08	4·37	4·59	4·78	4·94	5·20	5·40	5·96
20	2·95	3·58	3·96	4·23	4·45	4·62	4·77	5·01	5·20	5·71
30	2·89	3·49	3·85	4·10	4·30	4·46	4·60	4·82	5·00	5·47
60	2·83	3·40	3·74	3·98	4·16	4·31	4·44	4·65	4·81	5·24
120	2·80	3·36	3·68	3·92	4·10	4·24	4·36	4·56	4·71	5·13
∞	2·77	3·31	3·63	3·86	4·03	4·17	4·29	4·47	4·62	5·01

(b) 1% points

v	n 2	3	4	5	6	7	8	10	12	20
5	5·70	6·98	7·80	8·42	8·91	9·32	9·67	10·24	10·70	11·93
6	5·24	6·33	7·03	7·56	7·97	8·32	8·61	9·10	9·48	10·54
7	4·95	5·92	6·54	7·01	7·37	7·68	7·94	8·37	8·71	9·65
8	4·75	5·64	6·20	6·62	6·96	7·24	7·47	7·86	8·18	9·03
9	4·60	5·43	5·96	6·35	6·66	6·91	7·13	7·49	7·78	8·57
10	4·48	5·27	5·77	6·14	6·43	6·67	6·87	7·21	7·49	8·23
11	4·39	5·15	5·62	5·97	6·25	6·48	6·67	6·99	7·25	7·95
12	4·32	5·05	5·50	5·84	6·10	6·32	6·51	6·81	7·06	7·73
13	4·26	4·96	5·40	5·73	5·98	6·19	6·37	6·67	6·90	7·55
14	4·21	4·89	5·32	5·63	5·88	6·08	6·26	6·54	6·77	7·39
15	4·17	4·84	5·25	5·56	5·80	5·99	6·16	6·44	6·66	7·26
20	4·02	4·64	5·02	5·29	5·51	5·69	5·84	6·09	6·28	6·82
30	3·89	4·45	4·80	5·05	5·24	5·40	5·54	5·76	5·93	6·41
60	3·76	4·28	4·59	4·82	4·99	5·13	5·25	5·45	5·60	6·01
120	3·70	4·20	4·50	4·71	4·87	5·01	5·12	5·30	5·44	5·83
∞	3·64	4·12	4·40	4·60	4·76	4·88	4·99	5·16	5·29	5·65

Hints to the solution of selected exercises

General points about the exercises

Although it is not essential to work through every one of the exercises, they must not be thought of as optional, as they are closely bound up with the text. Some exercises continue the teaching of the section, such as Exercise 2.6.2., Exercise 2.6.4., Exercise 3.1.3., Exercise 3.4.3., Exercise 4.5.2., etc. Other exercises lead up to teaching to be given later; for example, Exercise 3.4.4., Exercise 5.2.2., Exercice 5.2.3., and Exercise 5.2.4. lead up to Section 5.7. Other exercises are numerical, and it is essential to work through these so that the full meaning and usefulness of the statistical techniques can be grasped.

This section contains the answers to almost all of the exercises. Some exercises are very straightforward, and merely involve plotting, etc. Some other exercises are easily done following hints given, for example, Exercise 2.6.5. For a few exercises, a reference is given to a readily available source where the answers are clearly set out, for example, Exercise 5.7.1. The answers are very brief, and in some cases merely vital calculations are given.

Some comments on teaching

Chapter 4 : In place of the sampling experiment described, sampling from Shewhart's normal bowl could be described. A description of this is given in Grant (1946, pp 85-100).

Chapter 5 : If desired, Section 5.4 could be done before Section 5.3, but inference about the mean comes more naturally to students. However, a sampling experiment to describe the X^2 distribution could be carried out at the begginning of the Chapter.

Chapter 7 : This chapter could be omitted if desired. However, if the idea of confidence intervals and significance tests is completely new to students, at least some of this chapter could be given in order to help clear up difficulties.

Chapter 8 : This chapter could be omitted in a short course. For further reading on the analysis of discrete data see Maxwell (1961).

Chapter 10 : In this chapter there are many applications of expectation and variance techniques, and of the t, X^2 and F tests discussed in Chapters 5 and 6. At least a very brief description of some of Madansky (1959) should be given, to emphasise that it is not *always* correct to apply linear regression techniques.

Chapter 12 : Students should be encouraged to consult Cameron (1951) with reference to Section 12.7 and Eisenhart (1947) with reference to Section 12.8; both of these are simple articles.

331

Miscellaneous exercises
See Fraser (1950) with regard to exercise 11.

Suitable articles in journals
Some journals contain articles which the students could be encouraged to read, or which
could be used in class for teaching or exercises.

In particular, Applied Statistics, Biometrics, and Industrial Quality Control are useful
in this connection.

Chapter 1

Page 32, Exercises 1.3
1.
$\quad$ (*i*) $\bar{x} = 128.47 \quad s^2 = 343.12$

$\quad$ (*ii*) $\bar{x} = 128.79 \quad s^2 = 337.45$ $\quad$ (In units of 5)

$\qquad \bar{x} = 128.71 \quad s^2 = 339.78$ $\quad$ (In units of 10)

$\qquad \bar{x} = 129.21 \quad s^2 = 342.68$ $\quad$ (In units of 15)

2. It is clear that truncated and Winsorized means will be biased estimates of the mean when
the distribution is skew.

Chapter 2

Page 48, Exercises 2.1.
1. $\quad$ (*i*) The real line, with rather ill-defined upper and lower bounds.

$\qquad$ (*ii*) A two-way table with, say six columns to represent the result of the first throw, and
six rows to represent the result of the second throw.

$\qquad$ (*iii*) 2,3, ..., 12.

$\quad$ Note especially that whereas every entry in (*ii*) is equally likely, every entry in (*iii*) is
not equally probable.

2. 0.05.

Page 51, Exercises 2.2
2. The events A and B *can* occur together, so that Addition Law II should have been used
(formula 2.13).

Page 54, Exercises 2.3
1. The events A and B are not independent. Pr $(A|B) = 1$.

2. Pr (all links hold) $= (1 - \theta)^n$

$\quad$ Pr (chain fails) $= 1 - (1 - \theta)^n$

3. $\qquad$ (*i*) $^{48}C_{13} \big/ ^{52}C_{13}$ $\qquad\qquad$ (*ii*) $^4C_1 \, ^{48}C_{12} \big/ ^{52}C_{13}$

Page 57, Exercises 2.4
1. The required probability is

$\qquad$ Pr {(c − 1) defectives in (n − 1) items} Pr {last item defective}

$\quad$ which is

$$^{n-1}C_{c-1}\,\theta^{c-1}(1 - \theta)^{n-c} \times \theta$$

2. The series to be summed is a geometric progression.

3. Let $B, \bar{B}$ be the events that B does and does not contract the disease respectively. The
probabilities are easily calculated from the following flow diagram.

For a family of four children, the probabilities that 0, 2, 2, 3 of the other children contract the disease are $(1 - \theta)^3$, $3\theta (1 - \theta)^4$, $6\theta^2 (1 - \theta)^4 + 3\theta^2 (1 - \theta)^2$, $\theta^3 (13 - 27\theta + 21\theta^2 - 6\theta^3)$

Page 62, Exercises 2.5

1. A table of results to this exercise will help the student to see the Poisson distribution as a limit of the binomial. See Biometrika Tables (3rd Edition) pp 210 and 218.

3. This is the geometric distribution, see Example 2.4

$$Pr\,(r) = (0.9)^{r-1}\,(0.1).$$

4. See exercise 2.4.2.

5. We assume that the probability of two events in the time interval $(t, t + \delta t)$ is of the second order, $0\,(\delta t)^2$. Then if there are x events to time $(t + \delta t)$, either there were x events to time t, and no event occured in the interval $(t, t + \delta t)$, or else there were $(x - 1)$ events to time t, and one event in $(t, t + \delta t)$. This argument leads to the equations given, and from these we obtain

$$\frac{dP_0\,(t)}{dt} = -\lambda P_0\,(t)$$

$$\frac{dP_x\,(t)}{dt} = \lambda P_{x-1}\,(t) - \lambda P_x\,(t), \qquad x = 1, 2, \ldots$$

These equations can easily be solved successively.
See Brownlee (1960) p. 137-9.

1. (i) 0.0913, (ii) 0.0478, (iii) 0.4962, (iv) 0.1872.
2. It is helpful for the student to see the pattern made by different distributions on probability papers.
3. See Birnbaum (1962) p. 153.
4. Following Exercise 2.5.5.

$$\Phi\,(t + \delta t) = \Phi\,(t)\,(1 - \lambda\delta t)$$
$$\Phi\,(t + \delta t) - \Phi\,(t) = -\lambda\Phi\,(t) \text{ and the result follows.}$$

Page 77, Exercises 2.7

1.

$z = 0$	x	
	0	1
0	0.18	0.12
y		
1	0.12	0.08

$z = 1$	x	
	0	1
0	0.245	0.105
y		
1	0.105	0.045

It is helpful to study the marginal and conditional distributions for this example.

2. In the following example, the marginal distributions for x and z, x and y, and y and z are identical and follow the pattern of independent probabilities.

$z = 0$	x	
	0	1
0	0.20	0.16
y		
1	0.16	0.08

$z = 1$	x	
	0	1
0	0.16	0.08
y		
1	0.08	0.08

The joint distribution of x, y, and z clearly shows statistical *dependence*.

3. The contours are ellipses. The mode is at:

$$x = \mu_x, \qquad y = \mu_y.$$

Chapter 3

Page 83, Exercise 3.1

2. $1/\theta$.

3. $E(x + y) = \int (x + y) f(x, y) \, dx \, dy$

$$= \int x f(x, y) \, dx \, dy + \int y f(x, y) \, dx \, dy$$

$$= E(x) + E(y).$$

4. $E(\bar{x}) = \dfrac{1}{n} E(x_i) = \mu$.

5. The constant is determined so that the integral of the p.d.f. over x and y is unity. Hence it can be shown that $k = 1$. The contours of equal probability density are straight lines. The marginal p.d.f. of x and the conditional p.d.f. of x when $y = \frac{1}{2}$ are both equal to $(x + \frac{1}{2})$

$$E(x) = E(x|y = \tfrac{1}{2}) = 7/12$$

$$E(xy) = 1/3$$

6. In the independence case, the integral factories.

Page 87, Exercises 3.2

1. For the geometric distribution $V(x) = (1 - \theta)/\theta^2$.

2. $V(x) = V(x|y = \tfrac{1}{2}) = 11/144$, $V(xy) = 1/18$.

3. The random variable

$$y = \sum_1^n x_i$$

has the binomial distribution. By (3.20)

$$V(y) = \Sigma V(x_i) = n \theta (1 - \theta).$$

Page 89, Exercises 3.3

		γ_1	γ_2
1.	Normal	0	0
	Exponential	2	6
	Rectangular	$3\sqrt{(128)}$	-1.2
	Binomial	$(1 - 2\theta)/\sqrt{n\theta(1 - \theta)}$	$(1 - 6\theta + 6\theta^2)/n\theta(1 - \theta)$

2. For the normal distribution,

$$V(S'^2) = 2\sigma^4/n.$$

Page 92, Exercise 3.4

1. $E(x_1) = E(x_2) = 1.00$, $E(x_1 x_2) = 1.20$, $C(x_1, x_2) = 0.20$.

2. $C(x, y) = -1/144$, $\rho = -1/11$.

3. $V(\Sigma a_i x_i) = E\{\Sigma a_i x_i - (E \Sigma a_i x_i)\}^2$

$$= E\{\Sigma (a_i x_i - a_i E(x_i))\}^2$$

The result follows on expanding the square.

4. This follows by a straighforward application of the result to Exercise 3.4.3; see also Exercise 5.2.3.

5. $V(y + x/a) = 2(1 + \rho)$

$$V(y - x/a) = 2(1 - \rho)$$

Since a variance must be positive or zero, $|\rho| \leqslant 1$.

6.

$$E(x) = \int_0^1 \int_0^{(1-x)} 2 x \, dx \, dy = 1/3$$

$$E(x^2) = \int_0^1 \int_0^{(1-x)} 2 x^2 \, dx \, dy = 1/6$$

Similarly $E(xy) = 1/12$, so that $C(x,y) = -1/36$, and $\rho = -1/2$.
Note. Another useful result can be obtained as an exercise.
Let x_i, $i = 1, 2, ..., n$ be any random variables, then define

$$y = \sum_1^n a_i x_i \qquad z = \sum_1^n b_i x_i.$$

It can easily be shown that

$$C(y,z) = \sum_{i < j} \sum a_i b_j\, C(x_i x_j) + \sum a_i b_i\, V(x_i).$$

Page 99, Exercise 3.5

1. Put $z = \log W$, then it looks reasonable to try the model $E(z) = a + bR$, $V(z) = \sigma^2$, where the variance is constant.
2. See Section 12.7.

Chapter 4

Page 104, Exercise 4.1

2. A model for Example 1.8 is discussed in the answers to Exercise 3.5.1:
   the main problems of inference arising are point and interval estimates for a, b, and σ^2, and the problem of studying how well the model fits. Problems of inference for Example 1.9 are discussed in Chapter 10, but it is helpful if the student can see beforehand that these problems arise. Models and problems of inference for Example 3.15 and Example 3.16 are obvious from the discussion in Section 3.5.

Page 109, Exercise 4.2

1. The tests described on pp. 106-7 could be carried out, and similar tests are easily devised. The student should see a need for some measure of agreement between theory and observed results; these problems are discussed in Chapter 8. See also Exercise 5.1.1.

   The parameters for $k = 11$ give a cycle of 50 terms only. The last digit in the cycles for all the remaining sets of parameters goes up by one each time. The starting value $u_1 = 1$ is also unsatisfactory. For $k = 201$, $u_1 = 643$ leads to a questionable portion of the series. The choice of a really good set of parameters is not a simple matter.

Page 116, Exercise 4.3

1. The distribution of the median has the greater dispersion.
2. It would be interesting to design a set of sampling trials to estimate the ratio of the variance of the mean to the variance of the median, for different variances of the 'wild' population. The truncated and Winsorised means could also be investigated. Such experiments lead to the general conclusion that even a small percentage of wild observations makes it profitable to consider a statistic such as the Winsorised mean.
3. If the sampling experiment is done as a class exercise the work of drawing these graphs (and other) can be divided among the class, and the results duplicated.
4. The integral is easily done by adding the exponents in the integral, and completing the square in x. Since z also has a normal distribution, this proves that any sum of normal variables is normally distributed.
5. If (x_i, s^2) is written out, it is seen to be a series of terms of zero expectation. Also, we have

$$C(\bar{x}, s^2) = \frac{1}{n} \sum_1^n C(x_i, s^2) = 0.$$

Now s^2 is invariant to a change in scale, so if we put
$$z_i = x_i + \theta,$$
$$C(z, s^2) = E\{(\bar{x} + \theta)\, s^2\} - E(\bar{x} + \theta)\, E(s^2)$$
$$= E(\bar{x} s^2) - E(\bar{x})\, E(s^2) = 0.$$

6. The standard errors of $\bar{x}$ and s^2 are respectively

$$\frac{\sigma}{\sqrt{n}} \text{ and } \sigma\sqrt{\left(\frac{2}{(n-1)}\right)}$$

It is particularly vital that the student be clear about the meaning of a sampling distribution.

Page 123, Exercise 4.4

1. The standard error of the mean of ten plots is $2.5/\sqrt{10} = 0.79$.
   If there are n plots, the width of 95% confidence intervals for the mean is $2 \times 1.96 \times 2.5/\sqrt{n}$
   Hence

$$\sqrt{n} = 2 \times 1.96 \times 2.5 \text{ or } n \simeq 96$$

2.
$$E(\hat{\theta}_1) = E(\bar{x}) = \theta$$
$$E(\hat{\theta}_2) = k\theta$$
$$E(\hat{\theta}_3) = \theta_3$$

The mean squared errors are respectively

$$\frac{\sigma^2}{n} \; ; \; \frac{k^2\sigma^2}{n} + (k-1)^2\theta^2 \; ; \text{ and } (\theta_3 - \theta)^2.$$

A graph of the three M.S.E.'s against θ should be given. The estimate $\hat{\theta}_3$ is useless, because θ is unknown.

Page 128, Exercise 4.5

1. The statistic

$$z = (\bar{x} - \bar{y})\Big/\sqrt{\left(\frac{\sigma_1^2}{n} \quad \frac{\sigma_2^2}{m}\right)}$$

is standard normal if $\mu_1 = \mu_2$, and large values of z are regarded as significant. A $100(1-\alpha)\%$ lower confidence bound for $\mu_1 - \mu_2$ is

$$(\bar{x} - \bar{y}) - \lambda_\alpha\sqrt{\left(\frac{\sigma_1^2}{n} + \frac{\sigma_2^2}{m}\right)}$$

where
$$\alpha = \int_{\lambda_\alpha}^{\infty} \sqrt{\left(\frac{1}{2\pi}\right)} \exp\left\{-\frac{1}{2}t^2\right\}dt$$

2. The statistic $\bar{x}\sqrt{n}/\sigma$ is compared with the standard normal distribution. But if $\mu = 0$, $\bar{x}\sqrt{n}/\sigma$ is $N(0,1)$.

Power	$= 1 - \Phi(1.96 - \mu_1) + \Phi(-1.96 - \mu_1)$			
μ_1	0.50	1.00	1.50	2.00
Power	0.0790	0.1700	0.3231	0.5160

The power for negative μ_1 is clearly equal to that for positive μ_1. A graph of the power should be drawn.

3. The power is $\{1 - \Phi(1.65 - 0.5\sqrt{n})\}$, and we want this to be 99%, or $\Phi(-2.33)$. Hence
$$1.65 - 0.5\sqrt{n} = -2.33, \; n \simeq 64.$$

Chapter 5

Page 130, Exercises 5.1

1. An interesting pattern emerges if u_r is plotted against u_{r+1}, in Exercise 4.2.1. This is best done for low values of M, and a large number of points is required.
2. This is one of the methods of generating approximately normally distributed random variables.

Page 132, Exercise 5.2

1. Range estimate = 0.0088.

$$s'^2 = 89 \ 10^{-6} \qquad s' = 0.0094$$
$$s^2 \ = 72 \ 10^{-6} \qquad s \ = 0.0085$$

2. The relevance of these results is explained in Section 5.7.
3. In normal samples $V(s^2) = 2\sigma^4/(n-1)$.
4. The result follows easily by using the second expression for s^2 in Exercise 5.2.2 above. The implications of this result and the result of Exercise 3.4.4 are discussed in Section 5.7, see also Scheffé (1959), chapter 10.

Page 135, Exercises 5.3
1. We use t on 8 d.f.: (a) −0.0115,; (b) −0.0131.
2. The percentage points are as follows;

approximate:	2.53	2.19	2.09	2.07	2.03
true:	2.57	2.23	2.13	2.09	2.04

In obtaining the approximate values, the deviation from normality has been ignored.
4. 39.56. However, Example 5.1 states that the mean of the modified process cannot be less than 40 lb,

Page 139, Exercises 5.4
1. (a) $264.2 \ 10^{-6}$, $32.9 \ 10^{-6}$ (b) $28.7 \ 10^{-6}$.
2. See Birnbaum (1962) pp. 191-7.

Page 141, Exercises 5.5
1. Standard error = $10^{-3} \ 2.83$, $t = 1.77$ on 8 d.f. The data do not show any evidence of a departure of the meter constants from unity.

Page 143, Exercises 5.6
1. $X^2 = 9$, on 8 d.f. There is no evidence that the variance differs from the previously established value of $(0.008)^2$.

Chapter 6

Page 146, Exercises 6.1
1. The standard error of the difference of two means is not equal to the sum of the separate standard errors.

Page 149, Exercises 6.2

1.

	Cut shoots	Rooted plants
Total	587	465
Mean	58.7	46.5
USS	34785.0	21997.0
Corr	34456.9	21622.5
CSS	328.1	374.5
Pooled CSS		702.6
Pooled est. var.		39.03
$V(\bar{x}_1 - \bar{x}_2)$		7.807
Standard error		2.794

Thus $t = 4.37$ on 18 d.f., which is significant beyond the 0.1% level. The data give very strong evidence that the concentration of chloramphenicol is greater in cut shoots than in rooted plants: 99% C.I. for the difference in concentration are 12.20 ± 8.07.

2. The pooled CSS is 73.980 on 48 d.f., giving $\sigma^2 = 1.54$, $\sigma = 1.24$, estimated standard error = 0.555.
The 1% t value on 48 d.f. is 2.68, so that the distance to the warning limits is
$$2.68 \times 0.555 = 1.49.$$
However, the CSS for samples 9 and 10 alone are 11.98 and 31.09, and sample 10 contains an observation which must be regarded as suspicious on any count. If samples 9 and 10 are cut out of the estimation of σ^2, we have 40 d.f., and $\sigma^2 = 0.77$, $\sigma = 0.88$, estimated standard error = 0.39

distance to limits = 2.70 × 0.39 = 1.06.

A normal plot of the residuals reveals some skewness.

3. $V = \{2a^2/\nu_x + 2\,(1-a)^2/\nu_y\}\sigma^4,\ a = \nu_x/(\nu_x + \nu_y)$

Page 152, Exercises 6.3

3. The estimated variances are 36.46 and 41.61 for cut shoots and rooted plants, respectively. The F-ratio is 1.14 on (5,5) d.f., and 90% confidence intervals for the ratio of the variance of rooted plants to the variance of cut shoots are (0.36,3.64).

4. Write

$$s^2/\sigma^2 = 1 + \left(\frac{s^2 - \sigma^2}{\sigma^2}\right)$$

and then expand this to obtain

$$E\,(\sigma^2/s^2)\ 1 + 2/\nu$$

$$E\,(\sigma^4/s^4)\ 1 + 6/\nu.$$

We then obtain

$$E\,(F) \simeq 1 + \frac{2}{\nu_2},\ V\,(F) \simeq 2\left(\frac{1}{\nu_1} + \frac{1}{\nu_2}\right)$$

These formulae help the student to appreciate the form of the F-distribution.

5. For several sample variances s_i^2, calculate the sample variance of log s_i^2, and compare the theoretical value of $2/\nu$ using the test of Section 5.6. This rests on the assumption that the distribution of log X^2 is nearly normal. The student should be cautioned that this method is rather sensitive to the normality of the original observations.

Page 154, Exercises 6.4

1. It is important to point out that the two-sample analysis is not strictly valid in this case. If the experiment is a paired comparisons design, some of the observations will be correlated. However, we obtain the following results.

Pooled CSS	1.19375
Est. variance	0.08527 (on 14 d.f.)
Est. var. mean	0.02132
Est. standard error	0.1460
t	1.80

This result is just significant at the 10% level. The standard error is about twice that of the paired samples analysis, as day to day variation has been included in the error.

The degrees of freedom are twice that of the paired samples analysis. With small samples, this would lead to a more powerful test than the paired samples analysis provided the standard errors of the two analyses were comparable.

2. It is important to avoid possible sources of bias, and the surest way to do this is to randomise.

3.

(a)	Total	90.0
	Mean	7.50
	USS	753.28
	Corr	675.00
	CSS	78.28
	Est. variance	7.1158 (on 11 d.f.)
	Est. var. mean	0.5930
	Est. standard error	0.7701
	95% C.I.	5.81, 9.19

(b) $2 \times 1.96 \times 3/\sqrt{n} = 2,\quad n \simeq 35$

Or, $2\,t_{n-1}\,(2\tfrac{1}{2}\%)/\sqrt{n} = 2,\quad n \simeq 36$

Page 159, Exercises 6.5

1. The mean difference is 0.588, with a standard error of 0.541. In two or three cases the pairing has worked out particularly badly.

2.

	Laboratory 1		Laboratory 2	
	A	B	A	B
Total	44.23	38.06	48.03	33.77
Mean	5.5287	4.7575	6.0037	4.2212
USS	250.2743	187.6130	291.5443	148.6771
Corr.	244.5388	181.0704	288.3577	142.5516
CSS	5.7355	6.5426	3.1866	6.1255
Est. σ^2	0.81	0.93	0.45	0.87

The most extreme F-ratio is 6.54/3.19, which is nowhere near significance on F (7,7)
This is a conservative test. Tests of the difference $(A - B)$ are as follows:

	Laboratory 1	Laboratory 2
Pooled CSS	12.2781	9.3121
Pooled est. σ^2	0.8770	0.6652
Est. $V(\bar{x}_A - \bar{x}_B)$	0.2192	0.1663
Standard error	0.4682	0.4078
Diff $(\bar{x}_A - \bar{x}_B)$	0.7713	1.7825
t (14 d.f.)	1.65	4.37

A one-sided test would be appropriate. The possibility of an interaction should be examined, by doing a test on

$$(\bar{x}_{1A} - \bar{x}_{1B}) - (\bar{x}_{2A} - \bar{x}_{2B})$$

This yields $t = 1.63$ on 28 d.f. (two-sided), which is not significant.

There are various other possible methods of answering the question, but interest lies chiefly in the treatment difference. There are also other t-tests which could be calculated.

Chapter 7

Page 165, Exercises 7.2

	True value	Normal approximation	
		with	without correction
1. $Pr (x \geqslant 9)$	0.00195	0.00383	0.00135
$Pr (x \leqslant 1)$	0.01953	0.02275	0.00982
$Pr (x \leqslant 2)$	0.0898	0.0913	0.0478
$Pr (3 \leqslant x \leqslant 7)$	0.8906	0.8860	0.9424

2.
$$Pr (x \leqslant r) = \Phi \left(\frac{r + \frac{1}{2} - \mu}{\sqrt{\mu}} \right)$$

Difficulties must arise if μ is too small, say
$$\mu - 3\sqrt{\mu} < 0$$
or
$$\mu < 9.$$

Page 168, Exercises 7.3
1. The significance level is 2.15%. A 97.8% C.I. is 0.11, 3.67.
2. $\Phi (- 8\frac{1}{2}/\sqrt{10}) = 0.0036.$

Page 173, Exercises 7.4
1.   S+   0 1 2 3 4 5 6 7 8 9 10 11 12 ... 21
        frequency 1 1 1 2 2 3 4 4 4 5 5 5 4 ... 1
2. We have S+ = 50, and the significance can be found either by the normal approximation (2.8%), or by exact calculation (0.2%).
        S+   55 54 53 52 51 50
        frequency 1 1 1 2 2 3
        Significance level $= 2 \times 10 \times 2^{-10} \simeq 0.002$

3. $(n (n + 1)/4) - 3\sqrt{[n (n + 1) (2n + 1)/24]} > 0$, or $n > 12$.
The calculation for $n = 10$ in Exercise 7.4.2 above, shows a large discrepancy between the true probability and the normal approximation.

Page 174, Exercises 7.5
1. $E (d_r) = m/(m + n)$, $V (d_r) = mn/(m + n)$,
                          $C (d_r, d_s) = - mn/(m + n)^2(m + n - 1)$.
2. There is one tied rank, which can be averaged. We find the smaller sum to be $62\frac{1}{2}$. By the normal approximation, the significance level is about 0.07%, which agrees very well with the earlier result.

Page 176, Exercises 7.6
1.   Significance level $= 7/128 = 0.0547$.
                          Power $= p^{10} + 10p^9 (1 - p) + 45p^8 (1 - p)^2$
                          where $p = Pr$ (positive observation) $= \Phi (\mu/\sigma)$

μ	0	0.40	0.80	1.20
p	0.50	0.6554	0.7881	0.8849
P	0.0547	0.2734	0.6416	0.9015

Chapter 8

Page 180, Exercises 8.2
2. We find

$$z = \sin^{-1} \sqrt{\theta} + \frac{(r - n\theta)}{2n\sqrt{\{\theta (1 - \theta)\}}} + \cdots$$

from which the result follows.
4. From Exercise 7.2.2 it is natural to try $m = 9$ first. One possibility is to try
                          $\Phi \{ 2\sqrt{} (r + \frac{1}{2}) - 2\sqrt{m} \}$,
and denote this approximation (2), and approximation (8.5) as (1).
We find the following values

	$Pr (r \leqslant 2)$	$Pr (r \leqslant 4)$
True	0.00623	0.0550
(1)	0.015005	0.0668
(2)	0.00225	0.0394

A further possibility is to examine approximations of the type
                          $\Phi \{ 2 (\sqrt{} (x + a) - \sqrt{} (m + b)) \}$.

Page 184, Exercises 8.3
1. See Table 8.7 on p. 196. $\hat{\mu} = 6.9150$. 95% C.I. are (6.56, 7.29). Compare result with Exercise 8.13, page 197.
3. C.I. are $\{x + \frac{1}{4} \lambda^2 \pm \lambda\sqrt{x}\}$. 95% C.I. are (20.23, 41.70).
4. 95% C.I. are (20.24, 42.82). But the ordinate $Pr (r = x)$ has been included in calculating both limits.
5. Approximately 222 observations would be required.

Page 186, Exercises 8.4
1. 99% C.I. are (0.9001, 0.9167). [Use $x = 7264$, $n = 8000$.]

Page 187, Exercises 8.5
1. A confidence interval for $\theta = k/(1 + k)$ can be obtained using (8.13) with $r = 30$, $n = 51$. This leads to (0.4857, 0.7661). Here $k = \mu_1/\mu_2$, and to obtain confidence intervals for the rate per hour, k must be multiplied by 26/75. In marginal cases, results conflicting with significance tests can occur, owing to continuity corrections, etc.
3. The value of the statistic (8.15) is 3.42.

Page 192, Execersises 8.6

1. For Exercise 2.1 X^2 is negligible and non significant. (In fact it is suspiciously small.) For Exercise 2.2 $X^2 = 220.9$, and very highly significant.

2.

	paralysis	no paralysis
Vac.	2	10
Not vac.	9	4

$X_1^2 = 3.68$

Page 194, Exercises 8.7

1.

		A	
		I	N
	I	25	9
B			
	N	1	5

The statistic (8.15) has the value 2.21.

Page 197, Exercises 8.8

1.

Total	136
Mean	27.2
USS	3958
Corr.	3699.2
CSS	258.8

$$X^2 = \frac{258.8}{27.2} = 9.51$$

on 3d.f.

2. Since the dispersion test deals with the overall frequency distribution, a trend would probably pass undetected. Various techniques could be used in the analysis. One simple device would be to use the square root transformation and then regression analysis.

3. The method of proof can be extended to discuss the variance of the statistic.

Page 201, Exercises 8.9

2. $\qquad CSS = 17.0, X^2 = 4.26$ on 3d.f.

Page 205, Exercises 8.10

1. From table 8.7, $\bar{x} = 6.915$. The expectations for $x = 0$ and 1 should be combined. We find X^2 on 12d.f. is 9.16.

2. Here

$$\hat{\lambda} = \Sigma x_i / \Sigma t_i,$$

and we calculate

$$X^2 = \Sigma \{x_i - \hat{\lambda} t_i\}^2 / \hat{\lambda} t_i.$$

Page 209, Exercises 8.11

1. (*i*) Expected values for the 3×3 table given in Exercise 8.11.1 are as follows:

58.23	464.27	1363.50
63.70	507.84	1491.46
25.07	199.89	587.04

$X_4^2 = 295.6$

(*ii*) Test the 2×3 table of total columns from Table 8.12 and Exercise 8.11.1. Expected values are

	α	β	γ	
A	3107.87	3180.78	1409.35	7698
B	1922.13	1967.22	871.65	4761
	5030	5148	2281	12459

$X_2^2 = 15.25$

(*iii*) Expected values for newspaper group α are:

22.50	333.71	1529.80	1886
37.50	556.29	2550.20	3144

$X_2^2 = 22.60$

For newspaper group β:

72.53	552.62	1437.85	2063
108.47	826.38	2150.15	3085

$\chi_2^2 = 5.41$

For newspaper group γ:

52.69	271.97	487.34	812
95.31	492.03	881.66	1469

$\chi_2^2 = 12.49$

(iv) Expected values are:

	Well to do	Middle class	Working class	Total
A	240.35	1873.99	5583.66	7698
B	148.65	1159.01	3453.34	4761

$\chi_2^2 = 0.33$

Page 214, Exercises 8.12

2.

	Table 8.17	Table 8.18
Total	145	132
USS	699	472
Corr	539.1	396
CSS	159.9	76
θ	0.5311	0.4286
$n\theta(1-\theta)$	1.7432	1.7143
χ^2	91.2	44.3
d.f.	38	43

4. Frequency distributions are as follows

Col	1	2	3	4	5	6	7	8	Total
0	1	0	2	1	4	1	3	1	13
1	3	4	3	7	8	8	4	1	38
2	5	2	4	4	2	3	2	5	27
3	5	5	3	5	3	4	4	1	30
4	0	3	6	1	1	1	2	3	17
5	4	3	0	0	0	1	1	0	9
6	0	1	0	0	0	0	2	3	6
7	0	0	0	0	0	0	0	3	3
8	0	0	0	0	0	0	0	1	1

χ^2 dispersion for the total column is 176 on 143 d.f., which is significant at about 3%.

The column totals are respectively 48, 56, 44, 34, 25, 35, 45, 73, showing very strong evidence of a difference in density along the columns.

A glance at the data shows evidence of patches of low and high density. One method of analysis would be to group the rows in twos or threes, and analyse the square root of the totals by analysis of variance. A number of methods of attack are possible.

5.

Litter size	4	5	6	7	Total
No. Males	106	283	658	1284	2331
No. Females	106	297	668	1334	2405
Total	212	580	1326	2618	4736
Proportion male	0.50000	0.4879	0.4962	0.4905	0.4922

The χ^2 for testing variation in the proportion male with litter size is only about 0.21 on 3 d.f., and is suspiciously small. There is no significant difference between the overall proportion male and 0.50. Dispersion tests are as follows, and they all indicate less variation than in a binomial distribution.

Litter size	4	5	6	7
Total	106	283	658	1284
USS	248	823	2228	4994
Corr	212	690.42	1959.11	4408.17
CSS	36	132.58	268.89	585.83
$n\theta(1-\theta)$	1.00	1.2493	1.4999	1.7494
χ^2	36	106.1	179.3	334.9
d.f.	52	115	220	373

Chapter 9

Page 221, Exercises 9.2

1. One criticism of rough methods of estimation is that they do not usually yield standard errors. Furthermore, any method of estimating a standard error which is devised would tend to be rather sensitive to the accuracy of the point estimates in question.

Page 223, Exercises 9.3

1.
$$\hat{\theta}_1 = (2y_1 - y_2 + y_3)/3$$
$$\hat{\theta}_2 = (2y_2 - y_1 + y_3)/3$$
$$\hat{\theta}_3 = (y_1 + y_2 + 2y_3)/3$$

The variances are equal at $2\sigma^2/3$.

Chapter 10

Page 227, Exercises 10.1

2. Put Y = yield strength, and X = Brinell hardness, and try $Y = \alpha + \beta X$, but where the observed variables are not Y and X, but
$$y = Y + u, \quad x = X + v,$$
and u, v are normally distributed errors.

Page 230, Exercises 10.2

2.

	y	yx	x	
Total	97.3	—	43.2	
Mean	9.73	—	4.32	
USS	950.11	425.78	199.64	$\hat{\beta} = 0.4178$
Corr	946.73	420.34	186.62	Reg. SS = 2.273
CSS	3.38	5.44	13.02	Dev. SS = 1.107

Page 234, Exercises 10.3

1.

Source	CSS	df	M.S.
Regression	2.273	1	2.273
Deviation	1.107	8	0.138
Total	3.380	9	2.411

$F = 16.5$ on (1,8) d.f., very highly significant. 99% C.I. for σ^2: 0.050, 0.826

95% C.I. for α: $9.73 \pm 2.31 \times 0.1747 = 9.33, 10.13$

95% C.I. for β: $0.4178 \pm 2.31 \times 0.1030 = 0.1799, 0.6557$

3. The residuals are functions of ϵ_i and x_i only, and not of α and β. Hence the Dev. SS is independent of α and β. Therefore put $\alpha = \beta = 0$, and use
$$E(\text{Dev. SS}) = E(\text{Total SS}) - E(\text{Regr. SS}).$$

Page 236, Exercises 10.4

1.
$$y = 133.96 + 37.54(x - 0.575)$$

Predicted mean = 138.65

Deviations mean square = 0.09

Estimated variance of prediction = 0.135

This estimated variance is worth very little since it is based on the doubtful assumption that the regression is still linear. Confidence intervals are obtained by using t on 2d.f.

Page 237, Exercises 10.5

1. $V(\hat{\beta}) = 0.0454 / (2 \times 532.63) = 0.00004262$
    95% C.I. for β are $0.3041 \pm 2.16 \times 0.0065 = 0.2901, 0.3181$
2. $(0.3087 - 0.2995) \pm 2.16 \sqrt{(0.000170)} = -0.0189, 0.0373$
3.

	y	yx	x
Total	99.3		36.4
mean	9.93		3.64
USS	988.95	365.64	140.74
Corr	986.05	361.45	132.50
CSS	2.90	4.19	8.24

Combined $\hat{\sigma}^2 = 0.1175$.
Est. $V(\hat{\beta}_1 - \hat{\beta}_2) = 0.0233$. (16d.f.)

There is no significant difference between the slopes. The common estimate of slope is 0.4530.

4. Use (6.11) and (6.12) with $CS_1(x, x)$ and $CS_2(x, x)$ instead of n_1 and n_2, $\hat{\beta}_1$ and $\hat{\beta}_2$ instead of $\bar{x}_1$ and $\bar{x}_2$, and the deviations mean squares instead of s_1^2 and s_2^2.

Page 239, Exercises 10.6

1. Use a scale for x, $-3, -1, +1, +3$. We then obtain $CS(y, y) = 17.80$ $CS(y, X) = 18.77$, $CS(X, X) = 20$ Reg. SS. $(y, x) = 17.62$, Dev. SS $(y, x) = 0.18$

Page 244, Exercises 10.8

1. The intercepts are $(\bar{y} - \hat{\beta}\bar{x})$. The difference in intercepts is 0.51. The variance of this is

$$\sigma^2 \left(\frac{1}{n_1} + \frac{1}{n_2} \right) + (\bar{x}_1 - \bar{x}_2)^2 \, V(\hat{\beta})$$

which is 0.0260, standard error 0.161. Hence

$$t = 0.51 / 0.161 = 3.17, \text{ on 16d.f.}$$

This example sets out the basic idea of the analysis of covariance.

3. The predicted volume is

$$\bar{z} + (Y_0 - \bar{y}) / \hat{\beta}$$

and the variance must be obtained by approximate methods. The reference gives a full solution.

Chapter 11

Page 249, Exercises 11.1

1.
$$\Sigma y_i - n\hat{a} - \hat{\beta}_1 \Sigma \exp \{\beta_2 \, x_i\} = 0$$

$$\Sigma y_i \exp \{\hat{\beta}_2 \, x_i\} - \hat{a} \Sigma \exp \{\hat{\beta}_2 \, x_i\} - \hat{\beta}_1 \Sigma \exp \{2\beta_2 \, x_i\} = 0$$

$$\Sigma y_i \, x_i \exp \{\hat{\beta}_2 \, x_i\} - \hat{a} \Sigma x_i \exp \{\hat{\beta}_2 \, x_i\} - \hat{\beta}_1 \Sigma x_i \exp \{2 \, \hat{\beta}_2 \, x_2\} = 0.$$

Page 252, Exercises 11.2

3. Compare this with Exercise 10.3.3. The quick method of solution can be used.

Page 258, Exercises 11.3

1. Use transformed scale for x_1 throughout.
$$CS(x_1, x_1) = 50, \ CS(x_2, x_2) = 1.7217, \ CS(x_1, y) = 18.99$$
$$CS(x_2, y) = 1.3047, \ CS(y, y) = 11.2074.$$

Anova			
Source	CSS	d.f.	M.S.
Regression on (x_1, x_2)	7.92	2	3.96
Regression on x_i (ign. x_2)	7.21	1	
Regression on x_2 (adj. x_1)	0.71	1	0.71
Deviations	3.28	47	0.07
Total	11.20	49	

Page 261, Exercises 11.4

$$\Sigma y_i = 50.77, \ CS\,(y, y) = 6.2217, \ CS\,(x_1, y) = -6.77$$
$$CS\,(x_2, y) = -1.31, \ CS\,(x_3, y) = -0.71, \ CS\,(x_4, y) = 0.03.$$

Anova			
Source	CSS	d.f.	M.S.
Due to x_1	5.7291	1	5.73
Due to x_2	0.2145	1	0.21
Due to x_3	0.0630	1	0.06
Due to x_4	0.0000	.1	0.00
Devs.	0.2151	7	0.03
Total	6.2217	11	

Chapter 12

Page 266, Exercises 12.1

1. $$E\,(W_1) = \mu \ \Sigma a_{1i} = 0$$

Hence
$$E\,(W_1{}^2) = V\,(W_1) = \Sigma a_{1i}^2 \ V\,(\bar{x}_i)$$
$$= \sigma^2 \ \Sigma a_{1i}^2 / n.$$

But W_1 is a linear combination of normal variables, and hence is normal. Therefore the distribution of

$$W_1{}^2 / (\sigma^2 \ \Sigma a_{1i}^2 / n)$$
is χ^2 on 1 d.f.

Now
$$C\,(W_i, W_j) = \Sigma\Sigma a_{ir} a_{js} \ C\,(\bar{x}_r, \bar{x}_s)$$
$$= \sigma^2 \ \Sigma\Sigma a_{ir} a_{jr} / n.$$

Therefore W_i, W_j are uncorrelated, and therefore also independent in this case, if
$$\Sigma\Sigma a_{ir} a_{jr} = 0.$$

This problem leads on to Section 12.5. Students are naturally led to ask for a proof that the sum of the CSS for $(g - 1)$ independent contrasts is equal to the total CSS between groups. Such a proof is easy by matrix theory, but I do not know of a simple algebraic proof.

Page 270, Exercises 12.2

1. Source	CSS	d.f.	E (M.S.)
Between groups	$\sum_{1}^{g} n_i \,(\bar{x}_i - \bar{x})^2$	$g - 1$	$\sigma^2 + \dfrac{1}{(g-1)} \Sigma n_i \,(\mu_i - \mu)^2$
Within groups	$\sum_{i=1}^{g} \sum_{j=1}^{n_i} (x_{ij} - \bar{x}_i)^2$	$N - g$	σ^2
Total	$\sum_{i=1}^{g} \sum_{j=1}^{n_i} (x_{ij} - \bar{x})^2$	$N - 1$	
where	$N = \sum_{1}^{g} n_i$		

Page 275, Exercises 12.3

2.

Anova

Source	CSS	d.f.	M.S.
Between groups	15.10	3	5.03
Within groups	21.59	28	0.77
Total	36.69	31	

When linear contrasts have been done (see Section 12.5) the 3 d.f. between groups can be separated; see Exercise 12.5.1. At this stage an analysis by LSD or studentised range should follow.

3. (i)
$$V\left(\overline{x}_0 - \overline{x}_i\right) = \sigma^2\left(\frac{1}{m} + \frac{1}{n}\right)$$

(ii)
$$m = N/(1 + \sqrt{t}), \; n = N/(t + \sqrt{t}), \; N = m + tn.$$

Thus we have $m = \sqrt{tn}$.

Page 285, Exercises 12.5

1. The between groups CSS can be split up as follows:

	CSS	d.f.
Between laboratories	0.01	1
Between treatments	13.04	1
Labs. × treatments	2.05	1
Between groups	15.10	3

Page 288, Exercises 12.6

1.
Potash per acre	36	54	72	108	155
Treatment total	23.55	24.16	23.23	22.54	22.35

These totals indicate a smooth response curve. This indicates that the next step would be to fit a curve by regression on potash level, and then estimate the optimum potash level. The analysis of variance is as follows:

Anova

Source	CSS	d.f.	M.S.
Blocks	0.0971	2	0.049
Treatments	0.7324	4	0.183
Residual	0.3495	8	0.044
Total	1.1790	14	

2.
Treatment code	00	01	10	11	20	21
Total	43.9	50.7	44.3	57.2	53.2	57.3

Anova

Source	CSS	d.f.	M.S.
Blocks	87.52	2	43.76
Treatment	59.43	5	11.89
Residual	33.26	10	3.33
Total	180.21	17	

The treatments CSS can be separated as indicated.

Contrast	Meaning	CSS	d.f.
1	Phosphate effect	31.47	1
2	Nitrogen linear effect	21.07	1
3	Nitrogen quadratic effect	0.12	1
4	Phosphate × Nitrogen linear	0.61	1
5	Phosphate × Nitrogen quadratic	6.16	1
	Total	59.43	5

The result for contrast 5 is surprising, especially since contrast 3 is so small. However, if the nitrogen levels are plotted out separately for the two levels of phosphate, we see clear but opposite quadratic effects, which cancel out on average. None of the residuals are large.

Page 294, Exercises 12.7

1.

	Anova		
Source	CSS	d.f.	M.S.
Regulators	2.745	7	0.392
Testing stations	0.200	3	0.067
Residual	0.670	21	0.032
Total	3.615	31	

Page, 297, Exercises 12.8

1. The Anova without transformation is as follows:

Source	CSS	d.f.	M.S.
Towns	71.5	3	23.8
Preparations	34.0	3	11.3
Residuals	198.5	9	22.0
Total	304.0	15	

However, the residual mean square should be just under 10 (see the top of p. 296), and this is suspiciously large when tested by X^2. If the arcsin transformation is used we have results as follows:

Source	CSS	d.f.	M.S.
Towns	0.0475	3	0.0158
Preparations	0.0226	3	0.0075
Residual	0.1329	9	0.0148
Total	0.2031	15	

The residual mean square is now in good agreement with the theoretical value of 1/40. In neither analysis do the residuals give any grounds for suspicion.

References

ANDERSON, E., 1954, 'Efficient and inefficient methods of measuring specific differences'. pp. 93–106 in *Statistics and Mathematics in Biology*, Ed., O. Kempthorne, Iowa State College Press, Amos, Iowa.

ANDERSON, S. L., and COX, D. R., 1950, 'The relation between strength and diameter of wool fibres'. *J. Textile Inst.*, **41**, 481–91.

ANSCOMBE, F. J., and TUKEY, J. W., 1963, 'The examination and analysis of residuals'. *Technometrics*, **5**, 141–60 (Corr. **5**, 536; **6**, 127).

ARMITAGE, P., FOX, W., ROSE, G. A., and TINKER, C. M., 1966, 'The variability of measurements of casual blood pressure'. *Cli. Sci.*, **30**, 337–44.

ATIQULLAH, M., 1963, 'On the randomisation distribution and power of the variance ratio test'. *J. R. Statist. Soc. B*, **25**, 334–47.

BAKER, A. G., 1957, 'Analysis and presentation of the results of factorial experiments'. *Applied Statistics*, **6**, 45–55.

BANERJEE, S. K., and NAIR, K. R., 1940, 'Tables of confidence intervals for the median in samples from any continuous population. *Sankhya*, **4**, 551–8.

BARNETT, V., and LEWIS, T., 1978, *Outliers in Statistical Data*, John Wiley, New York.

BARR, A., 1957, 'Differences between experienced interviewers'. *Applied Statistics*, **6**, 180–8.

BARTLETT, M. S., 1935, 'The effect of non-normality on the t-distribution'. *Proc. Camb. Phil. Soc.*, **31**, 223–31.

BERKSON, J., and ELVEBACK, L., 1960, 'Competing risks, with particular reference to the study of smoking and lung cancer'. *J. Amer. Statist. Ass.*, **55**, 415–28.

BOX, G. E. P., and COX, D. R., 1964, 'An analysis of transformations'. *J. R. Statist. Soc. B*, **26**, 211–52.

BROCKMEYER, E., HALSTROM, H. L., and JENSEN, A., 1948, 'The life and works of A. K. Erlang'. *Trans. Danish Acad. Technical Sci.*, **3**.

CAMERON, J. M., 1951, 'The use of components of variance in preparing schedules for sampling of baled wool'. *Biometrics*, **7**, 83–96.

CANE, V. R., and HORN, V., 1951, 'The timing of responses to spatial perception questions'. *Quart. J. Exp. Psych.*, **3**, 133–45.

CHERNOFF, H., and LIEBERMAN, G. J., 1954, 'Use of normal probability paper'. *J. Amer. Statist. Ass.*, **49**, 778–85.

COCHRAN, W. G., 1952, 'The χ^2-test of goodness of fit'. *Ann. Math. Statist.*, **23**, 315–45.

COCHRAN, W. G., 1954, 'Some methods for strengthening the common χ^2 tests'. *Biometrics*, **10**, 417–51.

COCHRAN, W. G., 1963, *Sampling Techniques*, John Wiley, New York.

COCHRAN, W. G., and COX, G. M., 1957, *Experimental Designs*, John Wiley, New York.

COMRIE, L. J. (Ed.), 1965, *Barlows Tables*, Spon. (Science paperbacks.)

COX, D. R., 1958, *Planning of Experiments*, John Wiley, New York.

COX, D. R., and LEWIS, P. A. W., 1966, *The Statistical Analysis of Series of Events*, Methuen, London.

CRAMÉR, H., 1946, *Mathematical Methods of Statistics*, Princeton Univ. Press, N.J.

DESMOND, D. J., 1954, 'Quality control on the setting of voltage regulators'. *Applied Statistics*, **3**, 65–73.

DOUGLAS, J. B., 1955, 'Fitting Neyman type A (two parameter) contagious distribution'. *Biometrics*, **11**, 149–73.

DREW, G. C., 1951, 'Variation in reflex blink rate during visual motor tasks'. *Quart. J. Exp. Psych.*, **3**, 73–88.

EISENHART, C., 1947, 'The assumptions underlying the analysis of variance'. *Biometrics*, **5**, 1–21.

EVANS, D. A., 1953, 'Experimental evidence concerning contagious distributions in ecology'. *Biometrika*, **40**, 186–211.

FELLER, W., 1957, *An Introduction to Probability Theory and its Applications*, vol. i, John Wiley, New York.

FELLER, W., 1966, *An Introduction to Probability Theory and its Applications*, vol. ii, John Wiley, New York.

FERTIG, J. W., and HELLER, A. N., 1950, 'The application of statistical techniques to sewage treatment processes'. *Biometrics*, **6**, 127–35.

FISHER, SIR R. A., 1958, *Statistical Methods for Research Workers*, Oliver & Boyd, Edinburgh.

FISHER, SIR R. A., and YATES, F., 1963, *Statistical Tables*, 6th Ed., Oliver & Boyd, Edinburgh.

FRASER, D. C., 1950, 'The relation between angle of display and performance in a prolonged visual task'. *Quart. J. Exp. Psych.*, **2**, 176–81.

FREEMAN, M. F., and TUKEY, J. W., 1950, 'Transformations related to the angular and square root'. *Ann. Math. Statist.*, **21**, 607–11.

GAYEN, A. K., 1949, 'The distribution of Student's t in random samples of any size drawn from non-normal universes'. *Biometrika*, **36**, 353–69.

GAYEN, A. K., 1950, 'Significance of difference between the means of two non-normal samples'. *Biometrika*, **37**, 399–408.

GEARY, R. C., 1936, 'The distribution of Student's ratio for non-normal samples'. *J. R. Statist. Soc. Suppl.*, **3**, 178–84.

GHURYE, S. G., 1949, 'On the use of Student's t-test in an asymmetrical population'. *Biometrika*, **36**, 426–30.

GIBSON, W. M., and JOWETT, G. H., 1957, 'Three-group regression analysis. Part II. Multiple regression analysis'. *Applied Statistics*, **6**, 189–97.

GNEDENKO, B. V., 1962, *The Theory of Probability*, Chelsea, New York.

GREENWOOD, M., and YULE, G. U., 1920, 'An enquiry into the nature of frequency distributions of multiple happenings'. *J. R. Statist. Soc.*, **83**, 255.

GRIDGEMAN, N. T., 1956, 'A tasting experiment'. *Applied Statistics*, **5**, 106–112.

HAMMERSLEY, J. M., and HANDSCOMB, D. C., 1964, *Monte Carlo Methods*, Methuen, London.

HASTINGS, C., HAYWARD, J. T., and WONG, J. P., 1955, *Approximations for Digital Computers*, Princeton Univ. Press, N.J.

JOWETT, G. H., and DAVIES, HILDA M., 1960, 'Practical experimentation as a teaching method in statistics'. *J. R. Statist. Soc. A*, **123**, 10–36.

KENDALL, M. G., and BABINGTON-SMITH, B., 1938, 'Randomness and random sampling numbers'. *J. R. Statist. Soc. A*, **101**, 157.

KENDALL, M. G., and BABINGTON-SMITH, B., 1939, 'Second paper on random sampling numbers'. *J. R. Statist. Soc. Suppl.*, **6**, 51.

KENDALL, M. G., and BABINGTON-SMITH, B., 1951, *Tracts for Computers XXIV. Tables of Random Sampling Numbers*, C.U.P., Cambridge.

KENDALL, M. G., and STUART, A., 1958, *The Advanced Theory of Statistics*, vol. I, Griffin, London. (Vol. II, 1961, Vol. III, 1966).

KURTZ, T. E., LINK, R. F., TUKEY, J. W., and WALLACE, D. L., 1965a, 'Short cut multiple comparisons for balanced single and double classifications. Part 2. Derivations and approximations'. *Biometrika*, **52**, 485–98.

KURTZ, T. E., LINK, R. F., TUKEY, J. W., and WALLACE, D. L., 1965b, 'Short cut multiple comparisons for balanced single and double classifications. Part I. Results'. *Technometrics*, **7**, 95–162.

LANER, S., MORRIS, P., and OLDFIELD, R. C., 1957, 'A random pattern screen'. *Quart. J. Exp. Psych.*, **9**, 105–8.

LEHMAN, E. H., 1961, 'Exact and approximate distributions for the Wilcoxon statistic with ties'. *J. Amer. Statist. Ass.*, **56**, 293–8.

LESLIE, P. H., 1951, 'The calculation of χ^2 for an $r \times c$ contingency table'. *Biometrics*, **7**, 283–6.

LINDLEY, D. V., 1964, *Introduction to Probability and Statistics*, Part 1, C.U.P., Cambridge.

LINDLEY, D. V., and MILLER, J. C. P., 1966, *Cambridge Elementary Statistical Tables*, C.U.P., Cambridge.

LOEVE, M., 1960, *Probability Theory*, Van Nostrand, Princeton, N.J.

LYDALL, H. F., 1959, 'The long-term trend in the size distribution of income'. *J. R. Statist. Soc. A*, **122**, 1–46.

MADANSKY, A., 1959, 'The fitting of straight lines when both variables are subject to error'. *J. Amer. Statist. Ass.*, **54**, 173–205 (Corr. **54**, 812).

MANN, H. B., and WHITNEY, D.R., 1947, 'On a test of whether one of two random variables is stochastically larger than the other'. *Ann Math. Statist.*, **18**, 50–60.

MCNEIL, D., 1977, *Interactive Data Analysis*, John Wiley, New York.

MILLER, R. G., JR, 1966, *Simultaneous Statistical Inference*, McGraw-Hill, New York.

MOOD, A. M., and GRAYBILL, F. A., 1963 *Introduction to the Theory of Statistics*, McGraw-Hill, New York.

MOSER, C. A., 1958, *Survey Methods in Social Investigations*, Heinemann.

MOSTELLER, F., and YOUTZ, C., 1961, 'Tables of the Freeman–Tukey transformations for the binomial and Poisson distributions'. *Biometrika*, **48**, 433–440.

OAKLAND, G. B., 1950, 'An application of sequential analysis to whitefish sampling'. *Biometrics*, **6**, 59–67.

OGILVIE, J. C., RYAN, J. E. B., COWAN, R. F., and QUERENGESSER, E. I., 1955, 'Interrelations and reproducibility of measures of light threshold and scotopic acuity'. *J. App. Psych.*, **7**, 519–22.

PAGE, E. S., 1959, 'Pseudo-random elements for computers'. *Applied Statistics*, **8**, 124–31.

PARZEN, E., 1960, *Modern Probability Theory and its Applications*, John Wiley, New York.

PEACH, P., 1961, 'Bias in pseudo-random numbers'. *J. Amer. Statist. Ass.*, **56**, 610–18.

PEARSON, E. S., 1956, 'Some aspects of the geometry of statistics. The use of visual presentation in understanding the theory and application of mathematical statistics'. *J. R. Statist. Soc. A*, **119**, 125–49.

PEARSON, E. S., and HARTLEY, H. O., 1966, *Biometrika Tables for Statisticians*, C.U.P., Cambridge.

PEARSON, E. S., and PLEASE, N. W., 1975, Relations between the shape of population distribution and the robustness of four simple test statistics. *Biometrika*, **62**, 223–41.

PRATT, J. W., 1959, 'Remarks on zeros and ties in the Wilcoxon signed-ranked procedures'. *J. Amer. Statist. Ass.*, **54**, 655–67.

SAVAGE, I. R., 1953, 'Bibliography of nonparametric statistics and related topics'. *J. Amer. Statist. Ass.*, **48**, 844–906.

SCHEFFE, H., 1959, *The Analysis of Variance*, John Wiley, New York.

SHAPIRO, S. S., and WILK, M. B., 1965, 'An analysis of variance test for normality'. *Biometrika*, **52**, 591–612.

SIEGEL, S., 1956, *Nonparametric Statistics for the Behavioural sciences*, McGraw-Hill, New York.

SKORY, J., 1952, 'Automatic machine method of calculating contingency'. *Biometrics*, **8**, 380–2.

SNOKE, L. R., 1956, 'Specific studies on soil-block procedure for bioassay of wood preservatives'. *Applied Microbiology*, **4**, 21–31.

STEVENS, S. S., 1951, *Handbook of Experimental Psychology*, John Wiley, New York, 1–49.

STUDENT, 1907, 'On the error of counting with a haemacytometer'. *Biometrika*, **5**, 351–60.

THOMPSON, W. E., 1959, 'ERNIE – a mathematical and statistical analysis'. *J. R. Statist. Soc. A*, **122**, 301–33.

TIKU, M. L., 1963, 'Approximation to Student's t distribution in terms of Hermite and Laguerre polynomials'. *J. Indian Math. Soc.*, **27**, 91–102.

TUKEY, J. W., 1949a, 'Comparing individual means in the analysis of variance', *Biometrics*, **5**, 99–114.

TUKEY, J. W., 1949b, 'One degree of freedom for non-additivity'. *Biometrics*. **5**, 232–42.

TUKEY, J. W., 1962, 'The future of data analysis'. *Ann. Math. Statist.*, **33**, 1–67.

WATERS, W. E., 1955, 'Sequential analysis of forest insect surveys'. *Forest Science*, **1**, 68–79.

WEHFRITZ, F. W., 1962, 'Calibration problems in chemistry and engineering'. *Trans. Amer. Soc. Quality Control*, September 1962.

WETHERILL, G. B., 1965, 'An approximation to the inverse normal function suitable for the generation of random normal deviates on electronic computers'. *Applied Statistics*, **14**, 201–5.

WETHERILL, G. B., 1960, 'The Wilcoxon test and non-null hypotheses'. *J. R. Statist. Soc. B*, **22**, 402–18.

WETHERILL, G. B., and O'NEILL, R., 1971, 'The present state of multiple comparison methods'. *J. R. Statist. Soc., B*, **33**, 218–50.

WOLD, H., 1954, *Tracts for Computers XXV. Random Normal Deviates*, C.U.P., Cambridge.

WU, S. M., 1963, 'Tool life testing by response surface methodology – Part 2'. *Trans. A.S.M.E. Paper* 63 – Prod – 7.

Author Index

Anderson, E., 37, 253
Anderson, S. L., and Cox, D. R., 36
Anscombe, F. J., and Tukey, J. W., 242
Armitage, P., *et al.*, 21
Atiqullah, M., 295

Baker, A. G., 218
Banerjee, S. K., and Nair, K. R., 168
Barnett, V., and Lewis, T., 242
Barr, A., 188
Bartlett, M. S., 143
Berkson, J., and Elveback, L., 36
Box, G. E. P., and Cox, D. R., 297
Brockmeyer, E., *et al.*, 62

Cameron, J. M., 43, 293
Cane, V. R., and Horn, V., 276
Chernoff, H., and Lieberman, G. J., 242
Cochran, W. G., 191, 200
Cochran, W. G., and Cox, G. M., 285
Comrie, L. J., 208
Cox, D. R., 286, 288
Cox, D. R., and Lewis, P. A. W., 183, 197
Cramer, H., 104, 113, 114, 135, 203

Desmond, D. J., 293, 294
Douglas, J. B., 180
Drew, G. C., 155

Eisenhart, C., 294
Evans, D. A., 180

Feller, W., 47, 139, 164
Fertig, J. W., and Heller, A. N., 153
Fisher, Sir R. A., 34
Fisher, Sir R. A., and Yates F., 152, 178
Fraser, D. C., 304
Freeman, M. F., and Tukey, J. W., 178, 179

Gayen, A. K., 143
Geary, R. C., 143
Ghurye, S. G., 143

Gibson, W. M., and Jowett, G. H., 245
Gnedenko, B. V., 47
Greenwood, M., and Yule, G. U., 62
Gridgeman, N. T., 210, 214

Hammersley, J. M., and Handscomb, D. C., 106
Hastings, C., Hayward, J. T., and Wong, J. P., 109

Jowett, G. H., and Davies, Hilda M., 26

Kendall, M. G., and Babington-Smith, B., 105
Kendall, M. G., and Stuart, A., 19, 73
Kurtz, T. E., *et al.*, 275

Laner, S., Morris, P., and Oldfield, R. C., 198
Lehman, E. H., 168
Leslie, P. H., 208
Lindley, D. V., 104
Lindley, D. V., and Miller, J. C. P., 178
Loeve, M., 47
Lydall, H. F., 26

McNeil, D., 37
Madansky, A., 224, 244
Mann, N. B., and Whitney, D. R., 174
Miller, R. G., 275
Mood, A. M., and Graybill, F. A., 135
Moser, C. A., 191
Mosteller, F., and Youtz, C., 178, 179

Oakland, G. B., 179
Ogilvie, J. C., *et al.*, 258

Page, E. S., 105
Parzen, E., 47, 63
Peach, P., 106
Pearson, E. S., 37

352

Pearson, E. S., and Hartley, H. O., 29, 118, 152
Pearson, E. S., and Please, N. W., 89, 143
Pratt, J. W., 168

Savage, I. R., 162
Scheffe, H., 143, 144, 145, 160, 294
Seigel, S., 162, 168, 170
Shapiro, S. S., and Wilk, M. B., 129, 295
Skory, J., 208
Snoke, L. R., 38

Stevens, S. S., 177

Thompson, W. E., 106
Tiku, M. L., 143
Tukey, J. W., 242, 275, 295

Waters, W. E., 179
Wehfritz, F. W., 244
Wetherill, G. B., 109, 176
Wetherill, G. B., and O'Neill, R., 275
Wold, H., 111
Wu, S. M., 261

Subject Index

Additive effects, 97, 295
Analysis of variance, 263–97, 314
  components of variance, 289–94
Asymptotic relative efficiency, 175–6

Baled wool example, 43–4, 98–9, 291–2
BASIC programs, 318–23
Binomial distribution
  expectation, 80
  variance, 87
  normal approximation to, 162–5
  transformations for, 178, 295–6
  methods of inference for, 185–6, 188–92
  dispersion test for, 199–201
  simulation of, 315
Blocking, see Randomized blocks

Central limit theorem, 71, 317
Chi-square distribution, 136–7
  confidence intervals for σ^2 using, 137–9
  effect of departures from assumption on, 143–5
  goodness of fit test, 201–5
  significance test for variance, 141–3
Computer programs, 318–23
Conclusions, how to give, 127, 142–3, 149, 154, 272–3, 275, 284
Conditional distribution, 76–7, 91
Conditional probability, 51–5
Confidence interval, 119–24
  one-sided, 121–2
  two-sided, 119–21
  simulation of, 316–7
Contingency tables
  2×2, 188–92, 204–5
  $r \times c$, 206–10
Continuity correction, 163, 204
Correlation,
  coefficient of, 92
  serial, 93
Covariance, 90–3
  of a linear function, 93
Cumulative distribution function, 62, 65–6
Cumulative frequency diagrams, 24–6

Cumulative sum chart, 33

Degrees of freedom, 136
  of s^2, 138
Dependence, see Independence
Dispersion, index of, 195–8
Distribution, probability
  definition, 58
  multivariate, 75–7
  marginal distribution, 76–7
  conditional distribution, 76–7
Distributions, continuous
  exponential, 67–8
  F-distribution, 150–2
  normal, 68–71
  t-distribution, 133–4
Distributions, discrete
  binomial, 55, 59–60
  discrete rectangular, 58
  geometric, 55, 60–1
  negative binomial, 179–80
  Poisson, 61–2
Dust, chamber experiment, 61

Estimate
  point estimate, 117
  unbiased, 118
Estimation, 117–24
  point estimation, 117–19
  interval estimation, 119–24
  for normal distribution, 130–9
Expectation, 79–84
  of many distributions, 80
  approximate methods, 135
Exponential distribution, 67–8, 315

F-distribution, 150–2
Frequency distributions
  one-dimensional, 19–26
  two-dimensional, 33–7
  marginal distributions, 34
Frequency table, 15, 16, 17, 112

Geometric distribution, 55, 60–1
Goodness of fit test, 201–5

Hardness of metals experiment, 94–8, 297
Heights of fathers and daughters example, 34, 52
Histograms, 21–4, 64, 317
 of mean and median, 113

Independence and dependence, 36, 54, 90–93
 of $\bar{x}$ and s^2 in normal samples, 114–16, 139, 315
Interaction, 97, 281–2
Interquartile range, 30

Kurtosis, 72
 coefficient of, 88

Least significant difference method, 272–3
Least squares, 221–3, 227–8, 247–8
 normal equations of, 248
 in analysis of variance, 269
Linear contrast, 266, 280–5

Marginal distribution, 76–7
Mean
 calculation of, 308–13
 truncated, 27, 32, 110–16
 Winsorized, 27, 32
 moments of, 83, 86, 132
 variance of with serial correlation, 93
Mean squared error, 124
Median, 26, 110–16
Model, statistical, 93–9, 101, 216–21
Moments, 87–9
Multiplicative congruential method, 105–6
Multivariate data, 19, 33–7

Negative binomial distribution, 179–80
Non-parametric tests, 162–76
Normal distribution, 68–71
 measures of departures from, 87–8
 standard normal, 69
 bivariate normal, 77
 simulation of, 316
Null hypothesis, 125

Order statistic, 241
Outliers, wild observations, 114, 116, 242

Paired samples, analysis of
 continuous variables, 152–6
 discrete variables, 193–5
Percentiles, 26–7
Poisson distribution
 expectation, 80
 variance, 84–5
 normal approximation, 178
 transformations for, 179
 methods of inference for, 182–4, 186–187

Poisson distribution – *continued*
 dispersion test for, 195–8
 simulation of, 315–6
Poisson process, 74
Population, 45
Power, of a significance test, 128, 175
Probability, 45–6, 47–57
 additional law, 48–51
 conditional, 51–5
 multiplication law, 54
Probability density function, 65
Probability paper (normal), 72, 73, 112
Pseudo random deviates, 105–10, 315

Quality control, 30–2

Radioactive emissions example, 17, 33–4, 195–7
Random digits, 58, 63
Random sample, 45
Random variable
 discrete, 58
 continuous, 63
 sums of, 180
Random variation, 15–19
Randomized blocks, 285–8
Randomness, tests of, 106–7
Range, 28–9, 31
 estimation of σ from, 118–19, 131
 range chart, 32, 110–11
 studentized, 273–4
Ranked data, analysis, 162–76
Regression
 linear, 224–39, 313–4
 multiple, 245–62
Regression coefficient, 225
 partial, 245
Residuals
 definition, 220
 analysis of, 239–43
 in analysis of variance, 270, 275, 279, 314
 in linear regression, 313
 in multiple regression, 257
Rounding-off errors, 67, 309

Sample, definition, 45
Sample space, 16
Sampling distribution, 102–4, 110–16
Sampling inspection, 60
Scatter diagram, 35–6, 40
 multivariate, 37, 255
 of $\bar{x}$ and s^2, 115
Seed germination example, 15–17, 49–50
Sign test, 165–9
Signed rank test, 169–73
Significance level, 125
Significance test, 124–8
 interpretation of, 126–7, 140

Significance test – *continued*
  one and two-sided, 127–8, 140–1
Simulation, 314–7
Skew distributions, 26–8, 30, 72
Skewness, coefficient of, 88
Standard deviation, 29
  calculation of, 308–13
  of populations, 84
  sampling distribution of, 110–12, 114–116
Standard error, 103, 123

t-distribution, 133–4
  effect of departures from assumptions on, 143–5
  effect of serial correlation on, 145
  confidence intervals using, 134–5
  moments of, 135
t-test
  one sample, 139–41
  two sample, 147–9, 313
  variances unequal, 160–1
Test statistic, 125, 165, 169, 174
Transformations
  for stabilizing variance, 179–81

Transformations – *continued*
  for simplifying model, 218
  of s^2, 152
  to analysis of variance, 295–7
  to normality, 71
  to symmetry, 30
  in simulation, 315–6

Variance, of random variables, 84–7
  for many distributions, 84, 87
  of a difference of two variables, 123
  of a sum of variables, 92, 93
  of $\bar{x}$ in normal samples, 86
  of s^2 in normal samples, 132
Variance, sample, 29
  calculation of, 308–13
  combination of, 147, 149–50
  distribution of, 138
  moments of, 132
  expectation of, with serial correlation, 144

Wilcoxon test,
  one sample, 169–73
  two sample, 173–5